3

£16-50

Molecular Biology
Biochemistry and Biophysics
14

Volker Neuhoff

Micromethods in Molecular Biology

With Contributions by

G. F. Bahr · P. Dörmer · J.-E. Edström
U. Leemann · G.M. Lehrer · F. Ruch
H.-G. Zimmer

With 275 Figures

Springer-Verlag Berlin · Heidelberg · New York 1973

Professor Dr. Volker Neuhoff

Max-Planck-Institut für Experimentelle
Medizin (Arbeitsgruppe Neurochemie)
3400 Göttingen
Hermann-Rein-Str. 3

ISBN 3-540-06319-6 Springer-Verlag Berlin Heidelberg New York
ISBN 0-387-06319-6 Springer-Verlag New York Heidelberg Berlin

Dedicated to my wife
in gratitude for so many years
of serene understanding

Preface

This book is based on practical experience and is therefore written as a practical manual. The fore-runners of the book were the manuals of the first and second EMBO-Courses on "Micromethods in Molecular Biology" which were held in Göttingen in the spring of 1970 and the autumn of 1971. This book may serve as a manual not only for the participants of the third EMBO-Course to be held in Göttingen in autumn 1973, but also for all experimenters who are interested in using micromethods. It must be emphasized from the outset that this book is conceived as a "cook book" and not as a monograph which attempts to revue the literature on micromethods critically.

The methods described here in detail are performed routinely in the authors' laboratories and include all the practical details necessary for the successful application of the micromethods. There are many other sensitive and excellent micromethods which are not included in this book, because the authors feel that in a "cook book" only methods for which they have personal experience and proficiency should be described. Some readers may feel that the title promises more than the present contents of this book; however, if sufficient interest is shown in this volume, it may be possible to remedy such deficiencies in future editions.

In general, micromethods are no more arduous than the equivalent method on the macro scale, and the saving in time is usually considerable. For instance, sometimes a procedure in the macro scale takes hours, and on the micro scale only minutes, yet the amount of information obtained is the same. Thus it is often advantageous to use micro methods even when there is sufficient material available for macro scale analysis.

Many existing macro scale methods can be made a hundred- or thousand-fold more sensitive by simply scaling down the dimensions of the analytical medium in use, but in some cases it may be necessary to change the conditions of separation when adapting a method. However, in changing from the normal to the micro scale, the biggest barrier is often the sceptism of the experimenter about any method which has been reduced to a micro scale. Once this has been overcome it is fascinating to see what possibilities exist for micro methods in one's own field of research.

Acknowledgments: Dipl. chem. F. Boschke, editor of Die Naturwissenschaften, Prof. Dr. G. Czihak, Institut für Genetik und Entwicklungsbiologie der Universität Salzburg as participant of the second EMBO-Course, Prof. Dr. H. Hydén, Institute of Neurobiology, Göteborg, and Prof. Dr. H.-G. Wittmann, Max-Planck-Institut für molekulare Genetik, Berlin, who have independently suggested that this book be written. I would also like to thank my coworkers, E. M. Adam, H.-H. Althaus, Dr. W. Behbehani, Dr. G. Briel, W. Dames, Dr. F.-H. Hubmann, F. Kiehl, M. Maier, S. Mesecke, E. Priggemeier, Dr. C.-D. Quentin,

I. URBAN, Dr. T. V. WAEHNELT, Dr. D. WOLFRUM, for their advice on specific methods, and especially Dr. R. RÜCHEL, and H. ROPTE for producing so many photographs under difficult technical conditions. My thanks are also due to Dr. J. HOBBS for translating the German text, Dr. B. LEONARD, Dr. SHIRLEY MORRIS, and Dr. N. N. OSBORNE for reading the text critically, and Mrs. I. von BISCHOFFS-HAUSEN for her assistance in typing the manuscript. Last, but by no means least, my thanks go also to the co-authors for their contributions and to Dr. K. F. SPRINGER and his staff for their part in preparing the book in the present form.

V. NEUHOFF

Contents

Chapter 11   **Quantitative Autoradiography at the Cellular Level** 347
PETER DÖRMER

Contributors

GUNTER F. BAHR   M. D., Prof. of Pathology, Armed Forces Institute of Pathology, Biophysics Branch, Washington D.C. 20306/USA

PETER DÖRMER   Priv.-Doz. Dr. med., Institut für Hämatologie der Gesellschaft für Strahlen- und Umweltforschung, Landwehrstr. 61, 8000 München 2/W. Germany

JAN-ERIK EDSTRÖM   Prof. Dr. med., Department of Histology, Karolinska Institutet Stockholm/Sweden

URSULA LEEMANN   Dr. s. c. nat., Department of General Botany, Swiss Federal Institute of Technology, 8006 Zürich/Switzerland

GERARD M. LEHRER   M. D., Prof. of Neurology and Director, Division of Neurochemistry, The Mount Sinai School of Medicine of the City University of New York, N.Y. 10029/USA

VOLKER NEUHOFF   Prof. Dr. med., Max-Planck-Institut für Experimentelle Medizin, Arbeitsgruppe Neurochemie, Hermann-Rein-Str. 3, 3400 Göttingen/W.Germany

FRITZ RUCH   Prof. Dr. s. c. nat., Department of General Botany, Swiss Federal Institute of Technology, 8006 Zürich/Switzerland

HANS-GEORG ZIMMER   Dr. rer. nat., CARL ZEISS, Labor für Mikroskopie, 7082 Oberkochen/W. Germany

Chapter 1

Micro-Electrophoresis on Polyacrylamide Gels

Polyacrylamide gels were introduced in 1959 by RAYMOND and WEINTRAUB, as supports for electrophoretic separations. The polyacrylamide gel is produced by polymerising acrylamide, with N,N-methylenebisacrylamide or ethylene di-acrylate as the cross-linking component. Catalytic redox systems, which yield free radicals, are used to initate copolymerisation (e.g. ammonium peroxydisulphate and N,N,N',N'-tetramethylethylene diamine). Electrophoresis on polyacrylamide gels is now in general use as a laboratory technique. Its popularity owes much to the transparency of the gel, its mechanical stability and inertness, its stability over a very wide range of pH and its insolubility in most of the solvents commonly used for electrophoresis. The gels can be prepared reliably and reproducibly from analytically pure starting materials, and possesses the decisive advantage that by varying the proportions of the starting materials, gels of different density and pore diameter can be prepared. Various other substances can also be copolymerised into these gels.

In carrier electrophoresis, the chemical and physical properties of the support influence the mobility of the components to be fractionated and the sharpness of separation. The capacity of the carrier material is limited, and it is inhomogeneous; this gives rise to adsorption, electroosmotic suction, and "wick" effects, which are difficult to control and which influence the fractionation adversely. In contrast, polyacrylamide gels are almost completely homogeneous (compare Fig. 1) and therefore adsorption and electroosmosis do not occur. Wick effects can also be minimized by choice of a suitable gel concentration, buffer system, and current strength. These advantage are particularly valuable when macromolecules are to be fractionated and characterised on such gels.

Continuous electrophoresis on polyacrylamide gels (RAYMOND and WEIN-TRAUB, 1959), as for all other methods of carrier electrophoresis, is based on a homogeneous buffer system of fixed pH. In contrast, disc electrophoresis, developed by ORNSTEIN (1964) and DAVIES (1964), employs a discontinuous separating system. Disc electrophoresis can be performed with different buffer systems, different pH values, and different pore sizes of the polyacrylamide gel used as carrier, yet it still maintains its amazingly high quality of separation. In practice, the degree of gel discontinuity can be adjusted for each separation problem. The term "disc" also indicates a characteristic which determines the quality of separation: the macro-molecules to be fractionated are concentrated from dilute solution into a sharply defined zone. Since discontinuous polyacrylamide gel electrophoresis is usually carried out in a glass tube with an inner diameter of 5–7 mm, the starting zone is

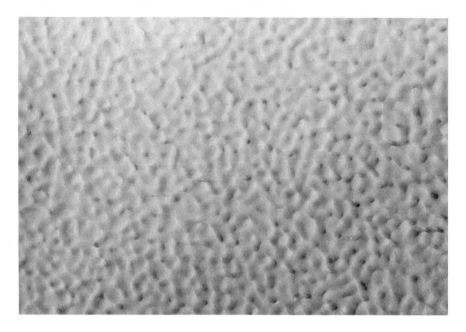

Fig. 1. Electronmicrograph of a 30 % polyacrylamide gel. Freeze-etching and carbon-platinum shadowing. (Preparation of RÜCHEL and AMELUNXEN.) Magnification 53 700 ×

actually in the form of a disc in which the mixture of molecules to be separated is highly concentrated; this is difficult to attain if fractionation by continuous electrophoresis is used.

The physical theory for the method of discontinuous electrophoresis, with numerous examples of its use, and practical advice on disc electrophoresis as normally performed, has been amply described by MAURER (1971). In this chapter, therefore, only micro-disc electrophoresis will be described, the application of which is always recommended if only very small quantities of the substances are available for investigation. For example, micro-electrophoresis can be used after a lengthy purification procedure in which only minimal quantities of, for example, a pure enzyme solution, are finally available for analysis. The introduction of micromethods is becoming more and more necessary in neurochemistry, since it is already evident that different anatomical regions of the brain differ in their metabolism; it is therefore necessary to use methods which enable changes in metabolism to be assessed in well defined anatomical regions and also in isolated nerve cells.

The first application on the microscale of polyacrylamide gel electrophoresis was carried out in 1964 when PUN and LOMBROZO fractionated brain proteins. In 1965, GROSSBACH used the 5 µl Drummond microcap for this technique, which was further refined in 1966 by HYDÉN, BJURSTAM and McEWEN, McEWEN and HYDÉN, who used 2 µl capillaries for the fractionation of brain proteins. In 1968, NEUHOFF introduced a gel mixture which had been specially developed for the micro fractionation of water-soluble brain proteins; these gels were later found to be suitable for many different fractionation problems. HYDÉN and LANGE (1972) used the micro-disc electrophoresis for the analysis of the changes in proteins in different brain areas as a function of intermittent training. GRIFFITH and LAVELLE (1971) have analysed changes in the developmental proteins in facial nerve nuclear regions by this method. ANSORG, DAMES and NEUHOFF (1971) have used micro-disc electrophoresis to study the effect of different extraction procedures on the pattern of brain proteins, and ALTHAUS et al. (1972) have used the method for the analysis of the effect of post tetanic potentiation of the monosynaptic reflexes in the spinal cord of cats on the water soluble proteins produced. GROSSBACH (1969, 1971) has used micro-disc electrophoresis for the analysis of chromosomal activity in the salivary glands of Camptochironomus. Glycoproteins of the alveolar surfactant of rat lung were analysed by REIFENRATH and ELLSSEL (1973), using micro-disc electrophoresis, and a modified PAS staining according to ZACHARIUS and ZELL (1969). 10 ng of glycoprotein can be detected and quantitatively determined in 5 µl gels. After recording the positions of the red stained glycoproteins, a second staining with amido black coloured the protein fractions blue and glycoproteins violet.

After a little practice, disc electrophoresis on the micro scale is hardly more difficult than the normal method. In addition to requiring smaller quantities of material, it has the added advantage of giving results agreeing with those obtained by the macro method in an appreciably shorter time. The lower limit for a single protein band in a 5 µl gel of 450 µm diameter is 10^{-9} g of albumin, if visualized with amido black 10 B. Micro interferometric determination of amounts of proteins in separated fractions in the nanogram range is described by HYDÉN and LANGE (1968). 0.1 to 0.5 µg of a protein mixture can be fractionated by electrophoresis in 5 µl capillaries; this means that 5000–10000 estimations can performed with 1 ml solution containing 1 mg protein. The duration of electrophoresis depends on the type of protein to be fractionated, but is generally between 20 and 60 min. With gels of small diameter, staining with amido black requires only 5 min, and the decolourising process about 30 min. Electrophoretic destaining is not necessary. To give an idea of the dimensions of a microgel, Fig. 2 shows a gel with a diameter of 0.45 mm near the head of a normal household match.

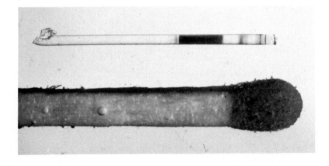

Fig. 2. Match stick and a 5 µl polyacrylamid gel. Magnification 5 ×

Micro-Disc Electrophoresis

Use of Capillaries

The Drummond Microcaps[1] introduced by GROSSBACH (1965) for micro-disc electrophoresis have proved to be extraordinarily well suited for this technique. They are obtainable in various size. 1 µl caps have an inner diameter of 0.24 mm and a length of 33 mm, 2 µl caps: inner diameter 0.28 mm, length 33 mm, 5 µl caps: inner diameter 0.45 mm, length 33 mm, 10 µl caps: inner diameter 0.56 mm, length 42 mm. The 5 µl caps are the most widely used for this method. There is sometimes a difference in length of a few millimetres between different batches of these capillaries. However, as the volume is always exact, the inner diameter of the capillaries is correspondingly larger or smaller.

For many purposes the capillaries can be used directly and filled with the gel mixture without pre-treatment. For some applications it is necessary to siliconize the capillaries. To do this, dimethyldichlorosilane is dissolved in benzene (2 % v/v); capillaries which have been cleaned in chromic acid are filled up to the top by capillary attraction by dipping one end in the siliconizing solution, the solution is completely removed by placing the end of the capillary on absorbant filter paper, and the capillary is then dried for 1 hr. at 80° C. When the capillaries have been siliconized, even if the gel contains no Triton X-100 it can be expelled by applying slight pressure from a water-filled syringe.

For electrophoresis on polyacrylamide gradients, even new capillaries must be carefully cleaned before being charged with polyacrylamide. For this purpose 200–300 microcaps are transferred to a suction flask which is half-filled with chromic acid. The capillaries are completely filled with the chromic acid by creating a vacuum by means of a waterpump. In order to fill the capillaries completely with the acid, it is necessary to release the vacuum repeatedly by opening the tap quickly. It is recommended that the capillaries should stay in the acid overnight. The acid is then poured off and the capillaries are transferred to a suitable sintered-glass filter-funnel over which a separating funnel fitted with a stopper is placed (see Fig. 3). The separating funnel is filled with distilled water, and the whole system is connected to a vacuum line via a filter funnel. On closing the funnel tap the pressure in the sintered funnel is reduced; when the tap connecting the separating funnel to the sintered glass filter is opened suddenly, water enters the glass funnel rapidly. By frequent repetition, the chromic acid is completely removed from the capillaries. The capillaries are then rinsed several times with absolute ethanol and finally with acetone. To ensure complete drying, the capillaries are left over night in drying cabinet at 37° C.

Capillaries which have already been used may be cleaned for re-use by the following procedure: when the gels have been pressed out of the capillaries after an electrophoresis run, the empty capillaries are collected in a beaker full of water; when a reasonable quantity of used capillaries has been collected, cleaning is performed as above. It is advantageous to precleanse the used capillaries in undiluted potassium hypochlorite (JAVELLE's solution) to dissolve any remaining

[1] Drummond Scientific Co., U.S.A.

polyacrylamide before transferring them to the chromic acid. Since the capillaries are relatively expensive cleaning is probably worth while.

JAVELLE's solution is also suitable for cleaning capillaries which are completely filled with polyacrylamide gel. It is advantageous to transfer the capillaries to a suction flask half filled with JAVELLE's solution and connected to a vacuum pump.

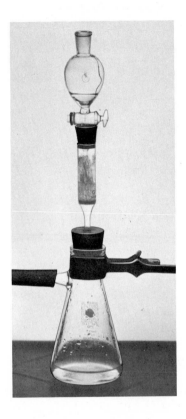

Fig. 3. Equipment for cleaning capillaries

The suction flask should have a perforated stopper with metal tube and a magnetic valve that is operated by a synchron motor, so that the vacuum is interrupted at short intervals and the gas formed by the action of the potassium hypochlorite is immediately removed from the capillaries. If this is not done, the gas bubbles will remain between the solution and the acrylamide in the capillaries, and stop the gel disolving.

The dissolution of a complete gel in a 5 μl capillary takes about 10 hrs. The dissolution time can be reduced to about 2 hrs. if the capillaries containing the gels are first dried for several days at approximately 100° C. In this case the JAVELLE's solution can enter the space between the dried gel and the capillary wall so increasing the surface of action.

Preparation of Gels

The polyacrylamide gels are made up from the following stock solutions:

Stock A: 860 mg Tris
       +8 ml H_2O
       +0.063 ml N,N,N',N'-Tetramethyl-ethylendiamine (TEMED)
       +3.6 N H_2SO_4 to pH 8.8 (ca. 0.45 ml)
       +H_2O ad 10 ml

Stock B: 2.85 g Tris
       +25 ml H_2O
       +8.7 M H_3PO_4 to pH 7
       +1 M H_3PO_4 to pH 6.7
       +H_2O ad 50 ml

Stock C for 20% Gel:
       20 g Acrylamide
       +200 mg N,N'-Methylene-bis-acrylamide
       +3.75 mg $K_3Fe(CN)_6$
       +H_2O ad 37.5 ml

Stock D: 70 mg Ammonium peroxydisulphate
       +25 ml 4% Triton X-100 in H_2O
       +25 ml H_2O

Stock E: 5.98 g Tris
       +50 ml H_2O
       +0.46 ml TEMED
       +8.7 M H_3PO_4 to pH 7
       +1 M H_3PO_4 to pH 6.7
       +H_2O ad 100 ml

Stock F: 200 mg Ammonium peroxydisulphate
       +5 ml 4% Triton X-100
       +5 ml H_2O

Electrodebuffer:
       3.0 g Tris
       +14.4 g Glycine
       +H_2O ad 500 ml
       pH 8.4

Bromophenol blue:
       100 mg/5 ml H_2O

Fluorescein:
       saturated solution in H_2O

Amido black 10 B:
       1.0% in 7.5% CH_3COOH

pH adjustment must be carried out *after* the addition of TEMED, since this reagent is alkaline. Potassium ferricyanide is used in solution C to retard the polymerisation and give a better quality gel. Triton X-100 is necessary to facilitate the removal of the gels from the capillaries by water pressure, but does not effect the separation quality of the gel. However, Triton X-100 interferes so strongly with SDS (sodium dodecyl sulphate) that Triton-containing gels are unsuitable for SDS electrophoresis.

Gels which are to be dissolved for radioactivity measurements are made up with ethylene diacrylate in place of bisacrylamide (CHOULES and ZIMM, 1965). Ethylene diacrylate does not markedly affect the resolution of the gel. Gel slices are cut according to the stained protein bands and are transferred to counting vials which contain ca. 1 ml conc. ammonia. When the slices are completely dissolved, the ammonia is allowed to evaporate. In order to disperse remaining traces of water, 0.5–1 ml of absolute ethanol (according to the quantity of water) is added, and finally the vial is filled with scintillation solution.

Apart from solution D, all solutions can be stored at 0–4° C for 2–3 months. Gels prepared from the stock solutions listed above were introduced for the fractionation of brain proteins (NEUHOFF, 1968), but they can also be used for many other fractionations. Proteins with molecular weights between 700 000 and 20 000 can be fractionated on these gels; however, the optimal pH and gel concentration should be determined for each protein mixture. All substances used for preparing the gels should be of the purest quality and should be stored in brown bottles at 4° C.

It is not always necessary to work with freshly recrystallized acrylamide or bisacrylamide. However, if preparations have been stored for a long time, or for the electrophoresis of particularly "sensitive" proteins, it is advisable to re-crystallize the acrylamide or bisacrylamide. 70 g of acrylamide are dissolved in a litre of chloroform (analytical grade) at 50° C, filtered warm, and recrystallized at −20° C. The crystals are filtered off, washed with ice-cold chloroform, and dried in air. 10 g of bisacrylamide are dissolved in a litre of acetone (analytical grade) at 50° C, filtered hot, and allowed to crystallize slowly. The solution is first cooled to room temperature, then to 4° C, and left at −20° C. The crystals are filtered off, washed with ice-cold acetone, and dried in air.

To prepare 20% gels, 0.5 ml of solution A is mixed carefully with 1.5 ml of solution C. When mixing, care must be taken that no air bubbles enter the liquid. To 1 ml of this mixture, 1 ml of solution D is added. 25% gels are prepared by adding 100 mg acrylamide to 2 ml of a 20% gel mixture. Gel concentrations lower than 20% are obtainable by replacing a corresponding aliquot of solution C by water. The smaller the capillary diameter, the higher the polyacrylamide concentration needed in order to achieve a similarly good fractionation with the same protein mixture. For instance, if a 20% gel in 5 μl capillaries gives a good separation of a particular mixture, the polyacrylamide concentration must be raised to 25–30% for fractionation in 2 μl or 1 μl capillaries.

The sharpness of the protein bands can be improved by adding 10 mg hydantoin to 2 ml of the gel mixture (NEUHOFF, 1968; NEUHOFF and LEZIUS, 1967, 1968). This is due to the ionic nature of the compound (see Fig. 4); by adding hydantoin the total ion concentration of the gel mixture is increased. The glycine of the electrode buffer can be completely replaced by hydantoin; the hydantoin/H_2SO_4 buffer system gives the same separation of a protein mixture as the glycine/H_2SO_4 buffer system.

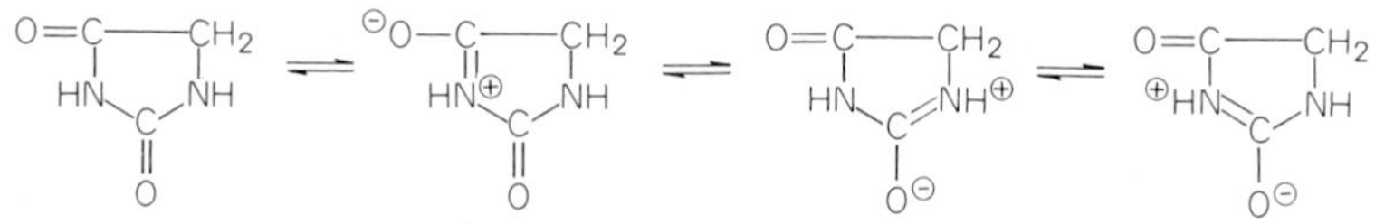

Fig. 4. Mesomeric forms of hydantoin

The capillaries are filled by capillary attraction to a half to two-thirds of their total volume by being dipped into the gel mixture. They are then pressed into a plasticine cushion approx. 2 mm thick, which effectively seals the bottom of the capillary. The cushion is made by smearing a small piece of plasticine on the bottom of a Petri dish (diameter ca. 3 cm). The capillary is then completely filled with water to the rim, using a fine pyrex glass capillary pipette. The diameter of the capillary pipette must be smaller than that of the gel capillary in order to allow air to escape during the filling procedure. The capillary pipette is attached to a rubber tube fitted at the other end with a plastic mouthpiece. Although the brown natural rubber tubes are less durable than silicone rubber tubes, they are more elastic and thus better suited for all types of micropipetting. The walls of the tubing should not be too thin, those with an outer diameter of 5.5 mm and a wall thickness of 1.5 mm being the best; the length of the tubing should be about 40 to 50 cm.

The introduction of the layer of water over the gel solution in the capillary must be carried out with the greatest care, since the quality of the subsequent electrophoresis is determined by the quality of the top of the gel. For this reason the experimenter should, if possible, sit in a relaxed position. Whether the elbows are loose and relaxed during manipulation of the capillary pipette, or whether they are supported by the bench-top, must be decided by personal experience. One hand holds the plasticine cushion in to which several gel capillaries have been inserted, the other holds the capillary pipette, which should have a capillary of 3 to 5 cm. The hands are then placed against each other, as shown in Fig. 5,

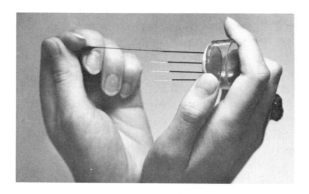

Fig. 5. Method used for filling the capillaries with water, sample etc.

and the capillary pipette, filled with water, is slowly and carefully introduced into the gel capillary. The experimenter should determine for himself whether this manipulation is more easily carried out against a bright or a dark background. A large magnifying lens may also be used. However, the use of a stereomicroscope is unnecessary; indeed, the relatively high magnification may be a hindrance.

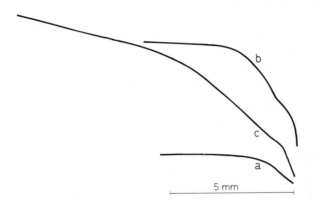

Fig. 6. Gel tip visualized by polymerisation of globulin with the gel, followed by staining with amido black and microdensitometry. The slope depends on the way in which the water is applied to the gel mixture in the capillary and can be illustrated by: *a* correct overlayering, *b* the upper 2–3 mm of the solution are mixed with water, *c* ca. 10 mm of the acrylamide solution are mixed with water

The capillary pipette is pushed down the wall of the gel capillary until the tip lies immediately above the gel solution, and the water is then allowed to flow slowly and evenly out of the capillary pipette. The pipette is withdrawn gradually as the tube fills with water. Under no circumstances should the capillary pipette be dipped into the gel solution, since the gel concentration at the interface will be altered on mixing with the water.

Fig. 6 shows a densitogram of protein at a gel surface after covering the gel with water correctly (a), and after mixing the top layer of the gel solution with water (b and c). The densitograms were recorded after staining with amido black, and the quality of the gel surface is indicated by mixing of a protein into the gel solution. If filling is carried out carefully, a short, steep gradient results, due to the inevitable diffusion between the gel phase and aqueous phase. This gradient is much less steep if the gel solution is diluted with water on filling the capillary incorrectly. The water layer should not contain buffer, because the ionic strength and the pH of the top of gel would be altered by diffusion, thus impairing the sharpness of the protein disc and the quality of the separation. In theory, the gel could be polymerized in the absence of the aqueous phase, but the extremely dense top of the gel layer formed under these conditions would lead to a poor separation of the proteins. Great care should be taken to avoid trapping air bubbles between the gel and the aqueous phase.

The gel mixture polymerises in the capillaries in ca. 45 min; the time of polymerisation depends on the concentrations of ammonium peroxydisulphate and TEMED. This redox catalytic system is superior to the use of riboflavin and UV for polymerisation (see MAURER, 1971) for the preparation of microgels. In microgels prepared by the riboflavin method, the sharpness of separation is generally poor. The polymerised gels must be allowed to stand for at least 10 hrs.

Fig. 7. Arrangement for storing the capillaries in a damp chamber

at room temperature before they are used; if the gels are used immediately after polymerisation the protein bands are poorly defined. Microgels prepared as described above can be used for up to two days after polymerisation without loss of quality. However, the tip of the gel should always covered with water; this can be achieved by storing them in a moist chamber as shown in Fig. 7.

Partial pre-fractionation may be necessary before applying the protein solution, as a concentrated protein solution may block the gel pores and allow only components with lower molecular weight to be separated in the gel. Partial fractionation can be effected by the use of Sephadex G-100 or G-200 superfine (HYDÉN et. al., 1966): The water over the gel is aspirated off, and replaced by buffer solution. 1–2 mm of a Sephadex suspension are then layered onto the gel column by means of a capillary. The Sephadex is sedimented onto the gel column by short centrifugation. One must accept that, under certain conditions, an undefined portion of the protein solution to be fractionated, penetrates into the gap between the glass wall of the capillary and the top of the gel. On subsequent staining, the proteins fractionated in the top of the gel are masked and can no longer be estimated by densitometry. It has been found that, by covering the tip of the gel with a 3–5 mm high layer of 25% saccharose solution in buffer (stock B, diluted 1 : 3), a discontinuous buffer system is produced which gives satisfactory fractionations.

Another way of circumventing the problems which arise on adding a concentrated protein solution to the gel is to cover the top of the gel with a layer, 5 mm thick, of a gel which polymerises rapidly, before adding the protein solution. A 5% gel seals off the capillary gap between the top of the gel and the glass wall, so ensuring a sharp buffer discontinuity and also giving some prefractionation of the protein. Such a gel layer also has the advantages of holding back some of the impurities in protein solutions, and of retarding proteins whose molecular weight is too high. Under some circumstances it is advisable to introduce a deeper layer (5–10 mm) of 5% gel over the main gel, particularly when dealing with concentrated protein solutions which contain a high proportion of impurities. All proteins which are fractionated in a 20% separation gel can pass through a 5% collection gel easily.

In practice, the 5% gel is introduced 1–2 hrs. before electrophoresis, and is layered over the 20% gel in the following way: 0.5 ml solution E and 1.5 ml solution C are mixed; 1 ml of this mixture is diluted with 1 ml water. 1 ml of this solution is mixed with 1 ml solution E, and 1 ml of this mixture is diluted with 0.8 ml water and 0.2 ml solution F. Polymerisation occurs within 5–10 min. The time for polymerisation can be prolonged by adding 0.9 ml, instead of 0.8 ml, water, and 0.1 ml instead of 0.2 ml solution F. The 5% gel solution should be applied immediately after the water covering the 20% gel has been removed. To do this, a capillary pipette is inserted into the 5 μl capillary to a point just over the gel, and the water is completely aspirated. Care should be taken not to damage the gel surface. The 5% mixture is now prepared and 5 mm of this solution is layered over the 20% gel. Once again, care must be taken not to damage the gel top, and not to allow air bubbles to form between the gel and the 5% gel solution. Finally, the 5% solution is covered with solution B, diluted 1 : 3. In this case, the extreme care necessary when covering the 20% separation gel is not required. Several capillaries can be covered with the 5% gel at the same time; the properties of the 5% gel do not deteriorate on storage for two days.

Electrophoresis

The optimum concentration of a solution for fractionation is 1–3 mg protein/ml. More concentrated solutions must be diluted, since the capacity of the gel for separation is easily exhausted by overloading. If the only solutions available for investigation are more dilute, so that the volume of solution which must be applied is greater than that above the gel column, a second or third

5 µl capillary can be attached to the column. To do this, a short piece of tightly fitting polyethylene tubing (inner diameter 1 mm) is pushed on to the gel capillary and the second 5 µl capillary pushed into it so that it sits directly on the lower capillary.

Normally, 0.1–1 µl of a protein solution, depending on the protein concentration, is used for electrophoresis (in a 5 µl capillary of 32 mm length, 1 mm corresponds to 0.156 µl). To load the column, the water above the 5% gel is removed, and the protein solution, free of air bubbles, is applied by means of a capillary pipette. Bubbles disturb the current considerably; it becomes uneven, subject to marked variation, and leads to a poor fractionation. Any free space in the capillary not filled by the protein solution, must be filled up with the buffer solution B, diluted 1 : 3.

After filling, the capillary is scratched with a diamond glass-writer at the level to which the plasticine penetrates the capillary, and this piece broken off. The bottom of the capillary must not be pulled, otherwise the whole gel can very easily be pulled out. If the capillary is broken so that the lower fragment is snapped away from the long upper end, a clean break across the gel is produced, resulting in an even current flow during electrophoresis. The capillary is then attached to a perforated rubber funnel cap, and the funnel is immediately filled with the electrode buffer. This must be carried out as quickly as possible, so that no buffer evaporates from the top of the capillary, and no air bubbles interrupt the flow of current at this site. The capillary with the funnel is then hung in the electrophoresis stand [2] (see Fig. 8) and the stand is adjusted so that the capillary dips into the lower electrode chamber. A small PETRI dish (diameter ca. 2 cm, height 2 cm) or a 5 ml beaker can serve as the lower chamber. The electrodes are made of platinum wire; the anode is connected to the lower vessel, containing buffer, and the cathode to the funnel of buffer above the gel.

An electronically controlled voltage generator (NEUHOFF, MÜHLBERG, MEIER, 1967), capable of supplying a constant potential and a steady current, is used [2]. The constant voltage is continuously adjustable between 0 and 450 V. The variation in the voltage applied should be less than $\pm 1\%$. The apparatus (see Fig. 8) is constructed so that a total current of 20 mA can be supplied, which is sufficient to carry out 10 or more separations simultaneously. The problem of the regulation and stability of very small currents in the milliamp range has been solved by the use of a noise diode switch; it is possible to regulate the current continuously, and to hold it constant over a range of 20 mA via a HELIPOT precision potentiometer. The voltage can be measured with the aid of a built-in very high resistance tube-voltmeter, so that conditions are obtained in which the current flowing through the capillary changes by a maximum of $\pm 2\%$ during measurement. This means that the current is not influenced appreciably by the measurement of the effective voltage, and is virtually independent of the resistance in the gel, which varies during electrophoresis. As well as monitoring the total current for 1 to 10 capillaries connected in parallel, the current flowing through each individual capillary can also be measured simultaneously. This is a great advantage for checking at the start of electrophoresis

---

[2] Obtainable from E. Schütt, Göttingen, Germany.

that the flow of current is even. It is thus possible to recognize and eliminate bad contacts occuring in the capillaries, for example, those due to air bubbles. A recorder socket allows continuous recording via a direct recorder, of all values measured. The polarity of all the electrophoresis columns can be simultaneously reversed by a switch on the voltage apparatus.

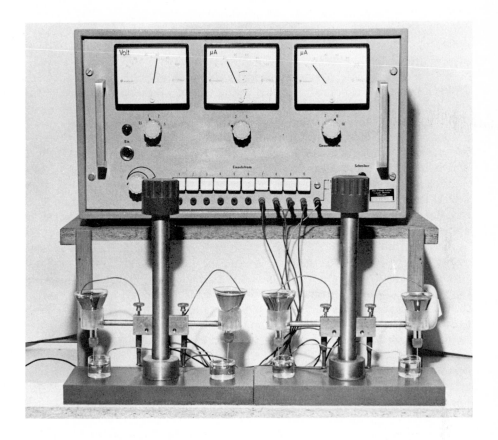

Fig. 8. Power supply and stands for capillary micro-electrophoresis

The constant voltage which needs to be applied depends on the acrylamide concentration in the gel, and on the type of protein in the mixture to be fractionated. As a rule, higher concentrations of gel require higher voltages for fractionation. 20% gels usually require 60–80 V. At the start of electrophoresis, 60–80 μA flows in the system described, then falls off slowly during the electrophoresis. The current is largely dependent on the ionic strength of the buffer employed; a current which is too low or unstable generally indicates that air bubbles are present in the capillary system.

If the voltage applied is too high, the proteins do not migrate evenly through the gel, owing to the formation of a temperature gradient across the gel. The heating in the middle of the gel results in an increase in protein mobility in this region. This may be recognized after staining, by protein bands which are convex in the direction of migration and are therefore unsuitable for microdensitometric estimation. In extreme cases application of higher voltages results in the heat denaturation of the protein, giving clearly visible brownish-coloured bands. If the voltage is correct, the narrow diameter of the gel is sufficient for dissipation of the heat generated, so that the individual capillaries do not need special cooling, nor need the electrophoresis be carried out in a cold room. When a separation has been carried out correctly, straight and undistorted bands of protein are normally obtained.

The progress of electrophoresis can be followed, on the voltage generator, for each individual capillary, or by following the migration of the proteins after staining with bromophenol blue. This can be achieved by mixing a small quantity of the dye with the electrode buffer; bromophenol blue binds tightly to the protein mixture during electrophoresis. At first, a somewhat diffuse zone of bromophenol blue is seen in the free space over the gel. This changes to a sharp blue band when the tip of the 5% collection gel is reached. The band migrates through the 5% gel and undergoes a further sharpening on reaching the separation gel. After the passage of the proteins into the separation gel the bromophenol blue bands spread out according to the protein positions. At the start of electrophoresis some sharp bands, coloured a pale blue, are easily recognizable. On further electrophoresis, the bromophenol blue is to a large extent detached from the bands, and migrates in the gel as a diffuse coloured cloud.

Bromophenol blue is not suitable as a marker for the buffer front, since it only migrates with the front at the commencement of electrophoresis, and is then left behind by it as the electrophoresis proceeds further. The buffer front can be indicated satisfactorily by using fluorescein, which can also be added to the electrode buffer in low concentration. It is recognizable as a distinct yellow band, clearly visible under UV light. This band migrates together with the bromophenol blue band at first, but leaves it behind as the electrophoresis proceeds. As a rule, electrophoresis is finished when the fluorescein band has migrated ca. 5–6 mm in the gel, a process which takes ca. 20–30 min at a potential of 60–80 V.

Immediately after electrophoresis the gels are expelled from the capillaries by water pressure, into a 1% amido black 10B solution in 7.5% acetic acid. These concentrations of acetic acid and dye have been found to be best for staining the protein bands. For expulsion of the gels a 2 ml syringe with a screw cone, to which a small tube (external diameter 1 mm) is fixed, is employed. The tip of the tube is blunted, and a short piece of tightly fitting polyethylene tubing is slipped over it (Fig. 9). The 5 μl capillary containing the gel is inserted into the empty end of this tubing, held firmly between thumb and index finger, and the gel is then pressed out. It is advisable not expell the gel too abruptly, but to regulate the pressure on the syringe plunger evenly.

Five minutes is generally sufficient for saturation of the gels and complete staining with amido black. If the staining time is curtailed, the protein bands are only stained at the periphery of the gel and cannot be estimated microdensitometrically. Staining for longer than 5 min does not result in any further

improvement in the colouration. The stained gels are then transferred to 7.5 %
acetic acid, and after about 30 min all the dye not bound to the proteins has
been washed out. By using this method for staining, the proteins are simultaneously
fixed and coloured. If the gels are first fixed in acetic acid, and then stained with
amido black, less colour is bound to the protein, so that bands which would
be only weakly stained may no longer be detectable.

Staining with Coomassie blue offers no decisive advantage over amido black
staining. Although there are some proteins which stain better with Coomassie
blue (0.1 % in 7.5 % acetic acid), in most cases staining with amido black is as

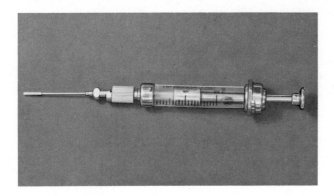

Fig. 9. Syringe with screw-cone for expelling gels from capillaries by water pressure

good and sometimes markedly better than with Coomassie blue. If gels are to
be stored, it is recommended to leave them in a 7.5 % acetic acid solution, which
contains sufficient amido black to give a weak blue colour, and to keep them in
the dark. If gels are left in acetic acid solution, slow leaching-out of the blue-
coloured protein bands takes place, and if exposed to light a brownish-black
discolouration of the proteins develops.

In principle quantitative dyeing is possible by this method so long as standardised conditions are maintained. It is important, however, that the dye-
binding capacity of the protein which it is intended to estimate be known. Since
different proteins have different dye-binding capacities, a separate standardisation
curve must be obtained for each individual protein before quantitative estimations can be made.

For manipulation of the gels, the forceps shown in Fig. 10, made from an
elastic spring steel, have been found to be the most suitable. The gels can be
damaged too easily if stronger forceps are used.

Gels can be photographed in a glass trough[3] filled with 7.5 % acetic acid
(Fig. 11a). The trough does not need to be covered with a sheet of glass. For routine evaluation, pherograms of the gels are recorded by means of a microdensitom-

---

[3] See footnote 2, p. 12.

eter. To do this the gels are transferred to a special glass chamber[4] (Fig. 11 b) whose 2 mm wide and 2 mm deep grooves are individually filled with 7.5% acetic acid solution. The floor of the trough is made of optical glass. After positioning the gels so that they lie as straight as possible in the middle of the trough, a cover glass is placed over it. Recording with the microdensitometer is generally performed using a gearing ratio of 1 : 20. The density of the grey wedge installed for the recording adjusts itself according to the colour intensity of the individual bands. Great care should be taken that the optical system of the densitometer is clean, and that the gel is well focussed in the light path. The Joyce Loebl double-beam microdensitometer has proved to be particularly suitable for evaluation of microgels; however, it is not possible to carry out estimations at several wavelengths using such an instrument.

$^1/_2$

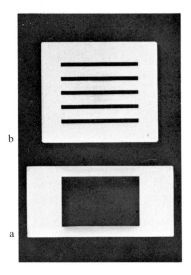

Fig. 10. Forceps made from elastic spring steel, used for handling microgels

b

a

Fig. 11a and b. Glass trough a, inner diameter 2.5×4cm, and glass chamber b for the recording of microgels

If proteins which are sensitive to oxidation, e.g. the brain-specific S 100 protein, are fractionated on gels made by the above method, artifacts may be formed during electrophoresis owing to excess ammonium disulphate, or potassium ferricyanate, or other side products of polymerisation reacting with the protein (ANSORG and NEUHOFF, 1971). This may be avoided by eliminating unused ammonium peroxydisulphate by a pre-electrophoresis run with gel buffer (stock solution A, diluted 1 : 8). In micro-disc electrophoresis, this pre-electrophoresis is performed before introducing the 5% gel, since this has a different pH to the 20% gel. For pre-electrophoresis, both electrode buffer vessels are filled with diluted stock solution A, and electrophoresis performed for one hour at about 60 V. The 5% gel is then introduced; however, since this also contains ammonium peroxydisulphate, very sensitive proteins may still be oxidized. A partial substitution of

---

[4] See footnote 2, p.12.

ammonium peroxydisulphate by sodium sulfite in the gel mixture is recommended by STEGEMANN (1967). The likelihood that a protein will be oxidised is greatest for those with low molecular weights, e.g. the S100 protein, which migrate close to the buffer front, since the persulphate also migrates here on electrophoresis. This is avoided by the electrophoresis of proteins on acrylamide gradient gels, since no 5% collecting gel is necessary and all interfering gel constituents can be removed completely by pre-electrophoresis.

For some investigations it may be necessary to determine the R_f value of a protein band in the gel, so that comparison with other fractionations may be made. If the separation is short, direct measurements of the distance of migration is practically impossible. To determine the R_f value, the gels are transferred to a 1% pyronine solution in 7.5% acetic acid immediately after electrophoresis. Pyronine forms a complex with the fluorescein band, which is preserved during the subsequent amido black staining (NEUHOFF, SCHILL und STERNBACH, 1970). This coloured complex is recorded as a peak during densitometry and can be used as a reference value for the determination of the R_f on the pherograms. This procedure has the disadvantage that the subsequent amido black staining is not so effective, owing to the denaturation of the proteins during the pyronine staining with acetic acid and rinsing out of uncomplexed pyronine. To avoid this, NOVOTNY (1972) did not add fluorescein to the electrophoresis buffer, but layered a small quantity of the fluorescein stock solution directly onto the 5% gel, and then applied the protein solution. By doing this, a sharp and intense fluorescein band is obtained at the buffer front. The fluorescein concentration at the front is so high that not all the fluorescein is washed out during amido black staining and subsequent rinsing in acetic acid. During the final incubation in the pyronine solution, the fluorescein-pyronine complex is formed. R_f-value determinations by this method are not suitable for very complex protein mixtures, but they may be used for the characterization of protein mixtures containing only a few components.

With care and practice, the reproducibility of micro-disc electrophoresis is astonishingly good. For the repeated fractionation of a standard albumin solution, reproducibility of ± 2–3% on gels of the same batch is attainable without further precautions. Even for a mixture of proteins as complicated as the water-soluble brain proteins, reproducibility of ± 5% is still attainable when they are estimated densitometrically and planimetrically. It is important to be extremely careful when covering the gels with water, and also to use the volumes of the protein solutions which have been indicated in this text.

The following procedure is recommended: After layering the 5% gel on the separating gel, the whole of the free space in the capillary is filled with buffer (stock solution B, diluted 1:3). A known volume of buffer is removed, and the capillary is filled to the tip with protein solution. The volume of buffer removed can be determined in millimetres, and can calculated from the length of the capillary and the total volume. However, it is difficult to avoid errors when estimating the volume in this way, particularly when the same volume of buffer must be removed and replaced by protein solution in a whole series of capillaries. For such experiments, the most satisfactory capillary pipettes are those shown in Fig. 12, which possess a distance marker in the form of a polyethylene tube or a piece of a 5 μl capillary fixed to the capillary pipette with Harvard glass wax [5]. This ensures that the capillary pipette can never be introduced further into the 5 μl capillary than the beginning of the distance marker, so that identical volumes of buffer can always be removed.

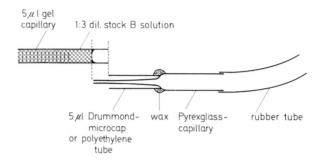

Fig. 12. Special capillary pipette for the application of defined volumes reproducibly

Isolation of Individual Protein Fractions

If individual proteins are to be eluted from the gel after fractionation, electrophoresis in a special elution capillary is generally used (Fig. 13). A 5 μl capillary is filled to a length of about 1 cm with the normal 20% gel solution; when covering the gel solution with water, it is not necessary to take the extreme care which is necessary for the production of good separating gels. After the polymerisation reaction, the water is aspirated off, and a short piece (ca. 5–7 mm) of capillary tubing, with inner diameter somewhat larger than the outer diameter of the 5 μl capillary, is fixed onto the upper end of the 5 μl capillary with Harvard glass wax [5]. The two capillaries are pushed together to give about 2 mm of overlap, and the space between the capillaries is filled with glass wax. The space in the capillaries is filled with solution B diluted 1:3. In order to obtain a more even flow of current and better elution of the proteins, it is recommended that sucrose, glycerol, or both, be added to the buffer, to a final concentration of 20% and 10% respectively.

So as to be able to elute undenatured protein from the gel, staining is carried out for a short time (ca. 1 min) with buffered 0.1% bromophenol blue solution at room temperature, or, if necessary, at 0° C. This staining is of course very weak, and only relatively intense protein bands can be located by this method. The stained gel is transferred to a drop of buffer on a paraffin-coated microscope slide, and the relevant bands are observed under a stereo microscope and excised with a razor blade. To do this, the gel is held with soft tweezers; after some practice, really thin slices of gel can be cut reliably. One or more gel slices from several experiments can then be transferred with the soft tweezers to the receiver capillary which has been fused onto the 5 μl capillary. After the elution system has been

[5] Richier a. Hoffmann, Harvard Dental Ges., W.-Berlin, Germany.

filled carefully to the top with buffer again, the capillary is inserted into the perforated rubber cap of the funnel. Care must be taken not to break off the elution capillary fused onto the electrophoresis capillary. After placing the capillaries in the electrophoresis stands and filling them with electrode buffer, electrophoresis

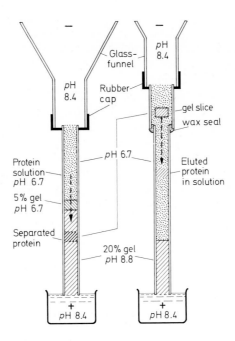

Fig. 13. Arrangement for the electrophoretic elution of proteins from isolated gel discs. (From NEUHOFF and SCHILL, 1968)

is carried out as described previously. It is advisable not to apply too high a voltage for elution. After electrophoresis has proceeded for a while, the eluted protein appears as a somewhat diffuse zone, coloured by bromophenol blue, in the free capillary space, above the 20% gel which has been polymerised so as to seal the capillary. When the protein is completely in the capillary space, the electrophoresis is stopped and the capillary is scratched with a glass writer immediately below the receiver capillary; the upper piece of the capillary is then broken off. To check for complete elution, the eluted slices of the gel can be restained with amido black 10 B. The eluted protein can be removed from the free capillary space with a fine capillary; since this protein has been in the electrode buffer solution which contains glycine, it may be necessary to purify it first by microdialysis (see chapter 12) before further analysis.

If individual protein bands can be eluted from the gel after complete amido black staining and fixing in acetic acid (e.g. for micro immune reactions (see chapter 4), or for radioactivity measurements), cutting up the gel under the stereomicroscope is simplified because the bands can be stained intensely. Before the slices of gel are transferred to the receiver capillary for elution, the acetic acid must be removed by brief washing (1–2 min) in physiological saline or in a suitable buffer. It is particularly important that the wash should not take too long when slices contain proteins of low molecular weight, so that as little material as possible is lost by diffusion. During washing, the amido black stays firmly bound to the protein. Elution is then carried out as described above. Because of the high pH of the electrode buffer, the amido black is washed off the protein, and appears as a blue cloud in the free capillary space. Then the protein is eluted from the gel slices, and can be visualized by its colouration with bromophenol blue. When this zone appears in the capillary the electrophoresis can be stopped and the protein may be removed from the capillary.

This method for electrophoretic elution is only effective with proteins up to a molecular weight of ca. 100000. Larger proteins are only incompletely eluted by this procedure; in order to elute such proteins, the gel slices are transferred to a beaker (5 ml or 10 ml) containing pure glycerine, and are then frozen at $-196°$ C. The glycerine infiltrates the small slices of gel relatively quickly and protects the proteins contained in them. The inner structure of the gel is broken to such an extent on freezing, that proteins of molecular weight larger than 100000 can be eluted. After the glycerol has frozen completely, it is thawed out at room temperature, the gel slices are removed, transferred to the elution capillary and eluted as described. The glycerine adhering to the slices has no adverse effect on elution. Subsequent staining of the slices after elution demonstrates that it is not always possible to obtain complete elution, even in this way. Experiments with RNA-polymerase from *E. coli* (see p. 28) show that the enzyme does not lose its enzymatic activity during elution by this method.

The same system of elution can be used to advantage if an isolated protein band is to be checked by a second electrophoresis, e.g. for its homogeneity. A 5 µl capillary filled with the appropriate separating and collecting gels is fitted with a short piece of receiver capillary as previously described. The isolated gel slice is placed in the capillary, and the second electrophoresis can be carried out in the appropriate gel immediately after the elution of the protein. At the end of the second electrophoresis, the protein band may be stained with amido black, or be re-eluted and analysed further.

During micro-crystallization, when the concentrations of aqueous solvents or solvent mixtures inside the capillary have to be changed slowly, a small gel layer as used in the elution procedure has been found to be very useful. For this, a 2 to 4 mm layer of a polyacrylamide gel of suitable concentration is used as a "dialysis membrane". The capillary containing the suspension for crystallisation is placed in a beaker filled with the solvent, which soaks into the capillary. It is sometimes necessary to remove unpolymerised material from the gel by an initial dialysis with water or a suitable buffer.

Microanalysis of DNA-Dependent DNA-Polymerase

The enzymatically active components of a preparation of DNA polymerase can be characterised by the use of two basically different methods. These are the *binding test*, which gives an indirect characterisation, and the *polymerisation test* which gives direct characterisation (NEUHOFF and LEZIUS, 1967, 1968).

Binding Test

For the binding test two parallel runs are made. One gel capillary is loaded only with the enzyme solution, while a second identical gel capillary is loaded with the same quantity of enzyme solution (Fig. 14) and in addition poly d(A-T) or DNA as primer, the substrates dATP and dTTP, as well as Mg^{++} and phosphate buffer. The capillaries are allowed to stand for 10 min at room temperature before the electrophoresis is started. For the electrophoresis of DNA polymerase from *Escherichia coli* (EC 2.7.7.7.) the electrophoresis system already described may be used (20 or 25 % separating gels with $Tris/H_2SO_4$, pH 8.8, 5 % upper gels with $Tris/H_3PO_4$, pH 6.7, and Tris/glycine, pH 8.4 as the electrode buffer). A distinct sharpening of the protein bands is obtained if the gel mixture contains hydantoin. The electrophoresis is carried out at room temperature with a constant potential of 60 V until the fluorescein front has migrated ca. 5 mm into the gel. Since the enzymatically-active protein components form a stable complex with the primer and substrates, the molecular weight of the complex is so high that the

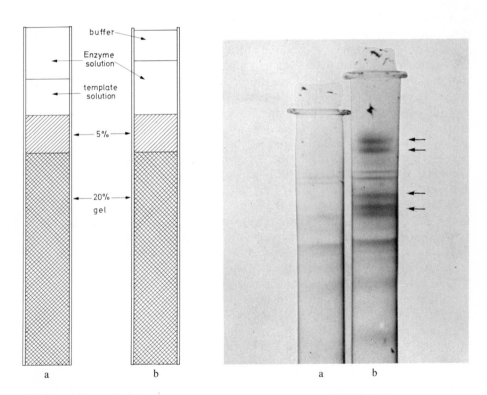

Fig. 14a and b. Binding test for DNA-dependent DNA- or RNA-polymerase. Left: scheme of the procedure, right: result of an experiment with DNA-polymerase. The points marked with an arrow in the control gel *b* are polymerases which are missing after fractionation of the same amount of enzyme through an layer of template *a*. (From NEUHOFF and LEZIUS, 1968)

components scarcely migrate into the gel, or do not penetrate the surface of the separating gel. On staining with amido black, the active protein components of DNA polymerase are either not detected in the gel or are found at a position near the cathode. By comparing such an analysis with one in which only the enzyme preparation was present, an unambigous differentiation between the enzymatically active and the inactive protein components of the fractionated solution can be made (see Fig. 14 a and b).

Using the same method, a series of binding tests can be carried out to test the influence of individual components, e.g. the effect of different substrates or ions on the formation of the enzyme-template complex (NEUHOFF and LEZIUS, 1968). To do this, the components are added singly and successively to the template solution on top of the gel, one gel capillary being used for each. Protein staining is carried out after the electrophoresis, the absence of one or more bands is noted, and the effect due to the presence or absence of a component on the binding of the primer can be determined.

Polymerisation Test

The polymerisation test is a direct method for characterising the polymerase, using the inclusion polymerisation step suggested by SIEPMANN and STEGEMANN (1967). 10–30 µl of the primer, poly d(A-T) or DNA (ca. 20 E_{260} units/ml), are added to 0.5 ml of a 20 or 25% gel mixture. The solution is mixed well, and the 5 µl capillaries are filled by the method already described. In this way the same concentration of primer is present throughout the polymerised gel. The 5% collecting gel is added, and the preparation of DNA polymerase is fractionated in the "primer" gel as before. The poly d(A-T) in the gel approximately doubles the time needed for separation, but does not affect the separation of the proteins as long as the concentration of primer in the gel is not too high. The primer remains evenly distributed in the gel during electrophoresis, and complex formation between the primer in the gel and the polymerase migrating through it does not occur.

If the concentration of primer in the gel is correct, staining with amido black shows that the proteins in the primer gel are fractioned in exactly the same way as when primer is absent. If the concentration of primer in the gel is too high, the fractionation of the proteins can be adversely affected, and the separation between individual components is decreased.

To obtain direct measurement of the polymerase activity, the gels are removed from the capillaries by water pressure, and then placed in a mixture of the substrates dATP and dTTP, Mg^{++}, and phosphate buffer (pH 7.4) and incubated at 37° C. This induces extensive synthesis of poly d(A-T). There is enough primer in the gel to form a complex with the protein which has polymerase activity, and the substrates can diffuse into the gel easily. After various intervals of time, a gel is removed and transferred to a solution of 1% pyronine in 7.5% acetic acid. After 5 min staining, it is cleared with 7.5% acetic acid. The newly-formed poly d(A-T) is seen as bright red bands of the pyronine complex at the positions of enzymatically active protein components of the polymerase preparation (see Fig. 15). Pyronine is preferable to toluidine blue for staining, since the former yields a stable complex which may be kept for a long time, whereas the coloured complex with toluidine blue disappears completely within a few hours. The pyronine staining is specific for the newly-formed DNA polymer, as can be shown by the simultaneous staining of a poly d(A-T) gel through which a similar preparation of polymerase has been fractionated. When staining with pyronine is carried out directly after the electrophoresis, only a homogeneous faint red background is detectable which colours the whole gel. This is due to the primer originally polymerised into the gel; apart from this, no other pyronine-staining fraction is seen.

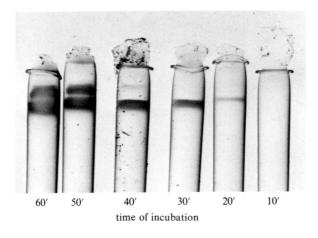

60'   50'   40'   30'   20'   10'

time of incubation

Fig. 15. Polymerisation test with DNA-dependent DNA-polymerase from *E. coli*. Staining with pyronine after incubation in the substrates for different time intervals of gels containing the template. (From NEUHOFF and LEZIUS, 1968.) Magnification 17 ×

The six gels shown in Fig. 15 are all derived from one batch of "primer gels" through which an identical quantity (0.6 µg protein) of a preparation of DNA-polymerase was fractionated simultaneously. All six gels were then incubated in the mixture of substrates, and gels were withdrawn at intervals of 10 min and stained with pyronine. The time dependence of the enzymatic synthesis of poly d(A-T) in the gel is clearly seen. No polymeric product is visible after 10 min incubation in the substrate mixture. After 20 min, the first band appears, and becomes more intense with time. After 40 min incubation, a new zone of polymer appears, which increases appreciably faster than the product formed earlier. Fig. 16 shows pherograms of these six gels as recorded by a Joyce-Loebl double-beam microdensitometer with an enlargement of 1:100 and a grey wedge of final density 1.6 D. In Fig. 17 the closed areas obtained planimetrically from these curves have been plotted against time. The pattern of the curves indicates a time-dependent enzymatic synthesis with a lag phase of 10 to 30 min.

For the determination of DNA-polymerase, a quantity of primer, ca. $2 \times 10^{-6} E_{260}$ units of poly d(A-T), at the site of an enzymatically active component in the gel (which itself can represent less than 10^{-8} g protein) is sufficient for the specific synthesis on incubation in the presence of the appropriate substrates. This simple method thus gives the same degree of sensitivity as the determination of activity using radioactive substrates. Determination using micro-disc electrophoresis on primer gels is superior to the normal radioactive procedure, however, in that it yields far more information. The isotope technique of activity determination gives information about the total enzymatic activity, but does not indicate how many components in the protein mixture are responsible

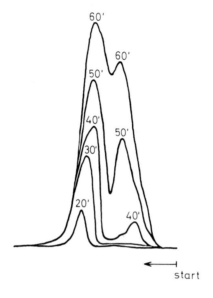

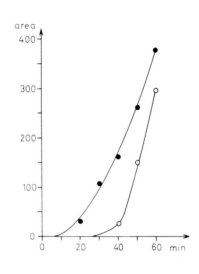

Fig. 16. Pherograms of the gels shown in Fig. 15

Fig. 17. The areas obtained planimetrically from the curves in Fig. 16 plotted against the incubation time

for this activity. With the micro-disc electrophoresis it is possible to determine in a single run not only the enzymatic activity, but also the specific activity of the individual components providing the procedure is carried out under standardised conditions. NEUHOFF, SCHILL and JACHERTS (1970) used the poly-merisation test to demonstration an RNA-dependent RNA-replicase from immunologically competent cells, using micro-disc electrophoresis. With RNA-polymerase, for example, there are also other means of demonstrating polymerase activity in a gel.

Microanalysis of DNA-Dependent RNA-Polymerase

It is well known that DNA-dependent RNA-polymerase is composed of 4 subunits ($2\alpha,\beta,\beta'$; Review, see BURGESS, 1971). In addition there are a series of factors which regulate transcription. Only the σ-factor, which is responsible for the specific initiation of transcription, has such a high affinity for the "Core" enzyme $(2\alpha,\beta,\beta')$ that it can be isolated with it during the purification of RNA-polymerase. On isolation of RNA-polymerase, a mixture of "Core" enzyme and "Holo" enzyme (enzyme$+\sigma$-factor) is always obtained, and the experiments which follows were performed on this mixture.

It has been shown that the subunits of the enzyme are capable of forming aggregates with different molecular weights ($\alpha\beta\sigma$, $\alpha\beta'$, σ, $\beta2\alpha$). The formation of the various aggregates depends predominantly on the ionic strength, but also on the pH and the degree of ageing of the enzyme. Normally the holoenzyme migrates with $S_{20\,w}=14.9$, and the core enzyme with $S_{20\,w}=12.5$. The holo-enzyme is capable of forming dimers ($S_{20\,w}=23$) while the core enzyme itself can aggregate to form tetramers and hexamers ($S_{20\,w}=44-48$). Concerning the function of the individual subunits, it is known that β' is responsible for binding to the DNA, and that the antibiotic rifampicin binds specifically only to σ (BURGESS, 1969, 1971; BURGESS and TRAVERS, 1970). A simple micro diffusion procedure for studying the binding of rifampicin or heparin, and substrates to the polymerase-template complex is described in chapter 4. LOUIS and FITT (1972) used micro disc electrophoresis for the analysis of subunits α and β of *Halobacterium cutirubrum* DNA-dependent RNA-polymerase.

When micro-disc electrophoresis on 20% gels was applied to pure RNA polymerase from *E. coli* (EC 2.7.7.6.) [prepared by the method of ZILLIG, FUCHS and MILETTE (1966), and which is thought to represent a mixture of core- and holoenzyme], it was established that 3 protein bands, sharply defined but lying quite close to one another, were poorly fractionated in the top of the gel. A good fractionation may be attained in 8% polyacrylamide gels with Tris/acetate buffer (NEUHOFF, SCHILL and STERNBACH, 1968, 1969a, b, 1970). In this case, the stock solution A for preparing gels is composed as follows: 10 ml buffer solution containing 0.01 M Tris, 0.01 M $MgCl_2$, 0.02 M NH_4Cl and 0.063 ml TEMED$+$ acetic acid to pH 7.4. The other stock solutions are as previously given. For the 8% gel, 0.5 ml of stock solution A $+0.6$ ml stock solution C $+0.9$ ml water are mixed thoroughly. 1 ml of this mixture is mixed with 1 ml of stock solution D. 10 mg of hydantoin, which is a significant help towards sharpening of the bands, is dissolved in this mixture. The 5 µl capillaries are filled to about $\frac{2}{3}$ full by capillary attraction, and covered with water. The further treatment of the gels and the introduction of the 5% gel is carried out as already described. The Triton X-100 contained in the gel does not affect the separation of the polymerase particles, as can be shown by using Triton-free gels.

The electrophoreses are begun at room temperature, with the usual electrode buffer (Tris/glycine pH 8.4) and a constant potential of 80 V, until the protein front, recognizable by the band of bromophenol blue, has reached the end of the 5% collecting gel. At this point the current, which was 90 to 110 µA, decreases to 40–50 µA. The potential must now be reduced to 10 V (in the normal technique of micro-disc electrophoresis the fractionation is carried out with constant voltage throughout). The current is about 5 µA, and decreases to 1–2 µA by the end of the electrophoresis. If the fractionation in the separating gel is carried out with higher voltages, the protein bands in the gel become distorted and bowl-shaped. During the fractionation, the protein bands can be recognized easily on inspection with a magnifying lens placed in front of a source of diffuse light; this is due to refraction of light, due to the high molecular weight of the separated material. The electrophoresis is complete when the fluorescein band has migrated 5–6 mm into the gel.

After the usual amido black staining, three protein bands are seen, clearly separated from one another (see Fig. 18, gel 1). To demonstrate that all three protein bands possess polymerase activity, the binding test, as described for DNA-polymerase, is carried out. All three protein species are able to complex with the template poly d(A-T) or with DNA (gel 2 in Fig. 18). No other protein

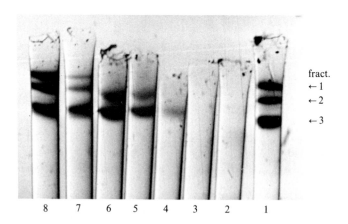

fract.
← 1
← 2

← 3

8   7   6   5   4   3   2   1

Fig. 18. Binding test of DNA-dependent RNA polymerase from *E. coli.* Staining with amido black. Gel 1: normal fractionation. Gel 2: Enzyme + poly d(A-T), Gel 3: Enzyme + poly d(A-T) + substrate, Gel 4–8: Enzyme + poly d(A-T) at a dilution of 1:5, 1:10, 1:50, 1:100, 1:1000. (From NEUHOFF, SCHILL and STERNBACH, 1968.) Magnification 14×

is detected in the gel on staining with amido black, thus indicating purity of the preparation of RNA-polymerase used. The three protein species possess the same affinity for the template in the presence of substrate (gel 3 in Fig. 18). A series of experiments in which a constant quantity of polymerase is fractionated in gels, which have previously been overlayered using identical volumes but different dilutions of poly d(A-T), show that the three protein species have different affinities for the template (gels 4–8 in Fig. 18). The microdensitometer curves of these gels are shown in Fig. 19. Using the enzyme (0.5 µg protein) and a concentrated poly d(A-T) mixture (0.3 µl of a solution of 24 E_{260} units/ml), no components of the enzyme are detectable in the gel after electrophoresis (gel 2 and 3). With increasing dilutions of the primer (1:5, 1:10, 1:50, 1:100, 1:1000) in the reaction mixture (gels 4–8 in Fig. 18), the three components appear again in the order of their molecular weights.

Fig. 20 illustrates the enzymic activity of all three protein species after fractionation on a "primer gel". After incubation for seven minutes at 37° C in the substrate mixture (0.03 M Tris acetate pH 7.9, 0.03 M magnesium acetate, 0.001 M ATP and 0.001 M UTP) the products of synthesis can be visualized by

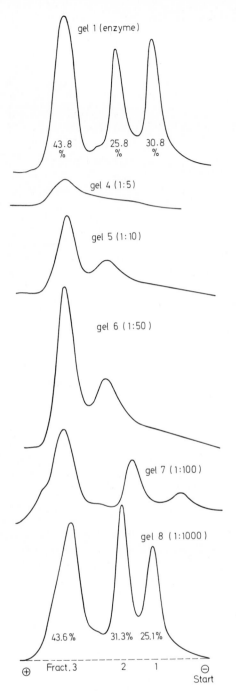

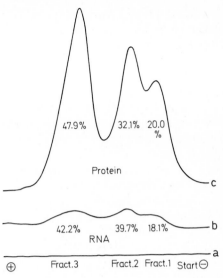

Fig. 20. Polymerisation test with DNA-dependent RNA polymerase from *E. coli*. Curve *a* after fractionation of the enzyme without incubation, followed immediately by staining with pyronine (control experiment for *b* and *c*). Curve *b* after fractionation of the enzyme, the gel was incubated for 7 min at 37° C in the substrate mixture and then stained with pyronine to detect the RNA synthesised. Curve *c* after staining as described for *b*, the same gel was stained with amido black to detect the protein bands. (From NEUHOFF, SCHILL and STERNBACH, 1968)

Fig. 19. Microdensitometer curves from the gels shown in Fig. 18

pyronine staining (curve b). If the same gel is stained subsequently with amido black, it is seen that the protein bands and the synthetic products occupy identical positions in the gel. After fractionation of the enzyme in a primer gel, without incubation in the substrate mixture and immediate pyronine staining (control experiment to b and c), no pyronine-positive band can be demonstrated (curve a).

Enzymatic polymerase activity can be assessed without pyronine staining if the enzyme preparation is fractionated on the primer gel and then incubated in a mixture containing ^{14}C-labelled substrates. Staining is then performed with amido black 10B for 5 min and all the radioactive substrates not incorporated into the synthetic product are removed by prolonged washing in 7.5% acetic acid. The acid should be changed frequently during the washing. Afterwards the gel slices corresponding to the protein bands can be cut out under the stereo-microscope and their radioactivity measured.

For radioactive labelling, it is advisable to use gels prepared with ethylene diacrylate as the cross-linking reagent, since these can be dissolved in ammonia. A piece of gel of identical size from a protein-free section of the gel is used as a blank. If the gel slices are counted directly in a toluene-based scintillation fluid without being dissolved, the counting yield is reduced considerably due to the shrinkage of the gel in toluene.

It is recommended that the products of synthesis be hydrolysed by incubation with a small volume (50–100 μl) of a solution of RNase or DNase if undissolved gel slices must be used for counting; 1 ml of absolute ethanol is added, and the volume made up with scintillator toluene, Aquasol (NEN), Bray's solution, or some other suitable scintillation medium. If radioactively-labelled proteins are to be estimated, the gel slices should be incubated in a solution of pronase (1 mg/ml in 10 mM Tris HCl pH 8).

Besides these methods, there are two other procedures for testing the enzymic activity of the individual polymerase species. In one, micro-disc electro-phoresis is carried out on normal 8% gels and the staining with bromophenol blue (0.01% in Tris acetate buffer pH 7.4) is performed at 0°C. The gel slices corresponding to the protein bands are cut out under the stereo microscope, and transferred directly to the polymerase test mixture with ^{14}C-labelled substrates as used in the procedure outlined above. After incubation, the mixture is con-centrated by the standard filter disc method commonly used in macroanalysis (ZILLIG, FUCHS and MILETTE, 1966).

The other method for the determination of enzymic activity is to elute the individual bands after micro-disc electrophoresis on normal gels and to visualise the protein bands with bromophenol blue. The proteins can then be removed from the capillary and processed as in the normal macro test. Table 1 shows that the enzymic activity of the RNase polymerase species is not affected on freezing; there is no difference in enzymic activity between frozen and non-frozen gel slices when their polymerase activity is measured directly. In contrast, the enzymic activity is appreciably higher if the enzyme is eluted from the slices electrophoretically and then used in the test.

If the three protein species of the polymerase are eluted electrophoretically and then fractionated on an 8% gel immediately afterwards in order to check

Table 1. Polymerase activity of the particles of DNA-dependent RNA polymerase from E. coli after electrophoretic fractionation. (From NEUHOFF, SCHILL and STERNBACH, 1969.) The protein-containing, isolated gel slices or alternatively the proteins eluted from them were added directly to the following test mixture: Total volume 250 µl with 0.03 M Tris/acetate, pH 7.9; 0.03 M magnesium acetate; 0.022 M ammonium chloride; 0.001 M ^{14}C ATP; 0.001 M ^{14}C UTP (specific activity 0.5 C/mole); 0.3 A_{260}-units poly d(A-T); 0.4 A_{260}-units calf thymus DNA as template (0.4 A_{260}-units), all four triphosphates (ATP, UTP, GTP, CTP, specific activity of each 0.5 C/mole) were added in 0.001 M concentration

Test material	Template	Fraction 1		Fraction 2		Fraction 3	
		cpm found	cpm found per gel slice	cpm found	cpm found per gel slice	cpm found	cpm found per gel slice
3 gel slices of each fraction directly in test mixture	p-d(A-T)	11038	3679	19376	6459	31228	10409
3 gel slices of each fraction directly in test mixture	DNA[a]	1759	586	2285	762	3843	1281
3 slices of each fraction, −170° C, then directly in test mixture	p-d(A-T)	13363	4454	18822	6274	22123	7374
2 slices of each fraction, −170° C, eluted and eluate in test mixture	p-d(A-T)	20224	10112	16732	8366	40320	20160
2 · 3 slices fraction 1, −170° C, eluted, refractionated, new bands in test mixture	p-d(A-T)	7476	1246	4683	781	1415	236
3 · 3 slices fraction 2, −170° C, eluted, refractionated, new bands in test mixture	p-d(A-T)	5989	665	9288	1032	4034	448
2 · 3 slides fraction 3, −170° C, eluted, refractionated, new bands in test mixture	p-d(A-T)	3622	604	6045	1008	3744 / 1147[b]	624 / 191
Gel slice without protein as control (mean of 4 estimations)	p-d(A-T)	206 ± 30					

[a] Calf-thymus DNA.
[b] This particle migrates faster than fraction 3 and is identical with a peak obtained after fractionation through a layer of NaCl or urea (compare Figs. 22–24).

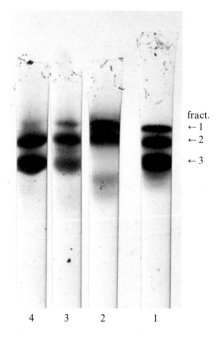

fract.
← 1
← 2

← 3

4   3   2   1

Fig. 21. Microphoto (magnification 16 ×) after staining with amido black. Gel 1: normal fractionation of 0.6 µl polymerase (1 mg protein/ml). Gel 2: the gel slices corresponding to the first fraction from three normal fractionations were cut out after bromophenol blue staining, frozen in glycerol at − 196° C for 10 min, then eluted electrophoretically and refractionated immediately on an 8 % gel. Gel 3: two gel slices containing fraction two were treated as described for gel 2. Gel 4: two gel slices containing fraction 3 were treated as described for gel 2. (From NEUHOFF, SCHILL and STERNBACH, 1969 b)

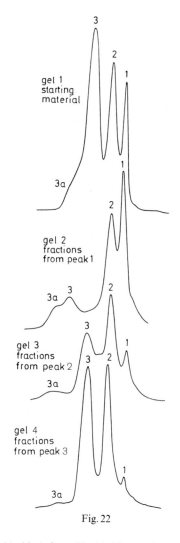

gel 1
starting
material

gel 2
fractions
from peak 1

gel 3
fractions
from peak 2

gel 4
fractions
from peak 3

Fig. 22

Fig. 22. Microdensitometer curves of the gels stained with amido black from Fig. 21. The numbers over the maxima indicate the numbers of the protein fractions

their homogeneity, it is seen that these components are interdependent (see Figs. 21 and 22). The first gel in Fig. 21 shows a normal fractionation. In gels 2, 3 and 4, two and three isolated gel slices of fractions 1, 2 and 3 were eluted electrophoretically and then (without modifying the eluted proteins) fractionated a second time in a closed capillary system containing an 8 % gel of identical composition to that used in the first electrophoresis. In these three gels all three enzyme components can be visualized by amido black staining. Fig. 22 shows the microdensitometric curves for these gels, from which the separation of the three fractions is clear. Furthermore, it can be seen from Figs. 21 and 22 that with this method of elution the components are not present in the same relative distribution as in the first fractionation. It may be seen from Table 1 that each of the three protein fractions derived from a single isolated fraction also possesses enzymatic activity.

The fact that the formation of the various aggregates of the polymerase depends on the ionic strength can be easily demonstrated by micro-disc electrophoresis. To do this, the 5% collecting gel, above an 8% separating gel in the capillary, is covered with a 3–6 mm layer of a solution of salt of the desired molarity, and the preparation of polymerase is applied on top of the salt. The polymerase passes through the salt solution before being fractionated in the gel, and aggregation due to the ionic strength is shown up in the gel on amido black staining. Fig. 23 shows the effect of different molarities of NaCl on the fractionation of a constant quantity of enzyme (0.5 μg protein); the 5% gels in a series of capillaries were covered with 0.9 μl (corresponding to 6 mm in a 5 μl capillary) NaCl of increasing concentration (0.1, 0.2, 0.4, 0.6 and 0.8 M) before the enzyme solution was applied. Fig. 23 shows that the protein which migrates most slowly, and thus has the highest molecular weight, diminishes on passage through the salt layer at a rate which is proportional to the concentration of NaCl. Simultaneously the quantity of the protein component with the next lowest molecular weight increases. With the high concentration of 0.8 M NaCl there is a marked increase in the protein with the fastest rate of migration, hence with the lowest molecular weight.

Similar results are also obtained if the enzyme is mixed into the NaCl solution, or if NH_4Cl is used instead of NaCl. However, under these conditions the variability in the results is considerably greater, and appreciably more protein is necessary if the polymerase is mixed with the salt solution in a test tube first.

Fig. 24 shows the effect of urea on the aggregation of the polymerase. Layers of about 6 mm (0.9 μl) of solutions of 1, 2, 4, and 6 M urea are applied to the tops of 5% collecting gels in different capillaries, and 0.4–0.5 μl of enzyme solution (corresponding to 0.4–0.5 μg protein) is layered carefully over each. Fig. 24 shows the microdensitometer curves for these gels after staining with amido black. It can be seen that by increasing the concentration of urea, the formation of the protein particle of lowest molecular weight is favoured, a result which differs significantly from those obtained when NaCl was used. The presence of a protein which migrates still further into the gel, and hence which has a lower molecular weight can be demonstrated. After migration through the 6 M urea layer, increasing the molarity of the urea even slightly above 6 M leads to complete destruction of the enzyme, and after electrophoresis no fractions can be recognized clearly.

Micro-disc electrophoresis of RNA polymerase on 8% gels which contain 6 M urea only in the gel is less satisfactory, because the experimental conditions are less defined. If the enzyme has to pass through a layer of urea before fractionation, there is sufficient time for it to form the various components while still in the aqueous phase, and these are then fractionated according to their shape and size. If the enzyme comes into contact with urea only when in the gel, the decomposition of the enzyme takes place at the same time as it is migrating through the stationary phase, and the results obtained are substantially more difficult to interpret, as shown by Fig. 25.

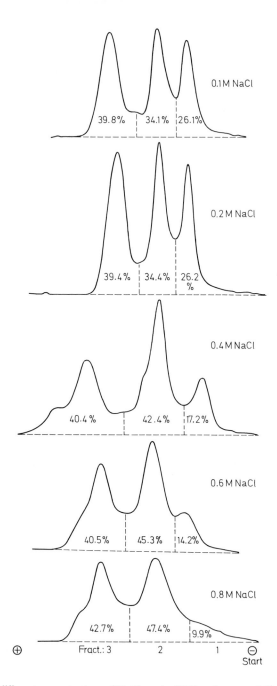

Fig. 23. Effect of different concentrations of NaCl on the RNA polymerase. Microdensitometer curves after fractionation on 8% gels and staining with amido black. Before being fractionated in the gel the enzyme migrates through a layer of NaCl solution, as described for the polymerase binding test. (From NEUHOFF, SCHILL and STERNBACH, 1968)

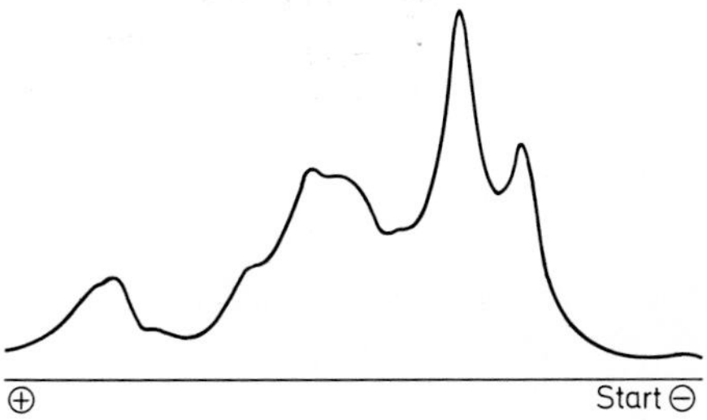

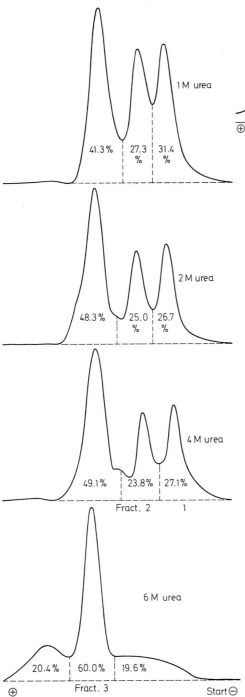

1 M urea

41.3%   27.3 %   31.4 %

2 M urea

48.3%   25.0 %   26.7 %

4 M urea

49.1%   23.8%   27.1%

Fract. 2   1

6 M urea

20.4%   60.0%   19.6%

Fract. 3

⊕   Start ⊖

Fig. 25. Microdensitometer curve for polymerase after fractionation in an 8% gel containing 6 molar urea followed by staining with amido black. Compare with Fig. 24

Fig. 24. Effect of urea on the three RNA polymerase species. The arrangement is the same as that described in Fig. 23

Quantitative Assay for Dehydrogenases

One of the advantages of the micromethod described (CREMER, DAMES and
NEUHOFF, 1972) is its high sensitivity for the determination of different dehydrogenases; it is at least 1 000 times more sensitive than a common photometric test
(LÖHR and WALLER, 1970). Furthermore, the tetrazolium method (BREWER and
SING, 1970; WILKINSON, 1970) has a higher sensitivity for staining bands of
enzymes than the common amido black staining procedure. The minimum
amount of pure glucose-6-phosphate dehydrogenase (Glc-6-P DH) from yeast,
which can be detected by amido black staining after fractionation on 20% polyacrylamide gels in 5 μl capillaries is approximately 1 ng (Fig. 26), as compared
to the ease of activity measurements, using amounts of less than 1 pg of yeast
Glc-6-P DH, with the tetrazolium assay.

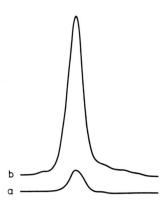

Fig. 26. Microdensitometer curves after fractionation of pure Glc-6-P DH from yeast on a 20% gel
in 5 μl capillaries. Curve a: 8×10^{-9} g after staining with amido black. Curve b: 8×10^{-12} g of the
same preparation, after electrophoresis and visualization under identical conditions, with incubation
for 30 min at 37° C in the Glc-6-P DH assay mixture

Because of the small diameter of the microgels, the time of diffusion of
substrates into the gel is so low that a linear increase in optical density can be
observed at the site of enzymic activity in the gel, after two or three minutes of
incubation in the assay mixture. This linear increase can be followed for a period
of between 15 and 80 min, depending on the enzyme concentration, showing
clearly that only negligible amounts of enzyme are lost from the gel during
incubation. It is therefore possible to compare a microgel to a "micro-cuvette"
with a volume from 0.003 to 0.015 μl, since a single protein band in a 5 μl gel
extends approximately 20 to 100 μm. In addition, the reactants are always in
excess of the small gel volume. Because of the coupled reactions of the tetrazolium
system, reduced pyridine nucleotides will be oxidised immediately and other
reaction products will diffuse into the incubation medium; e.g. 6-phospho-

gluconic acid in the Glc-6-P DH reaction. Therefore the products of the enzyme reaction will not cause an end-product inhibition. Only the blue formazan, which is the final product of the coupled reactions, remains as a precipitate at the site of enzymic activity in the gel. The method can be standardised in international units for each Glc-6-P DH variant by using enzyme preparations of known activity.

Another advantage of the gel electrophoresis procedure is the further purification of the enzyme. Thus the presence of other NADP reducing enzymes in an extract does not interfere with the assay for Glc-6-P DH; for example, 6-phosphogluconate dehydrogenase migrates much more slowly in the separation gel than Glc-6-P DH. Also substrates and inhibitors, unless they are tightly bound to an enzyme, are removed during electrophoresis. YOSHIDA (1966) has described a reversible partial inactivation of Glc-6-P DH by dilution which is independent of NADP. The micro-disc electrophoresis method avoids this effect because, even if very dilute solutions of enzyme are initially applied, the enzyme is concentrated in the 5% spacer gel. "Tetrazolium oxidase" (OEHLSCHLEGEL and STAHMANN, 1971) may interfere with the assay of Glc-6-P DH; however, it is possible to detect "tetrazolium oxidase" activity by incubating the gel in a solution of phenazine methosulphate and tetrazolium chloride in the light. A "tetrazolium oxidase" would produce an unstained band in a otherwise blue gel.

When enzymes are used as markers for genetic analysis it is often necessary to characterise an enzyme by the determination of several parameters. Comparison of a known protein with an unknown one by gel electrophoresis may give identical R_f-values for the two, but this does not prove conclusively that the two proteins are identical. In the case of Glc-6-P DH, the calculation of binding constants for the enzyme, even in the polyacrylamide micro-gel, may be advantageous. For the purpose of genetic analysis it may often be sufficient to detect differences in the binding properties of two variants. For further characterization of an enzyme after fractionation in microgels, the effect of activators or inhibitors can also be investigated, as was illustrated by the stimulation of Glc-6-P DH activity by Mg^{++} ions. The well-known cycling method described by LOWRY et al. (1961), LOWRY and PASSONNEAU (1972) is more sensitive than the method described here, and can also be used for the determination of Glc-6-P DH activities. However, for genetic problems [such as the investigation of gene activation and repression in the early stages of mammalian embryogenesis, or for the determination of linkage relationship by use of somatic cell hybride (HARRIS and WATKINS, 1965)], enzyme variants of Glc-6-P DH or other dehydrogenases which are used as genetic markers, have to be separated prior to identification (MILLER et al., 1971; RUDDLE et al., 1971; GRZESCHIK et al., 1972). Such work therefore requires methods which are not only sensitive but which can also separate variants of enzymes. The procedure described here meets these two requirements.

Assay Conditions

Micro-disc electrophoresis of the dehydrogenases is usually performed in 5 µl capillaries. The gel system is prepared with the stock solutions described previously. Other buffer systems, e.g. Tris/HCl, Tris/phosphate, triethanolamine/phosphate, or triethanolamine/sulphate can be used, provided the pH (8.8) is maintained. The system given can also be used for fractionations in 2 µl or even 1 µl Drummond microcaps, but in this case an increase in the concentration of acrylamide to 25 % is advantageous. The upper and lower electrode buffers are identical (Tris/glycine, pH 8.4), and each contains $NADP^+$ (0.1 mg/ml) and a trace of fluorescein. Electrophoresis is carried out at room temperature with a constant voltage of 50 V, and is stopped when the fluorescein marker has migrated a certain distance, depending on the type of separation (e.g. Glc-6-P DH variants), into the separating gel. Electrophoresis takes from 30 to 45 min depending on the migration distance of the sample into the separating gel. After electrophoresis the gels are extruded onto a watch glass by pressure from a syringe filled with triethanolamine buffer, and are then transferred to tetrazolium incubation mixture.

For testing the method, the assay is performed with pure Glc-6-P DH (Boehringer, Mannheim, 15304, EC 1.1.1.49.) from yeast. 0.3–1.5 µl of diluted solutions of the pure enzyme were subjected to electrophoresis. Pure Glc-6-P DH (1 mg/ml) can be dialyzed against 10 mM phosphate buffer, pH 7.0, containing 0.1 mg NADP/ml, for 24 hrs. at 4° C in micro-dialysis chambers (see chapter 12) to remove the ammonium sulphate from the Glc-6-P DH suspension. If very dilute solutions are used, the stock solution need not be dialysed.

Tissue samples are homogenized in ice cold 10 mM phosphate buffer pH 7.0 (1:10 w/v), and centrifuged at 0° C for 2 hrs. at 24000 g. The clear supernatant is divided into several portions and stored at −20° C. On storing very dilute solutions either at 0° C or −20° C, loss of activity was observed, so the final dilution of the supernatants with 10 mM phosphate buffer (pH 7.0) containing 0.1 mg NADP/ml, is made just prior to electrophoresis. For each sample, the optimal concentrations of, e.g. Glc-6-P DH, must be determined individually.

The incubation mixture used for testing Glc-6-P DH activity consists of 5 mg glucose-6-phosphate (as disodium salt, Boehringer, Mannheim), 4 mg NADP (as disodium salt, Boehringer, Mannheim), 6 mg p-nitro blue tetrazolium chloride (Serva, Heidelberg), and 1 mg phenazine methosulphate (Serva, Heidelberg) in 20 ml triethanolamine/HCl buffer (40 mM, pH 8.0). This amount is sufficient for the incubation of 20–40 microgels. The concentrations given are such that all reagents are present in excess. The blue formazan complex which is precipitated as the final product of the coupling reactions (WILKINSON, 1970) at the site of enzymic activity in the gel, can be recorded with a microdensitometer, and the peak areas measured by planimetry. The incubation mixture must be kept in the dark because the tetrazolium system is sensitive to light. Providing it is stored in the dark, the reaction mixture gives the same results when used immediately after preparation, or after 6 hrs. storage. Incubation is carried out at 25 or 37° C in the dark, and is terminated by transferring the gels to 7.5 % acetic acid. The colour of the tetrazolium band remains for several weeks providing it is stored in the dark.

For recording the enzymic activity in an individual gel at different time intervals, the chamber shown in Fig. 27 is used. This holds the gel and allows an excess of incubation mixture to bathe the gel during the recording. The formation

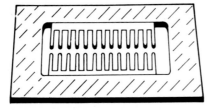

Fig. 27. Chamber for recording continuously the activity of an enzyme in a microgel. The two combs in the chamber are made from Perspex to allow the microgel to stay in a fixed position in sufficient excess of assay mixture during recording. The floor of the chamber is made from optical glass and, after being filled with assay mixture and gel, the chamber is closed with a cover glass

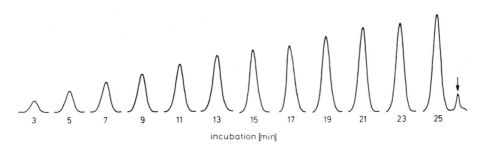

Fig. 28. Continuous registration of activity of Glc-6-P dehydrogenase in a microgel. After fractionating 0.8 µl (corresponding to 13.4 µg wet weight of an extract from rat kidney cortex) on a 20% polyacrylamide microgel (5 µl), the gel was placed in the chamber shown in Fig. 27, filled with the tetrazolium incubation mixture. At 2 min intervals the gel was measured with the microdensitometer in a dark room (grey wedge 1.6 d, ratio of record to sample travel 50 : 1). The peak areas representing the activity increase linearly with the time of incubation. The arrow indicates an artifact produced when the light beam of the microdensitometer is focussed on the gel during measurement

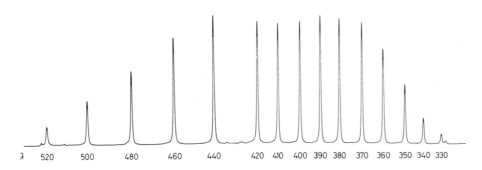

Fig. 29. Pherograms of the artificial blue bands produced in a blank gel by light of different wavelengths. A blank microgel was placed in a chamber filled with the standard tetrazolium incubation mixture and light of different wavelengths was focussed through a slit onto the gel with a UV microscope-photometer (Carl Zeiss). Each wavelength was used for 4 min at one gel locus and afterwards a pherogram was taken with the microdensitometer (grey wedge 1.6 d, ratio of record to sample travel 20 : 1). Artifacts could not be produced at wavelengths below 330 nm or above 520 nm, even if the exposure time was prolonged to 10 min

of the blue dye at the site of enzymic activity is recorded densitometrically, in the dark, at intervals of one minute. Fig. 28 demonstrates how the peak area increases with the time of incubation. When the light beam of the microdensitometer is focused for a longer period of time on one position in the gel, an artificial blue band is produced (Fig. 28 see arrow). As shown by Fig. 29, this blue band is only formed by light with a wavelength of between 320–520 nm (checked by a Carl Zeiss UV-microscope-photometer). This is probably due to the phenazine methosulfate which is present in the assay mixture and which absorbs light in this region, with a maximum of 383 nm. Although monochromatic light is not produced by the Joyce-Loebl microdensitometer, the formation of this artificial band is negligible if the instrument is used with a 5 V, 6.5 A lamp.

Since Mg^{++} ions stimulate the activity of Glc-6-P DH from yeast (GLASER and BROWN, 1955), it was necessary to check if the same effect could be observed after Glc-6-P DH had been fractionated on polyacrylamide microgels. 22 gels were each loaded with 1.5 µl of a solution diluted 1 : 100000 from a stock solution (1 mg/ml) of pure Glc-6-P DH. Electrophoresis was stopped after the fluorescein had migrated 10 mm into the separating gel. 11 of the gels were placed in an incubation mixture containing 10 mM $MgCl_2$, and the other 11 gels were incubated without $MgCl_2$, for a period of 30 min at 25° C in the dark. The mean values of the peak areas found by planimetry were $100 \pm 4.8\%$ (S.E.M.) for the gels incubated without of Mg^{++}, and $217 \pm 10\%$ for the series incubated with $MgCl_2$ in the assay mixture. Therefore, Mg^{++} ions have the same effect on fractionated Glc-6-P DH as on the normal yeast enzyme.

To investigate the influence of gel components (Triton X-100, N.N.N'.N'.,-tetramethyl-ethylendiamine (TEMED), $K_3Fe(CN)_6$, ammonium peroxide disulphate) on enzymic activity, these substances were added, in the same concentrations as in the gels, to a photometric Glc-6-P DH assay solution in a quartz cuvette. Glc-6-P DH activity was measured by following the production of $NADPH_2$, using its absorbance at 340 nm (LÖHR and WALLER, 1970). 1% (v/v) Triton X-100, (to facilitate removal of the gels from the capillaries), 0.63% (v/v) TEMED, and 0.11 mM $K_3Fe(CN)_6$ did not alter the enzymic activity significantly. Preincubation at 25° C, of the enzyme assay solution with ammonium persulphate (3 mM) for half an hour and for 1 hr. before the addition of glucose-6-phosphate to start the reaction resulted in decreases of 59% and 76% respectively in the activity as compared to controls. In spite of these results, ammonium persulphate does not markedly influence the activity of Glc-6-P DH in the gel system after electrophoresis, and pre-electrophoresis of the separating gel (4 hrs. at 60 V with 90 mM Tris/sulphate buffer, pH 8.8) to remove ammonium persulphate (ANSORG and NEUHOFF, 1971) did not result in higher values of enzyme activity. Replacing the 5% spacer gel by Sephadex G-100 superfine (Pharmacia, Sweden) (HYDÉN, BJURSTAM, McEWEN, 1966) in tris/phosphate buffer, pH 6.7 (to avoid the ammonium persulphate present in the 5% gel) decreases the quality of fractionation and therefore lowers the observed enzymic activity. The differences in the behaviour of ammonium persulphate in a cuvette assay and in the gel system may be explained by the facts that the persulphate concentration is markedly reduced during gel polymerisation and that the persulphate ions migrate in front of the enzyme during electrophoresis.

Enzyme Kinetics

Fig. 30 shows the result obtained with an extract from the kidney cortex of rat. 0.8 µl of the diluted extract, corresponding to 16 µg wet weight of tissue, were fractionated on 9 gels under identical conditions, followed by incubation at 37° C in the Glc-6-P DH assay mixture. The gels were transferred to 7.5% acetic acid at 2 min intervals. As shown in Fig. 30 there is a linear correlation

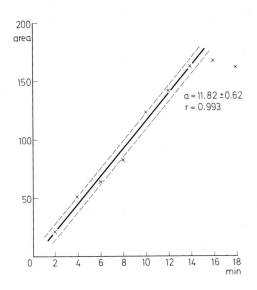

Fig. 30. Activity of Glc-6-P dehydrogenase from an extract of rat kidney, after electrophoresis and incubation for different periods.

9 gels (5 µl, 20% gel), 16 µg wet weight tissue, were each subjected to electrophoresis under identical conditions until the fluorescein marker had migrated 7 mm into the separating gel, incubated for the indicated periods in the assay mixture without $MgCl_2$, and traced under the microdensitometer after fixation in acetic acid (grey wedge 1.6 d). The peak areas given by planimetry in arbitrary units are plotted against incubation time. The peak areas increase linearly between 2 and 14 min of incubation. The dotted line indicates the standard deviation of observations.

Calculations: $y = (-2.57 \pm 5.56) + (11.82 \pm 0.62) \cdot x$
$\sigma = \pm 6.58, \quad r = 0.993$

($r_{\text{area time}} = 0.993$) between the area of the peaks and the time of incubation, from 2 to 14 min. The slope "a" of the curve is a measure of the increase in optical density per unit time, and can therefore be taken as a measure of enzymic activity. The equations of the curve were calculated according to the method of linear regression: $y = b + a x$. The calculated values of the regression lines, the standard deviation of the coefficients, the standard deviation of the observations, and the correlation coefficients r_{yx} for each curve are given in the legends to the figures.

As shown in Fig. 31, single gels can also be used for successive estimation of the enzyme activity; 0.3 to 1.5 µl of a dilute solution of Glc-6-P DH from yeast (corresponding to 3, 6, 9, 12 and 15 pg Glc-6-P DH per gel) were fractionated.

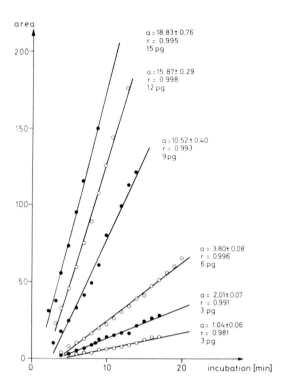

Fig. 31. Enzyme kinetics in microgels with different amounts of pure Glc-6-P dehydrogenase from yeast. 6 gels were loaded with different volumes of a dilute solution of the pure enzyme, corresponding to 3, 6, 9, 12 and 15 pg enzyme. Electrophoresis was performed until the fluorescein marker had migrated 10 mm and the gels were then incubated at 25° C in the chamber shown in Fig. 27. Pherograms were recorded (grey wedge 0.79 d) at 1 min intervals and the peak areas were plotted against time of incubation. The calculated values of the corresponding regression lines are:

$$3 \text{ pg:} \quad y = (-4.108 \pm 0.775) + (1.038 \pm 0.065) \cdot x$$
$$\sigma = 0.772, \quad r = 0.981$$

$$3 \text{ pg:} \quad y = (-6.949 \pm 0.880) + (2.009 \pm 0.078) \cdot x$$
$$\sigma = \pm 1.180, \quad r = 0.991$$

$$6 \text{ pg:} \quad y = (-13.809 \pm 1.135) + (3.805 \pm 0.088) \cdot x$$
$$\sigma = \pm 1.768, \quad r = 0.996$$

$$9 \text{ pg:} \quad y = (-28.05 \pm 3.63) + (10.52 \pm 0.40) \cdot x$$
$$\sigma = \pm 4.71, \quad r = 0.993$$

$$12 \text{ pg:} \quad y = (-33.17 \pm 2.51) + (15.89 \pm 0.29) \cdot x$$
$$\sigma = \pm 3.02, \quad r = 0.998$$

$$15 \text{ pg:} \quad y = (-18.88 \pm 4.30) + (18.83 \pm 0.76) \cdot x$$
$$\sigma = \pm 4.28, \quad r = 0.995$$

The incubation of each gel was followed densitometrically at one minute intervals at 25° C. As seen in Fig. 31, the enzymic activity is constant for each gel, as indicated by the linear increase of the areas under each peak with time. In another experiment it was found that when only 0.8 µl of Glc-6-P DH, corresponding to 0.64 pg (the stock solution was diluted 1 : 1 250 000), was fractionated on a 5 µl gel, a linear correlation ($r=0.996$) between peak area and time of incubation was observed for a period of up to 80 min.

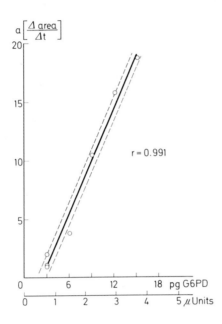

Fig. 32. Standard curve for assay of pure Glc-6-P dehydrogenase in microgels. The slopes "a" of the regression lines shown in Fig. 31, as measures of enzyme activity, are plotted against pg enzyme fractionated on each gel. The activity was calculated to be 242 U/mg protein under standard conditions as specified in the text (see lower scale of this figure). The dotted line indicates the standard deviation of observations.

$$\text{Calculation:} \quad y=(-3.58\pm0.77)+(2.30\pm0.14)\cdot x$$
$$\sigma=\pm1.07, \quad r=0.991$$

If the reaction rate, or the linear increment "a" from the experiments illustrated in Fig. 31, is plotted against the amount of enzyme subjected to electrophoresis (Fig. 32), a regression line ($r_{a\,pg}=0.991$) is obtained which can be used as a calibration curve for the quantitative determination of enzyme concentrations. The specific activity of the dialyzed Glc-6-P DH used in the experiments for Fig. 31 and 32 was 242 units/mg protein (LÖHR and WALLER, 1970). This value was used in calculating the lower scale in micro-units of enzyme as shown in Fig. 32.

Fig. 33 shows that the same assay conditions can be used for estimating the activity of Glc-6-P DH in biological materials. Extracts from rat kidney cortex,

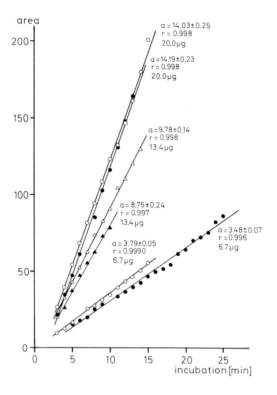

Fig. 33. Enzyme kinetics of Glc-6-P dehydrogenase from kidney cortex of rat.
Extract volumes prepared from kidney cortex of rat corresponding to 6.7, 13.4 and 20.0 µg wet weight
of tissue were fractionated on 20% polyacrylamide microgels until the fluorescein marker had migrated 7 mm into the separation gel. Incubation (in absence of $MgCl_2$) and registration at 1 min intervals were performed as in Fig. 31. The peak areas were plotted against time of incubation. The calculations of the corresponding regressionslines are:

6.7 µg: $y = (-3.975 \pm 1.213) + (3.485 \pm 0.073) \cdot x$
          $\sigma = \pm 2.014, \quad r = 0.996$
          $y = (-1.830 \pm 0.529) + (3.793 \pm 0.053) \cdot x$
          $\sigma = \pm 0.695, \quad r = 0.999$

13.4 µg: $y = (-6.476 \pm 1.666) + (8.756 \pm 0.242) \cdot x$
          $\sigma = \pm 1.566, \quad r = 0.997$
          $y = (-5.604 \pm 1.338) + (9.787 \pm 0.146) \cdot x$
          $\sigma = \pm 1.744, \quad r = 0.998$

20.0 µg: $y = (-23.164 \pm 2.000) + (14.191 \pm 0.233) \cdot x$
          $\sigma = \pm 2.434, \quad r = 0.998$
          $y = (-16.464 \pm 2.461) + (14.030 \pm 0.252) \cdot x$
          $\sigma = \pm 3.406, \quad r = 0.998$

corresponding to 6.7, 13.4 and 20.0 µg wet weight of tissue, were fractionated
on 20% polyacrylamide microgels and assayed for Glc-6-P DH. In each of the
6 experiments good linear correlation was observed between time of incubation
and the corresponding peak area. Fig. 34 re-illustrates that the slope "a" is
proportional to the amount of extract, with a correlation coefficient of 0.996 and

a standard deviation of $\pm 4.3\%$. This curve can also be taken as a standard graph for the quantitative determination of Glc-6-P DH activity from rat kidney cortex.

The method described is also suitable for the separation of Glc-6-P DH variants by assaying their enzymic activities in a "one step" procedure. Fig. 35 shows three microdensitometric tracings of the pattern of Glc-6-P DH activity from a mixture from human and rat kidney cortexes. The human and rat extracts were diluted to 20 µg wet weight per µl and mixed in ratios of 2:5, 1:1 and 5:2.

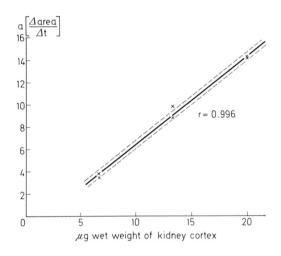

Fig. 34. Standard curve for assay in microgels of Glc-6-P dehydrogenase from kidney cortex of rat. The slopes "a" from the curves shown in Fig. 33 are plotted against the corresponding µg wet weights of rat kidney cortex. The dotted line represents the standard deviation of observations.

$$\text{Calculations:} \quad y = (-1.464 \pm 0.483) + (0.784 \pm 0.034) \cdot x$$
$$\sigma = \pm 0.448, \quad r = 0.996$$

0.8 µl of these mixtures, corresponding to 5.4 µg human tissue with 10.6 µg rat tissue, 8 µg with 8 µg, and 10.6 µg with 5.4 µg, respectively, were subjected to electrophoresis until the fluorescein marker had migrated 12 mm into the separation gel. The gels were then incubated for 15 min at 37° C. Fig. 35 demonstrates that the human and rat Glc-6-P DH pattern differs not only in the R_f values, but also in the activity of the enzyme bands. On the right side of Fig. 35 the peak areas are plotted against µg wet weight of tissue. The obvious difference between the two slopes reflects the difference between the two Glc-6-P DH species.

Another example of the separation of Glc-6-P DH species is shown in Fig. 36. Extracts of cells from a human myeloma (MATSUOKA et al., 1967) and from mouse fibroblasts (L 929, GIBCO) were diluted to give one symmetrical band for each of the Glc-6-P DH species after fractionation and incubation. Then 0.75 µl of both Glc-6-P DH species were mixed, fractionated on a single gel, and incubated for 30 min at 25° C. Normal times of electrophoresis (30–45 min) gave a poor

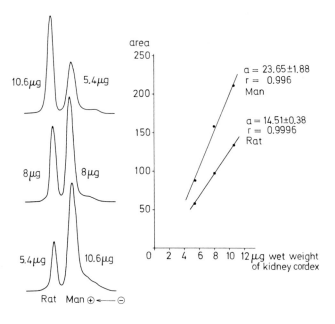

Fig. 35. Separation and enzyme assay of Glc-6-P dehydrogenase variants from rat and man. Three microgels were loaded with mixtures of extracts of human and rat kidney cortex corresponding to the wet weight of tissue given on the left side of the figure. Electrophoresis was carried out until the fluorescein marker had migrated a distance of 10 mm, then incubation at 37° C for 15 min; the gels were recorded in acetic acid with the microdensitometer (grey wedge 1.6 d, ratio of record to sample travel 50 : 1). It is obvious that under these experimental conditions good separation of both variants is observed, with the enzyme from the rat having the longer migration distance. The peak area representing the activity is larger for the enzyme from man than for that from rat. On the right side of the figure the corresponding regression lines are shown, with the slope "a" being greater for the enzyme from man than for that from rat.

Calculations:

human Glc-6-P DH: $y = (-36.90 \pm 1.562) + (23.65 \pm 1.89) \cdot x$
$\sigma = \pm 6.94$, $r = 0.996$

rat Glc-6-P DH: $y = (-19.82 \pm 3.22) + (14.51 \pm 0.38) \cdot x$
$\sigma = 1.43$, $r = 0.9996$

separation of these species of Glc-6-P DH (Fig. 36a). However, when the time was increased to 2 hrs., two clearly separated variant bands could be seen (Fig. 36c). The peak nearest the origin is that of human myeloma Glc-6-P DH. This is shown in Fig. 36b, where a 10-fold excess of the myeloma extract compared to the mouse fibroblast extract was fractionated. Fig. 36b also demonstrates that the resolving quality of the microgel system allows a clear separation of both species even under these unfavourable conditions.

As Figs. 37 and 38 show, the same method is also suitable for the detection of LDH isoenzymes. Extracts of guinea pig liver, corresponding to 6.0, 3.0, 2.3, 1.9, 1.5, 1.1, 0.8, 0.4, and 0.2 µg of wet tissue weight, were fractionated on 25% polyacrylamide gels in 2 µl Drummond microcaps. The gels corresponding to 6 and 3 µg were incubated for 10 min, while the other gels were incubated for

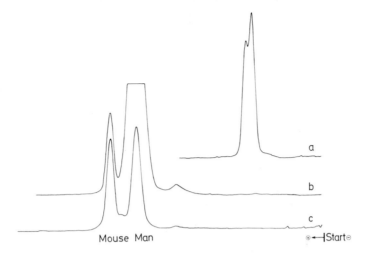

Fig. 36. Separation and assay of variants of Glc-6-P dehydrogenase from tissue cultures of mouse and man.

Extracts of tissue culture cells from a myeloma of man and from mouse fibroblasts were diluted so as to have similar activities for Glc-6-P dehydrogenase, and 0.75 μl portions were subjected to simultaneous electrophoresis on 20% polyacrylamide microgels for 40 min (curve *a*) and for 2 h (curve *b* and *c*). After 30 min incubation in the assay medium at 25° C, the gels were recorded in acetic acid. 40 min electrophoresis is too short for a good separation of both species (curve *a*), but separation can be demonstrated clearly after prolonged electrophoresis (curve *c*). In curve *b* the extract from the human myeloma was ten times more concentrated than the extract from mouse fibroblast, but even under these conditions prolonging the time of electrophoresis leads to a clear separation of both species. This figure also demonstrates that not only is the time of electrophoresis important for a good fractionation of enzyme species, but that it is also necessary to keep the amounts of extracts to be separated in the right concentrations. (Figs. 27–36 from CREMER, DAMES and NEUHOFF, 1972)

30 min, at 37° C, in the incubation mixture containing lactate as the substrate. After fixation in 7.5% acetic acid, pherograms were obtained with the microdensitometer as shown in Fig. 38.

A further demonstration of the sensitivity of these methods is the determination of Glc-6-P DH activity in a single mouse ovum. Single ova from untreated female mice were obtained by the method of BRINSTER (1963), using cold (4° C) 0.15 M saline solution instead of culture medium for extruding the ova from the oviducts. Cumulus cells were removed and the ova were washed three times by transferring them with a fine glass pipette into cold 0.15 M saline solution to remove oviductal fluid and debris. Single ova were transferred to 5 μl Drummond microcaps, 3 μl hypotonic phosphate buffer (10 mM, pH 7.0) was added; the microcaps were previously heat sealed at one end. The microcaps were frozen at −70° C to disrupt the ovum. Centrifugation was performed in a Heraeus-Christ haematocrit centrifuge equipped with a special rotor for capillary centrifugation (see chapter 5). The clear supernatant was used for the Glc-6-P DH assay. The extract from one single mouse ovum (3 μl) was fractionated on a 20% polyacrylamide gel in a 5 μl capillary. For the application of the extract, a piece of a 5 μl Drummond microcap was connected to the gel capillary by

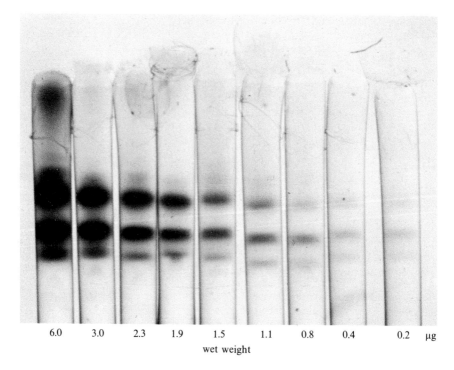

6.0   3.0   2.3   1.9   1.5   1.1   0.8   0.4   0.2   µg

wet weight

Fig. 37. Detection of isoenzymes of LDH from guinea pig liver. Extracts, corresponding to 6.0, 3.0, 2.3, 1.9, 1.5, 1.1, 0.8, 0.4 and 0.2 µg wet weight of liver tissue were fractionated on 25 % gels in 2 µl capillaries. The micrograph was prepared after incubation in the tetrazolium mixture, containing lactate as the substrate, and fixation in acetic acid. (Magnification 34 ×)

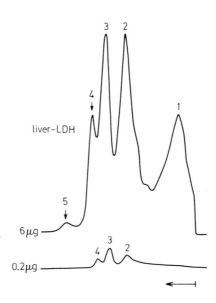

Fig. 38. Microdensitometer curves from two gels shown in Fig. 38

means of polyethylene tubing. Electrophoresis was stopped when the fluorescein marker had migrated 10 mm into the separation gel. The gel was then incubated for 20 min at 37° C in the Glc-6-P DH assay mixture. The incubation was stopped by transferring the gel to 7.5% acetic acid. Fig. 39 shows the microdensitometer tracing of this gel taken with a grey wedge of 1.6 d. In practice, it is quite possible to perform several determinations of Glc-6-P DH activity with the extract of one mouse ovum if longer periods of incubation are used or grey wedges of lower density are used for the measurement. Studies of enzyme kinetics are also possible if continuous recording at one or two minute intervals is performed as described for pure Glc-6-P DH and the kidney extracts. The sensitivity of the system can be increased by using 2 µl or even 1 µl capillaries for the fractionation of extracts.

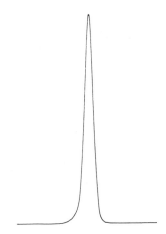

Fig. 39. Activity of Glc-6-P dehydrogenase of a single ovum of a mouse.
The extract of a single ovum of a mouse was subjected to electrophoresis on a 20% polyacrylamide microgel in a 5 µl capillary. After incubation for 20 min at 37° C in the assay mixture, the gel was recorded in acetic acid with the microdensitometer (grey wedge 1.6 d, ratio of record to sample travel 50:1). The activity represented by the peak area is sufficient to allow several electrophoreses and enzyme assays. (From CREMER, DAMES and NEUHOFF, 1972)

Michaelis-Menten Kinetics and the Determination of Enzyme Activity

For the determination of the K_m-value of Glc-6-P DH from yeast, using microgels with glucose-6-phosphate (Glc-6-P) as substrate, 1.5 µl portions (corresponding to 50 pg Glc-6-P DH) of freshly diluted Glc-6-P DH were subjected to electrophoresis in 18 gels. These were then incubated in pairs in the assay mixture, for 20 min at 25° C, and were evaluated by microdensitometry after fixation in 7.5% acetic acid. The concentration of Glc-6-P in the incubation mixture was 1.64×10^{-5} to 6.6×10^{-4} M. The result is shown in Fig. 40 (closed

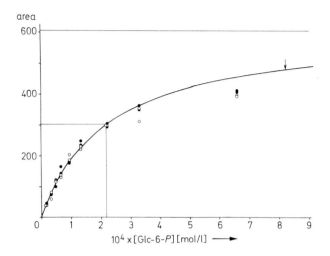

Fig. 40. Determination of the K_m value for Glc-6-P dehydrogenase from yeast and Glc-6-P in micro-gels. The pure enzyme was fractionated on 20% polyacrylamide microgels in 5 µl capillaries in two series of experiments. Each gel was loaded with 50 pg of enzyme and subjected to electrophoresis until the fluorescein marker had migrated 10 mm in the separation gel. The open circles represent the result of an experiment with 18 gels, and the closed circles an experiment with 19 gels. The gels were incubated for 20 min at 25° C in assay mixtures with concentrations of Glc-6-P as indicated on the abcissa of the figure. The peak areas of the individual experiments are plotted against concentration of Glc-6-P. The curve (full line) was calculated from the values of the regression line shown in Fig. 41. The lines parallel to the abcissa represent the peak areas corresponding to V and V/2 respectively. The concentration of substrate corresponding to V/2 represents the K_m value for Glc-6-P under these experimental conditions. The arrow indicates the concentration of Glc-6-P in the assay mixture used for all other experiments described

circles), where the peak areas of the pherograms are plotted against Glc-6-P concentration in the assay mixture. Fig. 41 shows the corresponding LINE-WEAVER-BURK diagram, with a linear regression line showing a correlation coefficient $r_{1/area,\ 1/[Glc-6-P]} = 0.998$. From this it is possible to calculate a binding constant of Glc-6-P DH for glucose-6-phosphate, of 2.2×10^{-4} M in the micro-gel. Another experiment, carried out under identical conditions (open circles in Fig. 40), gave a K_m-value of 2.7×10^{-4} M ($r_{1/area,\ 1/[Glc-6-P]} = 0.991$).

Fig. 40 also shows that for higher concentrations of Glc-6-P, the observed values of enzymic activity lie significantly below the hyperbola calculated from the regression line in Fig. 41. When the Glc-6-P concentration was increased to 24.6×10^{-4} M, a corresponding increase of enzyme activity was not observed. From these results it would seem that the increased precipitation of formazan at the site of enzymic activity diminishes this activity (compare Fig. 30). This is borne out by the fact that a linear increase between peak area and time of incuba-tion can be observed for much longer periods of time when small amounts of Glc-6-P DH are fractionated. One must also take into consideration the fact that, under unfavourable experimental conditions, more NADPH than formazan may be formed in the coupled reactions of the tetrazolium system, and that the

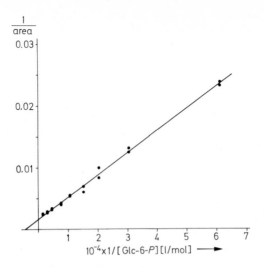

Fig. 41. LINEWEAVER-BURK diagram for Glc-6-P dehydrogenase and Glc-6-P determined after fractionation of Glc-6-P dehydrogenase on microgels.
The values shown in Fig. 40 with closed circles are plotted according to Lineweaver and Burk. The resulting regression line is calculated according to $y = b + a \cdot x$ to be:

$$y = (0.001651 \pm 0.000151) + (0.003595 \pm 0.000061) \cdot x$$

The standard deviation of observations $\sigma = 0.00046$, the correlation coefficient $r = 0.998$. From these data the K_m value is calculated ($= a/b$) to be $2.2 \cdot 10^{-4}$ M. (From CREMER, DAMES and NEUHOFF, 1972)

NADPH can then diffuse into the incubation medium. The K_m-value of 2.3×10^{-4} M found for yeast Glc-6-P DH in the gel system with glucose-6-phosphate as substrate is higher than those calculated by GLASER and BROWN (1955), LOWRY et al. (1961), and ROSE (1961). This may be due to the embedding of the enzyme in the gel matrix, since NEUHOFF, SCHILL and STERNBACH (1969 b) have shown that the activity of DNA-dependent RNA-polymerase from E. coli is lower in the gel than after elution from the gel (compare p. 29).

Micro Isoelectric Focusing

Since the development of ampholytes (KOLIN, 1954; SVENSSON, 1961) it is possible to fractionate single protein components from mixtures of different proteins according to their isoelectric points. Isoelectric focusing (IEF) was initially carried out in saccharose gradients (VESTERBERG and SVENSSON, 1966), and later in polyacrylamide gels (DALE and LATNER, 1968; WRIGLEY, 1968). In a comprehensive review, HAGLUND (1971) described the historical development, the theoretical foundations, and the applications on the macroscale of this very important and powerful method for protein fractionation. The micro version of this method, described by RILEY and COLEMAN (1968), and CATSIMPOOLAS (1968), requires 10 µg of a single protein or 0.2 to 0.4 mg of a protein mixture

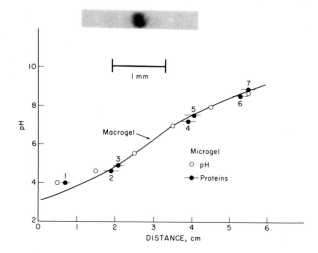

Fig. 42. Comparison of isoelectric focusing in 7.5% acrylamide gel on the macro and micro scales. The gels contained LKB Ampholine (pH 3–10). The line shows a pH gradient representative of those obtained in the macrogel. The open circles show the average pH value of each cm slice of the micro gel, and are plotted as half the "migration distance" of the gel slice. The closed circles represent the average migration distances of specific standard proteins, and are plotted against their isoelectric points. The lines through the closed circles are equal to three times the standard error of the mean. The numbers refer to the following proteins: 1 α-casein, 2 ovalbumin, 3 bovine serum albumin, 4 hemoglobin, 5 chymotrypsin, 6 α-chymotrypsinogen-A, 7 RNase. Upper inset: A photograph of a segment of gel containing 5×10^{-9} gram of ovalbumin after electrofocusing. (From GAINER, 1973)

for a single fractionation. GROSSBACH (1972), GAINER (1973), and QUENTIN and NEUHOFF (1972), working independently, refined the IEF methods further. GAINER described the IEF of proteins in the 10^{-9}–10^{-10} gram region, performing the IEF in glass capillary tubes (Corning 7740) of length 7 cm and inner diameter 0.58 mm, using 7.5% or 5.5% polyacrylamide gels with LKB Ampholine in the pH range 3–10. The purified proteins he used to characterize the microgels were α-casein, ovalbumin, chymotrypsin, α-chymotrypsinogen-A, bovine serum albumin, human hemoglobin and bovine pancreatic ribonuclease. Fig. 42 shows the comparison of a series of IEF experiments on the macro and micro scales, and demonstrates the outstanding agreement between the two methods. GROSSBACH (1972) has performed IEF in capillaries of 100 μm diameter using bovine serum albumin and β-lactoglobulin in the pico-gram range. QUENTIN and NEUHOFF (1972) used micro isoelectric focusing for the detection of isoenzymes of lactate dehydrogenase (LDH) in different regions of the brain of a rabbit. IEF is performed in 5 μl Drummond microcaps; the amounts of acrylamide, bisacrylamide, TEMED, ammonium peroxydisulphate, 40% Ampholine of pH 3–10 are given in Table 2. The concentration of acrylamide, T, =6.4% (%T=gm acrylamide+gm bis-acrylamide/100 ml solution), the crosslinking C=2.5% (%C= 100× bis-acrylamide/T), the concentration of ampholine is 3.8% in the final gel mixture. The most efficient fractionation of LDH isoenzymes was obtained with these proportions of gel constituents. If a concentration of ampholine of

Table 2. Gel mixture for isoelectric focusing in 5 μl capillaries

4 g Acrylamide + 100 mg Bisacrylamide in 7.5 ml H_2O	0.25 ml
0.063 ml/10 ml H_2O TEMED	0.25 ml
60% Saccharose	0.40 ml
40% Ampholine pH 3–10	0.20 ml
10 mg/10 ml H_2O Ammoniumperoxide disulphate	1.00 ml
	2.10 ml

1–2% is used, as is general for macroscale IEF, the resolving power of the micro gels is much worse than with 3.8% ampholine. According to DOERR and CHRAMBACH (1971), 12% sucrose is added to the gel mixture to stabilise the pH gradient in the gels. This concentration of sucrose is important, since higher or lower concentrations reduce the quality of separation. Furthermore, lower concentrations of ammonium peroxydisulphate lead to lower stability and lower resolution of the gels. Triton X-100, which is normally used in the gel mixtures for micro-disc electrophoresis to aid removal of the gels from the 5 μl capillary, was not necessary in the case of gels containing ampholines.

To carry out IEF on a micro-scale, about $\frac{3}{4}$ of the 5 μl capillary is filled by immersion in the gel mixture. The capillary is then pushed into a cushion of plasticine 2 mm thick. The remainder of the capillary is filled with water. Polymerisation starts after about 15 min, but in practice the gels are usually stored overnight at room temperature in a moist atmosphere. After removing the water layer on top of the gel, 0.6 μl of the diluted protein sample is applied to the column and the remaining space between the sample and the rim of the 5 μl capillary is filled with a solution containing 4% ampholine pH 3–10 and 12% sucrose. After removing the plasticine, the capillary is attached to the equipment described for micro-electrophoresis.

The electrolytes used are very dilute sulfuric acid, pH 2.95, at the anode, and dilute NaOH, pH 10.4, at the cathode. Both electrolytes contain 12% sucrose to reduce the endo-osmotic effects during electrophoresis. The anode must always be at the top of the capillaries. An IEF of LDH isoenzymes could also be obtained if the cathode was at the top of the capillary which was loaded with the sample. However, depending on the length of time of electrophoresis, a reversal of the direction of migration, back towards the cathode, occurs for the faster migrating components. GAINER (1973) also found that only samples loaded on the anodic ends of the capillaries could be electrofocused successfully on the microscale. IEF is carried out at a constant voltage of 200 V at room temperature; the current at the beginning is about 80 to 100 μA per gel and when it reaches 5 μA (about 50 min later) the electrophoresis is terminated. The gels are then extruded onto a watch glass by means of water pressure. The cathodic side of the gel is marked by dipping it for a short while into a 1% coomassie blue solution in water and the gel is then transferred to the incubation mixture to measure the enzymically active proteins.

The incubation mixture for assaying the LDH activity consists of: 1 mg phenazine methosulphate, 2 mg NAD, 3 mg p-nitro blue tetrazolium chloride, and 100 mg Na-lactate dissolved in 20 ml 0.05 M Tris/HCl buffer, pH 8.0. Incubation is carried out at 37° C in the dark, and is terminated by transferring the gels to 7.5% acetic acid. The blue precipitate of formazan at the sites of enzymic activity in the gel is then measured with a microdensitometer and the peak areas found by planimetry.

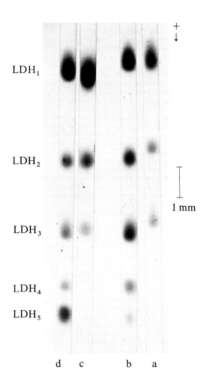

Fig. 43 a–d. Photograph (magnification 13 ×) showing LDH isoenzymes in microgels (6.4% polyacryl-amide and 4% ampholine, pH 3–10) after isoelectric focusing. The gels were incubated for 90 min in the tetrazolium mixture with lactate as substrate. *a* 60 µg wet weight tissue of rabbit cerebellum, *b* 60 µg frontal lobe, *c* 60 µg of lamina quadrigemina posterior and *d* 12 µg of lamina quadrigemina anterior

Fig. 43 shows 5 µl microgels of diluted extracts from the cerebellum, frontal lobe, the lamina quadrigemina posterior and anterior of rabbit, after IEF and incubation for 90 min in the tetrazolium mixture with lactate as the substrate. 5 to 50 mg of each tissue sample was homogenised in 0.01 M phosphate buffer, pH 7.0, in a ratio of 1:10 (w/v) at 0° C for two minutes, and centrifuged in a Servall RC-B for 60 min at 24000 g. The clear supernatants were stored at −20° C. Extracts were diluted just prior to electrophoresis with a solution of sucrose to give a final concentration of 12% sucrose in the sample. Incubation in the tetrazolium mixture for 90 min was used to detect those LDH isoenzymes which are present in very low concentrations in the tissue extracts. It was

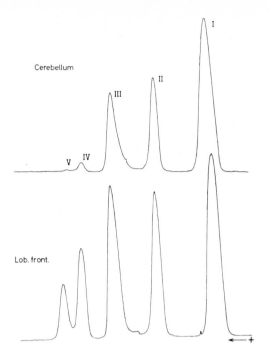

Fig. 44. Microdensitometer curves from the gels *a* and *b* from Fig. 43

necessary to use extracts with a higher protein concentration for the detection of the LDH 5 component from several brain regions. As for Glc-6-P DH, there is a linear correlation between enzymic activity and time of incubation, which depends on the concentration and the activity in the gel. For higher concentrations of enzyme the formation of the tetrazolium band reaches a plateau after a relatively short time of incubation (see Fig. 30). This means that, during the incubation time of 90 min, the pronounced peak of the LDH 1 isoenzyme can reach the plateau while components with lower activity are still in the linear range.

It can be seen from Figs. 43–45 that there are five isoenzymes of LDH in the frontal lobe (gel b) and that the isoenzyme LDH5 is almost completely missing from the cerebellum (gel a). The isoenzyme LDH4 is present, but only in a very low concentration, in the cerebellum. Furthermore, the concentrations of the isoenzymes LDH2 and LDH3 are lower in the cerebellum than in the frontal lobe. Gels c and d show the results for extracts from the lamina quadrigemina posterior and anterior respectively. In these regions again there are striking variations in the behavior of the five isoenzymes of LDH. In order to obtain similar peak areas for LDH1 from both parts of the lamina quadrigemina, 60 μg of wet weight from the posterior part and only 12 μg from the anterior region were fractionated. Even then, the amount of LDH5 isoenzyme in the posterior area was negligible compared to its prominence in the anterior part.

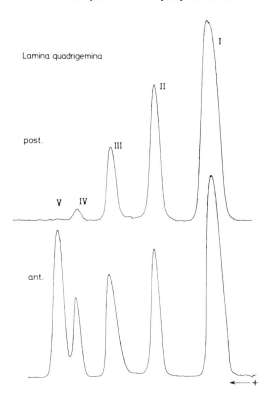

Fig. 45. Microdensitometer curves from the gels *c* and *d* from Fig. 43

The differences are clearly demonstrated by the microdensitometer recordings of the gels, as shown in Figs. 44 and 45.

In Table 3 the percentage mean values of the LDH isoenzyme activities are listed for all organs and brain regions analysed. The microdensitometer tracings were evaluated by planimetry, and the total sum was taken as 100%. For the analysis of the muscles from heart and quadriceps only 0.6 μg wet weight of tissue was necessary for fractionation, but for the demonstration of all five isoenzymes in the brain, expecially for LDH 5, volumes corresponding to 10 μg and 60 μg of wet weight were necessary. This clearly reflected the differences in the activities of the individual isoenzymes of LDH. The results obtained for LDH by the micro-procedure are in agreement with the macroanalysis of heart muscle, skeletal muscle, liver, spleen, and kidney from rabbit (PLAGEMANN *et al.*, 1960; WIELAND *et al.*, 1959), and for medulla, pons, cerebellum, frontal lobe, hippocampus, thalamus, and retina (BONAVITA and GUARNERI, 1963; PASANTES-MORALES *et al.*, 1972). Thus, the determination of LDH isoenzyme activities in 15 defined regions of the brain and in five other organs demonstrates the suitability of 5 μl capillaries for isoelectric focusing. The method is very sensitive, needing less than 1 μg wet weight of some tissues for an analysis; furthermore, it requires only about $2\frac{1}{2}$ hrs.

Table 3. Percentage composition of LDH isoenzyme in different regions of the brain and other organs in rabbit

	LDH^1	LDH^2	LDH^3	LDH^4	LDH^5
Lamina quadrigemina post	59.84	24.83	13.36	1.71	0.26
Hypothalamus	59.49	21.72	16.97	1.46	0.37
Cerebellum	51.17	21.91	24.47	1.92	0.53
Medulla	58.39	21.30	15.54	3.76	1.00
Hypophysis	49.32	35.40	12.38	1.74	1.16
Pons	48.73	24.47	19.47	6.96	0.63
Thalamus [*]	40.00	20.36	26.55	9.82	3.27
Bulbus olfactorius	31.44	20.11	26.63	13.26	8.56
Hippocampus [*]	30.25	16.33	20.75	19.14	13.52
Frontal Lobe	34.93	20.60	26.00	11.05	7.41
Retina	22.09	14.34	27.23	17.31	19.04
Lamina quadrigemina ant. [*]	38.17	15.94	16.19	8.48	21.21
N. opticus	18.77	15.15	22.07	17.40	26.61
Chiasma opt.	24.90	12.70	14.31	18.95	29.13
Tractus opt.	24.09	11.25	18.28	18.93	27.45
Heart [+]	100.00				
Spleen [*]	30.89	28.66	27.39	10.19	2.87
Kidney	55.20	24.26	14.11	6.44	
Liver [*]	0.94	0.94	11.92	46.08	40.13
Muscle [+]					100.00

Wet weight of tissue used for one IEF: [+] 0.6 μg, [*] 12 μg, 60 μg.

The peak areas of the pherograms were measured by planimetry and the total sum taken to be 100%. Each value represents the mean of three different animals. (From QUENTIN and NEUHOFF, 1972.)

for a complete analysis. The tetrazolium method used for the detection of the isoenzymes of LDH has a higher sensitivity than amido black staining for protein bands.

Comparison of the fractionation of the isoenzymes of LDH by IEF with that obtained by micro-disc electrophoresis (cf. Fig. 37) shows that better separation of the individual bands can be attained by IEF. However, the limit of detection using the tetrazolium incubation method is about 6–8 times higher for gel mixtures without ampholine (i.e. in the micro-disc electrophoresis method) compared with that for the ampholine-containing gels of IEF. BISPINK and DAMES (1973) have recently found that the enzymic activity of LDH isoenzymes in ampholine gels is increased when glycerine is added to a final concentration of 25–50% (v/v) to the tissue extract to be fractionated, and when the sample in the capillary is overlayered with a 15–30% glycerine/water solution. The IEF method may also be standardized in international units for each isoenzyme of LDH by using preparations of isoenzymes of known activities, and performing the measurements at defined time intervals during the incubation as described for the determination of the kinetics of Glc-6-P DH.

The isoenzymes of LDH are built from tetramers of the H and M subunits (APELLA and MARKERT, 1961; CAHN et al., 1962; KOWALEWSKI, 1972). The distance between the peak maxima after IEF should be constant if the pH gradient within the gel is linear. Different authors (CONWAY-JACOBS, 1971; DOERR and

CHRAMBACH, 1971) have shown that, although the pH gradient is linear at acid pH's, between pH 8 and pH 10 the gradient is not linear and normally reaches a plateau. This could explain why the distances between the peaks of isoenzymes become shorter on the cathodic (alkaline) side of the gel. It might be posible to obtain a better linear pH gradient by using a somewhat longer gel than the 5 µl capillaries used for these experiments.

Micro-Electrophoresis on Gradient Gels

Gradient gel electrophoresis has proved to be useful for various separations on the macro scale (Review see MAURER, 1971). Columns with continuous gradients or (by layering of individual acrylamide solutions with different concentrations) of discontinuous gradients can be produced by an appropriate mixing apparatus. Microcapillaries with such gradients can be produced either in large quantities, using a specially adapted mixing apparatus (DAMES, MAURER, NEUHOFF, 1973), or in single capillaries by using capillary attraction (RÜCHEL et al., 1973). The capillaries are filled by dipping them into the appropriate solutions. In this way the gel components are continuously mixed, to give a linear gradient of concentration in the polymerised gel. The length and steepness of the gradient can be varied, as can the range of concentration. The gradient may be shown by the polymerisation of protein and amido black staining into the gel. This procedure can replace micro-disc electrophoresis for some separations since it gives a sharper separation of the proteins, and the manipulation required is simpler.

In micro-disc electrophoresis, as in macro gradient electrophoresis, it is always necessary to add the sample to the gel by a fine capillary pipette. While this can be performed safely and easily after some practice, it does demand skill on the part of the experimentor. For gradient gel electrophoresis, a further advantage over micro-disc electrophoresis is that it is not necessary to add the 5% spacer gel.

By varying the method of producing the gradient, the application of the sample can also be performed by capillary attraction, thus avoiding the use of micropipettes. This procedure appears to be very suitable for routine laboratory analyses, something which cannot be said, without further qualification, for micro-disc electrophoresis. The limits of detectability for individual protein bands in a gradient gel, the sample volumes to be fractionated, and the quantities of protein contained in them, are identical with those for micro-disc electrophoresis.

Preparation of Gradient Gels

Stock solutions used for the production of the gradients.

Stock A_1: 0.86 g Tris
        8 ml H_2O
        0.063 ml TEMED
        3.6 N H_2SO_4 to pH 8.8
        H_2O to 10 ml

Stock A_2: 3.39 g Tris
            6 ml 1 N H_2SO_4
            0.063 ml TEMED
            6 N H_2SO_4 to pH 8.8 (approx. 0.6 ml)
            H_2O to 10 ml

Stock C: 20 g Acrylamide
            400 mg bisacrylamide
            3.75 mg $K_3Fe(CN)_6$
            H_2O to 37.5 ml

Stock D: 35 mg ammonium peroxydisulphate
            25 ml 4% Triton X-100 in H_2O + 25 ml H_2O

All stock solutions may be stored for prolonged periods at 4° C, apart from solution D, which must be renewed frequently. If additional sharpening of the protein bands in the gels is desired, the stock solution A_2 is used for preparing the gel. The higher viscosity of this solution leads to a decrease in the length of the resulting gradients as compared to gradients prepared with stock solution A_1.

The capillaries used for gradient gels must be cleaned before use as described on p. 4. Before filling, the middle of the capillary is marked with a felt-tip pen. The capillary, held by forceps, is first dipped into solution D, and filled up to this mark by capillary attraction. It is advisable to use the forefinger of the hand which hold the forceps to close the capillary tip if no suction is used. This prevents the formation of air bubbles during transportation of the capillary between the different solutions. Immediately afterwards it is held in a mixture of stock solutions A and C (one part A + 3 parts C v/v) and filled to the tip of the capillary by capillary attraction. Should a mistake be made at this point, the capillary can be emptied by applying filter paper to one end, and can then be refilled. The filled capillary is placed vertically in a plasticine cushion. However, for further use it is simpler to seal the capillary at the bottom by placing it in a small beaker with a 2–4 mm layer of 50% sucrose solution. Care must be taken that the capillary stands vertically. The adhesive force between the concentrated sucrose solution and the glass walls of the capillary and the beaker holds the capillary firmly to the side of the beaker (Fig. 46). This method avoids the need to cut off the lower piece of the capillary with the plasticine stopper. However, only those capillaries which are completely full can be placed in the sucrose solution, otherwise the sucrose enters the capillary and hinders the formation of an even gradient. It is easy to fill ca. 10 to 20 capillaries simultaneously by this simple method. To do this, capillaries of equal length are fixed together by means of a strip of sticky tape. They can then be held in the solution D, and subsequently in A/C, simultaneously, and sealed at the bottom by placing in the 50% sucrose solution.

For routine work it is advisable to color the freshly-prepared solution A/C with a crystal of brompenol blue, so allowing the formation of the gradient to be checked. The dye does not influence either the polymerisation of the gel or the fractionation of the proteins. During electrophoresis the dye migrates with the buffer front, though sometimes a small amount of the dye remains in the concentrated region of the gradient, appearing slightly yellow after staining with amido black in acetic acid.

Fig. 46   Fig. 47

Fig. 46. 5 µl capillaries filled with a polyacrylamide gradient, in a small beaker with a 3 mm layer of 50% sucrose

Fig. 47. Formation of gradients of polyacrylamide in 5 µl capillaries. In the left capillary, a "lancet" gel gradient is formed and visualized with bromophenol blue in the concentrated solution A/C directly after filling by capillary attraction. The right capillary shows that after 5 min horizontal diffusion, the gradient has become even across the whole diameter of the capillary. Magnification 2 ×

Using this method of filling, the formation of polyacrylamide gradients in capillaries takes place in two stages, which can be observed conveniently if the solution A/C is coloured with bromophenol blue. When solutions A/C enter the capillary, a lancet-shaped gradient is formed (Fig. 47, left capillary) due to the more rapid flow in the centre of the capillary. After standing for 10 to 15 min in a vertical position the gel gradient is even across the whole diameter of the capillary, due to horizontal diffusion (right capillary in Fig. 47). This can be recognized by the more intense colouration from top to bottom of the tube. The gel mixtures must therefore be made up in such a way that no polymerisation takes place during the first 10 to 15 min. Since the concentration of ammonium peroxydisulphate diminishes continuously from top to bottom on the formation of the gradient, the polymerisation also takes place continuously from top to bottom of the capillary. This has the great advantage that the escape of the heat generated on polymerisation presents no problem, and no convection currents disturb the formation of the continuous gradient.

Using the stock solutions given, the gels polymerise in ca. 45 min. As for the gels used for micro-disc electrophoresis, it is advisable to store gradient gels overnight in a moist chamber (see Fig. 7) before use. After polymerisation, about 10–12 mm of the aqueous solution remains in the end of the capillary above the soft gel tip. This liquid is withdrawn, leaving a sufficient length of capillary for application of the sample.

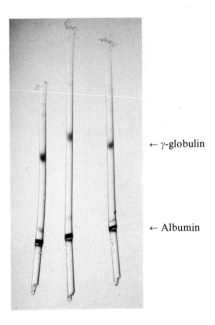

← γ-globulin

← Albumin

Fig. 48. Microphoto (magnification 4×) after fractionation of γ-globulin and albumin on gradient gels of different lengths. The time of electrophoresis was 16 hrs. The albumin remains in the part of the gel with the highest concentration of acrylamide. The gel segment shown in the lower photograph was formed on prolonged storage of the capillaries

The highest gel density (approximately 40%) is found at about 5 to 6 mm above the lower end of the capillary. After expulsion of the gels from the capillaries this region appears as a bulb-shaped thickening (see Figs. 49 and 52). The gel concentration in this region is so high that even after 16 hrs. of electrophoresis, albumin (MW 67000) has not migrated through it (Fig. 48). This high density at the lower end of the gel also enables the gels to be pushed out with a tightly fitting steel wire without damage to the gel above the dense section (i.e. the region in which the proteins are fractionated). This is particularly important if gels must be prepared without Triton X-100, as is necessary for SDS microgels. On longer storage of the polymerised gels the remaining 5–6 mm of the mixture in the lower part of the capillary also polymerises (see Fig. 48); this is unimportant how-ever, because fractionation is carried out only in the section of gel above the high-density region.

If the stock solutions described above are used, the following relationship exists between the usable length of a gel gradient and the diameter of the capillary

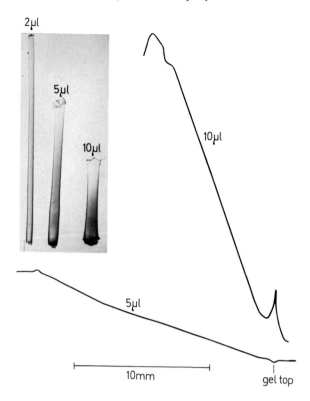

Fig. 49. Relationship between the slope and length of a polyacrylamide gradient, and the inner diameter of capillaries. This was obtained by mixing globulin in the polyacrylamide solution, then staining with amido black after polymerization, and microdensitometry of the gels. (Magnification of the gel photo 6,6×)

in which it is prepared. In a 5 µl capillary of 0.45 mm diameter the gradient has a length of ca. 15 mm. In 2 µl capillaries with a diameter of 0.28 mm, the gradient is appreciably longer (ca. 20 mm), so that under certain conditions there may be too little space over the tip of the gel for the sample. In 1 or 2 µl capillaries a gradient ca. 10 mm long was found by practical experience to be the most suitable. To obtain this length, only about 8 mm each of the solutions D and A/C should be allowed into the capillary. In 10 µl capillaries with a diameter of 0.56 mm, the gradient produced by the same method of filling is appreciably shorter (ca. 8 mm) than in 5 µl capillaries. With increasing capillary diameter the gel gradient thus becomes shorter and steeper (Fig. 49).

The length, and so the steepness of a gradient in a capillary, can also be varied by adding sucrose to the stock solution D. Fig. 50 shows the relationship between the concentration of sucrose and the length of the gradient in 10 µl and 5 µl capillaries. To show this, globulin (0.5 mg/ml) was dissolved in the stock solution A_2/C, and microdensitometer curves of the gels were recorded after staining with amido black. If the concentration of sucrose in solution D is above 25%, no

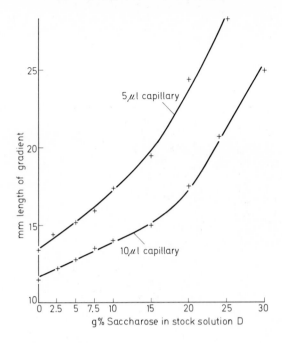

Fig. 50. Influence of the concentration of sucrose added to the gradient mixture, on the length of polyacrylamide gradients in 5 and 10 µl capillaries respectively

gradients are formed; instead, the whole gel polymerises homogeneously throughout its length and corresponds to a normal gel from the micro-disc electrophoresis
preparations. The gel gradients can also be varied by varying the concentration
of acrylamide in solution A/C. This means that the gradient can be adapted for
any separation problem.

Gradient gel electrophoresis can be further simplified so that the sample to
be separated can also be applied to the gels by capillary attraction. First the solution A/C is drawn up to about the middle of the 5 µl capillary. Solution D is then
added, followed by a few mm of the pH 6.7 Tris/H_3PO_4 buffer containing 20%
sucrose. Finally a small volume of the diluted sample is drawn up by capillary
attraction. The capillary is immediately inverted and stored upright in a moist
chamber until polymerisation is complete. If electrophoresis is carried out immediately, so little liquid evaporates from the top that filling the capillary presents no
problem. For longer storage in the moist chamber, e. g. overnight in the refrigerator,
the chamber has to be completely saturated with water. This can be done easily
by lining the beaker with moistened filter paper in such a way that it dips into the
water layer in the petri dish. As electrophoresis of serum proteins has shown,
this procedure gives fractionations which agree with those obtained by the
methods described previously. This method would therefore seem to be particularly useful for routine analyses, e.g. of samples of serum.

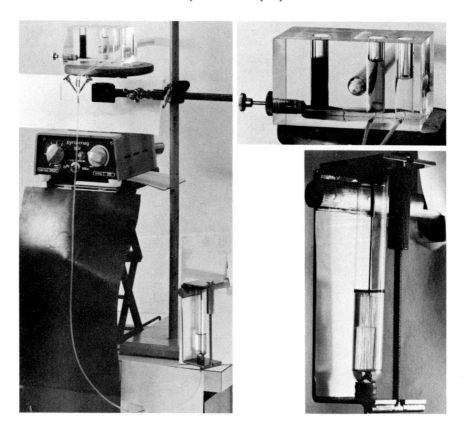

Fig. 51. Apparatus for filling a batch of capillaries with a gradient gel simultaneously and under indentical conditions

DAMES, MAURER and NEUHOFF (1973) have developed the apparatus shown in Fig. 51, which allows up to 180 capillaries of the same or of differing diameters to be filled simultaneously with a gradient gel under identical conditions. This method necessitates the use of a mixing chamber suited to the volumes necessary for filling the capillaries. However, in this method it is very important to neutralise the capillary attraction first. This is achieved by standing the bundle of capillaries on a nylon mesh and holding them upright in a polythene tube by means of a steel carrier. The tube is filled slowly from below with the gel buffer solution containing 0.2 % Triton X-100. The tube is then filled up slowly from below with the gel mixture. The gradient is produced by two communicating vessels, one containing a 2.5 %, and the other a 40 %, solution of acrylamide. Both solutions are mixed with the gel buffer (Tris/H_3PO_4, pH 8.8) and contain TEMED and peroxydisulphate. To increase the stability of the gradient a 10 % solution of sucrose is also added to both solutions together with 0.3 mM potassium ferricyanide to retard polymerisation. The system is made up so that the polymerisation starts after ca. 15 to 20 min, proceeds slowly from top to bottom, and is complete after ca. 45 min. The solutions are mixed thoroughly by a magnetic stirrer while they are evenly siphoned off. Since the gel mixture pushes the aqueous phase before it as it enters the capillary, the gel is always covered with an aqueous layer. In order to send the acrylamide from the mixing chamber into the capillary, a 50% solution of sucrose is finally siphoned through the system. This means that the tubing is cleaned of the gel mixture, which would otherwise polymerise in it, and also that the sucrose solution reaches the nylon mesh on which the capillaries stand. This prevents the capillaries polymerising on to the nylon mesh, and also forms a reliable seal for the bottom of the capillaries. With this method, the steepness of the gradient, its ionic concentration, etc. can be varied widely, so as to suit any specific separation.

SMEDS and BJÖRKMAN (1972), working independently from DAMES et al. (1973), reported an apparatus for a micro scale method for the preparation of rods of polyacrylamide gels having a continuous gradient of concentration. With this equipment the gradient solution is first prepared in a gradient tube, and then two 5, 50, or 100 µl Drummond Microcaps are filled by dipping them into the solution by means of an excentric plate connected to a motor with stepwise speed control.

Protein Fractionation

Fractionation by electrophoresis on polyacrylamide gradient gels can either be carried out with the discontinuous buffer system described by ORNSTEIN (1964) and DAVIES (1964) or by the procedure described by ALLEN, MOORE and DILWORTH (1969). In the first method, a 4 mm layer of $Tris/H_3PO_4$ buffer, pH 6.7 (stock solution B for the micro-disc electrophoresis), is introduced immediately after removal of the liquid over the gel tip. In this method the buffer contains 20% sucrose to raise the viscosity, and is used instead of a 5% spacer gel. The sample is pipetted on top, and electrophoresis is carried out using the same electrode buffer (Tris/glycine pH 8.4) as for micro-disc electrophoresis.

For electrophoresis using the discontinuous system described by ALLEN et al. (1969), the sample, dissolved in diluted (1 : 20) Tris/sulphate buffer (stock A), is applied immediately above the top of the gel. The remaining empty capillary space is filled with a buffer ("cap-buffer") of the same composition and pH as the gel buffer (i.e. 1 : 8 diluted stock solution A without TEMED) instead of the "cap-gel". The usual Tris/glycine buffer, pH 8.4, is used as electrode buffer for the electrophoresis. When the sample is applied in diluted buffer it forms a zone of decreased conductivity between zones of higher conductivity, so that a correspondingly smaller drop in potential occurs with this kind of discontinuity in the buffer system. For this reason, the sample is first of all concentrated in the gel and then fractionated. The glycine/sulphate which initially migrates through the gel after the sample has the effect of sharpening the protein bands in the gel as it overtakes them. Since the polyacrylamide gel gradient retards the migrating protein bands at regions in the gel of corresponding density according to their molecular weight, a clear separation of individual components may be obtained.

Due to the method for making the gel gradients, the ionic strength in the gel also has a gradient, since the stock solution D contains no buffer. Even if gel buffer is added to stock solution D, so that it cannot form a buffer gradient in the gel, this has no effect on the quality of the protein separation but merely increases the duration of the electrophoresis. If bromophenol blue is added to the electrode buffer, the fractionation of the proteins in the gel may be observed directly, and the time appropriate for ending the electrophoresis can be determined.

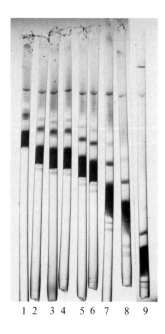

Fig. 52. Fractionation of diluted human serum on polyacrylamide gradients (5 µl capillaries) for different times of electrophoresis.
1: 10 min, 2: 20 min, 3: 25 min, 4: 30 min, 5: 60 min, 6: 90 min, 7: 2 hrs., 8: 4 hrs., 9: 8 hrs.

Since the mobility of the protein bands in gradient electrophoresis does not change appreciably once they have reached the point in the gel corresponding to their molecular weight, increasing the time of electrophoresis has little effect on the sharpness of the bands. When the glycine/sulphate front has passed the fractionated protein bands, continuation of electrophoresis corresponds essentially to the conditions described for continuous zone electrophoresis. Thus, because of the favourable conditions in a gel gradient, the diffusion of the individual bands is avoided and the sharpness of the bands is preserved, as is shown in Fig. 52. Samples (0.5 µl) of human serum diluted 1 : 40 were fractionated on gradient gels in 5 µl capillaries produced as described above. The best fractionation is obtained after 20 to 30 min (gel 2–4) with a constant potential of 60 V. Even after electrophoresis for 2 to 4 hrs. (gel 7 and 8), although the pattern of the bands changes, their relative positions are unaltered. After 8 hrs. electrophoresis (gel 9) at 60 V, the albumin band has stopped in the gel region of highest density, whereas the

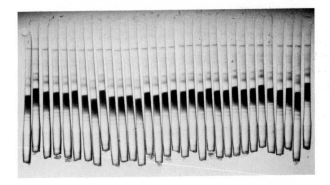

Fig. 53. Fractionation of diluted human serum on 32 polyacrylamide gradient gels (5 µl) under identical conditions

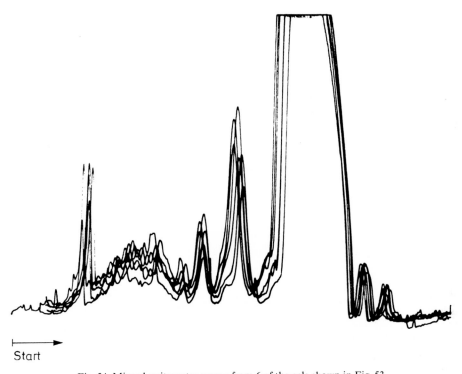

Start

Fig. 54. Microdensitometer curves from 6 of the gels shown in Fig. 53

very high molecular weight macroglobulin has hardly migrated any further in the gel (see also Fig. 48). Between these bands the normal serum proteins can still be recognized as diffuse bands. This diffusion is due to the long time of electrophoresis. If a similar experiment is carried out with appreciably steeper gradients, the relative positions of the bands are largely unaltered even when electrophoresis is continued for up to 72 hrs.

Fig. 53 shows that the production of the gradient gels, and the subsequent fractionation of proteins carried out on them, is reproducible. 0.5 µl samples of human serum diluted 1 : 40 were fractionated on 32 identical gradient gels prepared in 5 µl capillaries with the discontinuous buffer system described by ALLEN et al. (1969). Microdensitometer curves were recorded for 6 gels, and then super-imposed on one another, to demonstrate the good reproducibility (Fig. 54). The reproducibility of gradient formation can be verified further by mixing globulin into the A/C solution, followed by staining of the polymerised gels with amido black. After monitoring the gels with a microdensitometer, the gradient of the stain can be measured. In a series of 33 gels the standard deviation was found to be $\pm 6\%$.

Electrophoresis on gradient gels can also be used to determine the molecular weights of proteins, as is illustrated in Fig. 55. A preparation of 100% pure crystalline human albumin (Fluka A. G., Switzerland), containing several polymeric forms of albumin, was used for this fractionation on a 5 µl gradient gel. After staining with amido black, relative mobilities were determined by microdensitometry as will be described for SDS gradient gel electrophoresis. Besides the albumin peak, four polymeric forms can be detected. In this case the molecular weights ranged between 67000 and 335000. For determining the molecular weights of other proteins, the acrylamide gradient (e.g. slope, length, acrylamide concentration) must be adapted accordingly. A straight line with a correlation coefficient $r_{\lg MW\ mm} = -1.000$ is given when the migration distance is plotted against the logarithm of the respective molecular weights. This finding is in good agreement with results obtained by KOPPERSCHLÄGER et al. (1969) on linear polyacrylamide gradient gels on the macro scale.

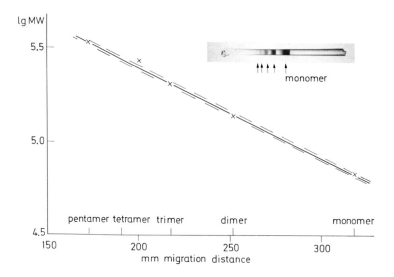

Fig. 55. Fractionation of pure human albumin and its polymeric forms in a 5 µl polyacrylamide gradient gel. (Magnification 3 ×).
The migration distance of the different polymers is plotted against the logarithms of their molecular weights

Fractionation of RNA

Electrophoresis on polyacrylamide gels has frequently been employed for the fractionation of RNA (review see MAURER, 1972). EGYHAZI et al. (1969) and RINGBORG et al. (1968) have isolated RNA from the salivary glands of *Chyronomus tentans* and fractionated nanogram quantities. For this they used agarose polyacrylamide composite gels (PEACOCK and DINGMAN, 1968), and were able to separate RNA components between 28 S and 4 S on the micro scale, with a separation distance of ca. 0.3 mm. NAUMOVA (1971) has frationated RNAs of 28 S to 4 S of differing origins on 2.5% polyacrylamide agarose composite gels in capillaries (diameter 0.2 mm).

WOLFRUM et al. (1973) employ continuous polyacrylamide gel gradients in 5 μl capillaries for the fractionation of RNA. The following stock solutions are necessary:

1) Acrylamide/bis acrylamide (recryst.)
   $T=30\%$ $C=4\%$
   2.88 g acrylamide
   0.12 g bisacrylamide
   to 10 ml H_2O
2a) Gel buffer Tris/H_2SO_4, 1.0 M, pH 9.0
   12.10 g Tris
   ca. 20.66 ml 1 N H_2SO_4 per 100 ml solution
2b) same buffer as above, half concentration (0.5 M)
3) Ammonium peroxydisulphate (PODS) (0.12% w/v in 2% w/v) Triton X-100
4) N,N,N′,N′-tetramethylethylenediamine (TEMED)
   0.24% (v/v) in H_2O
5) Sucrose (RNase free) from Schwarz-Mann, Orangeburg, U.S.A.
   60% (w/v)
6a) Electrode buffer (modified according to MAURER and ALLEN, 1972)
   Tris-borate, 0.065 M, pH 9.0
   7.68 g Tris
   1.925 g boric acid
   0.001 g bromophenol red
   to 1000 ml H_2O
6b) 7.86 g Tris
   1.925 g boric acid
   1.86 g EDTA (Ethylene diamine tetra-acetic acid, disodium salt)
   to 1000 ml H_2O
7) Solution A/C   1 part soln. No. 1
                   1 part soln. No. 2b   give a 7.5% gel on mixing 1:1 (v/v)
8) Solution D   1 part soln. No. 3
                   1 part soln. No. 2b
9) cap buffer: 0.02% Pyronine G in 1:4 diluted solution No. 2b (0.125 M).

10)   Sucrose/PODS   7 parts soln. No. 5
                     2 parts soln. No. 3
                     1 part soln. No. 2a
11)   Sample solution (10 mM Tris/H_2SO_4, pH 9.0)
       1 part soln. No. 2b
       9 parts H_2O
      10 parts soln. No. 5

As with gradient gels for fractionation of proteins, variations may be made over a wide range in composition for gels for fractionation of RNA. The characteristics of the gel can thus be fitted to the separation required, or to the type of sample to be separated. For example, a short (6–8 mm) steep gradient with a maximum concentration at the lower end of 40%, which can be produced from the stock solutions described for fractionation of proteins, is particularly well suited for fractionation of RNA smaller than 14 S (Fig. 56). If a gel system developed for the fractionation of RNA is used for the separation of albumin polymers, the linearity of the acrylamide gradient can be demonstrated by plotting the logarithms of the molecular weights against the relative migration distances, to give a correlation coefficient $r_{lg\,MW\,mm}$ equal to 0.999.

The following is a procedure suitable for the fractionation of RNA species from 28 S to 4 S. Cleaned 5 μl capillaries (see p. 4) are filled with 15 mm of solution D by capillary attraction, and 8–10 mm of solution A/C is then introduced. The gels are placed immediately in a plasticine cushion, or in a 5 ml beaker with a 2–3 mm layer of a sucrose/peroxydisulphate solution at pH 9 (soln. 10; see Fig. 46). The addition of ammonium peroxydisulphate to this solution ensures that the gel which is present in the lower end of the capillary is completely polymerised, thus giving a gel concentration of 15% at this point. This ca. 3–4 mm section of high gel concentration retards fast-migrating components, e.g. 4S RNA, during long fractionations, and is also a good "handle", since the low concentration gradient gel bound to it is very fragile. If a higher gel concentration at the lower end is required, e.g. to give sharper separation of the 4S RNA, the filled capillaries are placed to polymerise in a sucrose solution without ammonium persulfate. After polymerisation, the 2–3 mm sucrose phase is washed from the lower end and replaced by a polymerising acrylamide solution of reasonable concentration (e.g. 30–40%). With the mixture described here, a total length of 12–14 mm gel is obtained. If continuous gradients are to be formed the capillaries must stand upright during polymerisation, and during the sucking up of mixtures into the capillaries. The polymerised gels may be kept for up to 5 days at room temperature in a snap-top tube in solution 2b diluted 1:4 (corresponding to $\frac{1}{2}$ the buffer concentration in the gel).

Before applying the sample the upper aqueous phase is carefully sucked out without damaging the soft gel tip, and the free space in the capillary is rinsed with distilled water. Next, either 1 mm (= 0.16 μl) of a solution containing 1 mg/ml RNA (1 mg RNA is equivalent to 24 E_{260}-units) or up to 10 mm of a 0.1 mg/ml solution is applied. Higher concentrations of RNA cannot be used, since the capacity of the microgel is overloaded and no separation is obtained as shown by Fig. 57 (gel 3 and 4). However, the separation of the RNA components with lower molec-

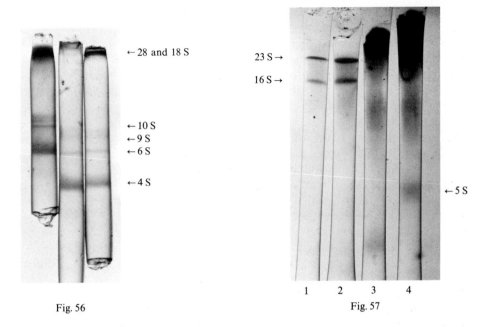

Fig. 56. Microphoto (magnification 14 ×) taken after fractionation of tumor RNA (MOPC 70e, POTTER and KUFF, 1964) on polyacrylamide gradient gels. The highest gel concentration was 40%. Methylene blue was used for staining. The left gel is shorter, so has a steeper gradient. (The RNA preparation was a gift from Dr. B. WEIMANN)

Fig. 57. Fractionation of RNA from E. coli, on polyacrylamide gradient gels (5 µl). Gel 1: 0.08 µg, gel 2: 0.16 µg, gel 3: 0.32 µg, gel 4: 0.64 µg. Methylene blue was used for staining. Magnification 13 ×

ular weights can be obtained with overloaded gels. In the original electrophoresis system of ALLEN, MOORE and DILWORTH (1969), a zone of low ionic strength is necessary, so the RNA to be fractionated must be dissolved in the sample solution (soln. No. 11). The "cap-gel" described by MAURER and ALLEN (1972) is replaced by a 4–6 mm layer of gel buffer (soln. 9) as "cap-buffer". This solution contains 0.02% Pyronine G, which binds to the RNA and indicates the position of the main bands of the RNA during electrophoresis. Since the duration of electrophoresis has a decisive influence on the quality of separation, observation of the separation during electrophoresis is an appreciable advantage. If individual components of the RNA must be eluted from the gel for further analysis, the Pyronine staining serves as a marker for the appropriate sectioning of the gel without further staining. Depending on the quality and the type of the RNA preparation, the reproducibility of fractionation of RNA in micro gradient gels is increased if an electrode buffer containing up to 5 mM EDTA (stock solution 6b) is used. In this case, bromophenol red cannot be used as the front marker, and also the pyronine does not stain the RNA sufficiently during electrophoresis.

If free space still remains in the capillary after charging it with the sample and the gel buffer, this is filled with electrophoresis buffer. If the free capillary space is not sufficient, e.g. when using dilute solutions of RNA, the capillary space can be increased by attaching a further section of capillary (details see p. 11). The apparatus described on p. 13 is used for electrophoresis. A separation takes ca. 30–60 min at a constant potential of 40–60 V. The cathode is at the gel tip end. During this time the moving boundary of the electrophoresis buffer, made visible by the addition of fluorescein or bromophenol red (FELGENHAUER, personal communication), migrates to the lower third of the gel. The initial current of 30–40 μA falls to about 10 μA. The gels are expelled from the capillaries with water pressure after electrophoresis. To protect the very fragile gel tip, a piece of polyethylene tubing, ca. 2 mm long, is fitted on to the upper end of the capillary and filled with Triton X-100 (5 %). The gel is then pressed out of the lower capillary end directly under the surface of the dye solution, using a water syringe fitted to a polyethylene tube (see Fig. 9). The lower end of the gel is sufficiently concentrated to be expelled from the capillary using a tightly-fitting steel wire.

Several dye solutions are suitable for detecting RNA: 0.2 % methylene blue in 10 % acetic acid, 0.2 % toluidine blue in 10 % acetic acid (KONINGS and BLOEMENDAL, 1969) chromalaun staining (GROSSBACH and WEINSTEIN, 1968) "stains all" (DAHLBERG, DINGMANN and PEACOCK, 1969). Methylene blue is particularly well suited for staining 4 S or 5 S RNA (PEACOCK and DINGMAN, 1967). A disadvantage common to all staining methods is that, owing to the small diameter of the gel, low molecular weight RNA diffuses out very rapidly (1–2 hrs.). This means that all densitometric estimations must be performed immediately after staining. Quantitative estimations are difficult with this method, but may be carried out using the UV-Scanning technique. For densitometric evaluations it is advisable to use acetic acid weakly coloured with the dye. For longer storage the gels are frozen at $-20°$ C in this solution.

Using the electrophoresis system given above, gradient gels containing 0.2 % SDS (without Triton X-100), can be used for the fractionation of RNA (LOENING, 1969) if the electrode buffer also contains 0.2 % SDS and 1 mM EDTA. SDS and EDTA are used to inhibit nucleases (compare Figs. 58 and 59, c, d). The buffer system according to LOENING (1969) gives a good separation of RNAs of high molecular weight (Fig. 59 d). RNA of low molecular weight is better separated by the buffer system of ALLEN et al. (1969) or as described above (Fig. 59 c). In this case it is necessary to add 0.2 % SDS to all solutions used, and 1 mM EDTA also to the electrode buffer. If gels containing SDS are stained directly, without previous soaking, the whole gel is stained homogeneously throughout. To destain such gels, even with micro gels, a washing time of more than 12 hrs. is necessary, during which time most of the RNA has diffused out of the gel. However, good staining of micro gels containing SDS is possible if the dye solution contains 30 % (v/v) formamide and 0.2 % (w/v) methylene blue or toluidine blue in 10 % acetic acid. The gels can be transferred to this solution directly after fractionation of the RNA, and are stained for one hour. Destaining is performed in 7.5 % acetic acid.

The ubiquity of ribonucleases demands sterile equipment. The capillary pipettes used are sterilised by the high temperature at which they are produced. Breaking off the long drawn-out tip immediately before use should be carried out

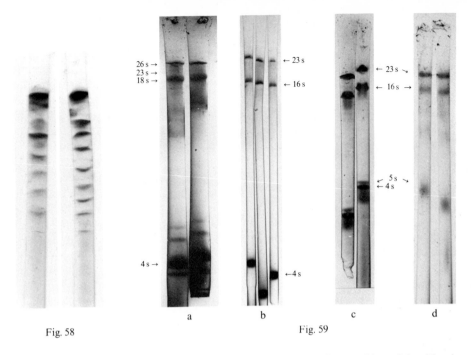

26 s →
23 s →
18 s →

← 23 s

← 16 s

← 23 s
← 16 s →

← 5 s
← 4 s

4 s →

←4 s

a   b   c   d

Fig. 58   Fig. 59

Fig. 58. RNA from *E. coli* after storage at 4° C for two days. Toluidine blue used for staining. Magnification 13 ×

Fig. 59 a–d. Fractionation of RNA from various species on 10 µl (*a*) and 5 µl gels respectively. Toluidine blue staining. For details see text

with a flame-sterilized pair of tweezers. The 5 µl capillaries used are sterilised sufficiently by cleaning them in JAVELLE's solution. All other vessels employed should preferably be siliconised after sterilisation. The cleaned glass tubes necessary for the preparation of RNA are sealed of one end, and filled with a 2% solution of dimethyldichlorosilane in benzene. The glass tubes are kept at 80° C for ca. 2 hrs. and are then ready for use. Fig. 58 shows a preparation of RNA from *E. coli*, which was stored for 2 days at 4° C. The numerous bands, which are well fractionated in the gradient gel, are due to the action of RNase. MCPHIE, HOUNSELL and GRATZER (1966) have described a similar result on RNase digestion of ribosomal RNA from yeast. Fig. 59 shows the separation of RNAs which were isolated from various species. The gels (10 µl) in Fig. 59a show the separation of RNA extracted from *Chlorella pyrenoidose* (kindly supplied by Dr. SYMANK). In *b* a mixture of RNA from *E. coli* and tRNA from yeast was fractionated. In Fig. 59c and d RNA from *E. coli* was fractionated on gradient gels containing SDS (0.2%). In *c* the stock solution 6a was used as electrode buffer with the addition of 0.2% SDS and 1 mM EDTA. In *d* a electrode buffer according to LOENING (1969) was used, but with a pH of 9.0.

Micro-Electrophoresis in SDS Gradient Gels

SDS gel electrophoresis cannot be performed well with the normal micro-disc electrophoresis system since the Triton X-100 contained in the gel mixtures is not compatible with SDS. The Triton X-100 cannot be omitted, because normal micro-disc gels, unlike gradient gels, cannot be pushed out of the capillaries with a steel wire without damaging them to such an extent that meaningful micro-densitometer curves cannot be recorded. If samples containing SDS are fractionated on Triton-containing gels, good separation of the protein bands cannot be achieved.

GAINER (1971) described micro-disc electrophoresis with SDS gels in capillaries 7 cm long, with an inner diameter of 0.58 mm, and homogeneous polyacrylamide gel concentrations of 5, 7.5, 10 or 15%. In his system SDS is not contained in the gel but only in the electrode buffer. In order to expell the gels (which do not contain Triton) by water pressure from the capillaries, the capillaries are first treated with column coat (Canalco). After the electrophoresis the capillaries are frozen at $-30°$ C, and the gels then pressed out of the capillaries while they are just thawing.

The polyacrylamide gradient gels which do not contain Triton X-100 have the great advantage that, owing to the high acrylamide concentration at the lower end, they can be pushed from the capillaries with a steel wire without damaging the section where separation has taken place. For the production of SDS gels, the solutions previously described can be used. The presence of SDS in the gel is unnecessary, because SDS migrates faster than any protein molecule, and sufficient SDS is contained in the electrode buffer to ensure stability to the SDS-protein complex. The production of the gradient gels is carried out as described above. The SDS concentration in the sample to be fractionated should, if possible, be not higher than 1.0 to 2.0%. Samples which contain more SDS fractionate badly. The electrode buffer remains unchanged apart from the addition of 0.1% SDS. Fig. 60 illustrates the fractionation of purified myelin (review, see WAEHNELDT and NEUHOFF, 1972) from mouse brain on an SDS polyacrylamide gradient gel in a 5 µl capillary. The pattern of proteins obtained after staining with amido black is similar to those reported else where. The additional protein component in myelin, described by AGRAWAL et al. (1972), and migrating between the proteolipid protein and the two basic proteins, was also found in the microgels.

As Fig. 60 shows, brain proteins can be fractionated with a discontinuous SDS electrophoresis system, as SDS can solubilise but not dissociate brain proteins and glycoproteins (KATZMAN, 1972). Conversely, if water-soluble proteins are to be fractionated after SDS treatment, the electrophoresis system must be selected with care. Fig. 61 shows the fractionation (RÜCHEL et al., 1973) of a solution of human albumin which contains four polymeric forms in addition to the monomer. The albumin solution was incubated with 10 mg SDS per 1 mg protein for 5 min at 100° C and 0.3 µl used for electrophoresis. Fractionation was performed on gradient gels with the discontinuous system described above. The Tris/glycine electrode buffer and the "cap buffer" contained 0.1% SDS. After staining with amido black (Fig. 61 b) at least 12 bands are demonstrable. The same

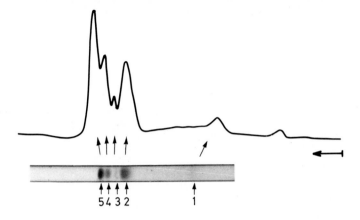

Fig. 60. Fractionation of purified myelin protein of mouse brain on SDS polyacrylamide gradient gel in a 5 µl capillary.
1: Wolfgram's protein, 2: proteolipid protein, 3: protein band DM-20 of AGRAWAL et al., 4 and 5: basic proteins

SDS-treated albumin solution fractionated, not with the discontinuous electrophoresis system but with a continuous system, displays the anticipated pattern of fractionation of albumin and its polymers (Fig. 61a). In this case 0.1 M sodium phosphate buffer pH 8.1+0.1% SDS was used as gel and electrode buffer.

The unexpected result obtained with discontinuous electrophoresis of the albumin solution could be due to different charging conditions of the albumin species with SDS. The result in Fig. 61c and d supports this. Discontinuous electrophoresis was first performed as in b, the electrode buffer was then changed for a Tris/glycine buffer of the same type but without SDS, and the electrophoresis continued for a further 30 min c or 60 min d at 60 V. Subsequently the gels were stained for 5 min in 0.5% toluidine blue in 10% acetic acid, and destained for 20 min in 7.5% acetic acid. This causes a complex to form from SDS and toluidine blue, which is dark violet and remains stable on storage for several hours in 7.5% acetic acid. The sulphate contained in the gel buffer also forms a complex with toluidine blue; this differs unmistakably from the intensely coloured SDS-toluidine complex by its pale blue colouration and the turbidity of the gel. Fig. 61c shows an intensive SDS peak at the migration front with toluidine blue staining. The rest of the gel, with toluidine blue staining alone, appears empty. If the same gel is stained with amido black 10B subsequent to the toluidine blue staining, the fractionated protein is also displayed. A comparison of Fig. 61b and c shows fewer bands after 30 min of SDS-free electrophoresis and still fewer after 60 min. Its appearance is now largely identical with that of fractionation with continuous electrophoresis. SDS is no longer detectable in these gels. On electrophoresis with Tris/glycine buffer it was not only removed from the gel, but also separated from the albumin polymers. This was also shown by autoradiography of the gels under the same experimental conditions but with ^{35}S-SDS.

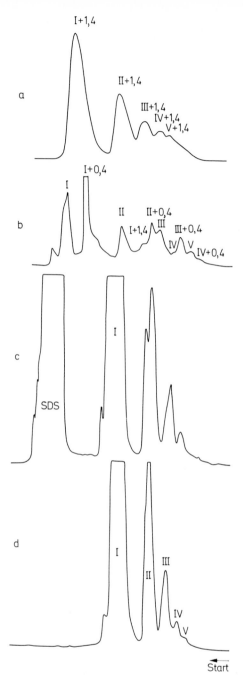

Fig. 61. Microdensitometer curves after fractionation of SDS-treated human albumin in gradient gels and amido black staining. *a* continuous electrophoresis system with 0.1% SDS in sodium phosphate buffer, *b* discontinuous system with 0.1% SDS in Tris/glycine electrode buffer, *c* initially discontinuous electrophoresis as in *b*, then replacement of the SDS-containing electrode buffer by SDS-free buffer and further electrophoresis for 30 min. Then, first toluidine blue staining to demonstrate SDS, and finally amido black staining of the same gel, *d* same experimental and staining conditions as in *c*, but with the second SDS-free electrophoresis carried out for 60 min at 60 V

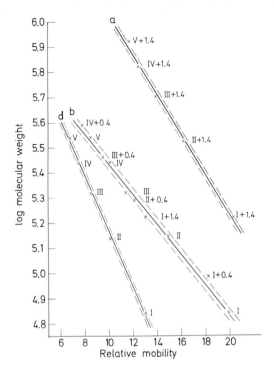

Fig. 62. Evaluation of the curves from Fig. 61 *a*, *b* and *d*. A perpendicular was dropped from each peak maximum to the baseline, and the gel top used as origin for measurement of the relative mobility. The logarithm of the molecular weight is plotted against the relative mobility. I = albumin monomer, II = dimer, III = trimer, IV = tetramer, V = pentamer. For calculation of the molecular weights in curve *a*, a SDS charging of the albumin (MW 69000) of 1.4 g SDS per g albumin was taken as a basis; for curve *d*, freedom from SDS was assumed. In curve *b* the states of SDS charging of the various forms are assumed to be I + 0.4 and I + 1.4 respectively, II + 0.4, III + 0.4 and V + 0.4, the figures referring to 0.4 g SDS or 1.4 g SDS per g albumin. The correlation coefficients $r_{\log MW\,rel.\,mobility}$ are -0.998 for curve *a*, -0.995 for curve *b*, and -0.999 for curve *d*. The dotted line indicates the standard deviation of observations

From these findings an explanation may be derived for the multiplicity of bands in Fig. 61 *b*. REYNOLD and TANFORD (1970) have shown that a saturated complex with the stoichiometry of 0.4 g of SDS per g of protein is formed between 5 and 8×10^{-4} M SDS monomer, and a second complex saturated at 1.4 g of SDS per g of protein is observed above 8×10^{-4} M SDS monomer. Under the experimental conditions chosen here, SDS was always present in the excess necessary to ensure maximal charging of the protein with SDS. Fig. 61a shows the corresponding normal pattern of fractionation of albumin polymers in a continuous electrophoresis. In contrast, in discontinuous electrophoresis, the glycine/sulphate front migrates over the SDS-charged protein, part of the protein is completely stripped of SDS, and another part is partially stripped, so that each of the original bands may present three different states of charging with SDS. Correspondingly, a straight line is obtained as in Fig. 62 *b* if the logarithm of the molecular weight corresponding to the charging state is plotted against the linear relative migration. In Fig. 62 *a* charging with 1.4 g SDS/g protein is taken as the basis for the graphic representation. A straight line is obtained here as well. After complete electrophoretic removal of SDS (Fig. 61 *d*) the evaluation in Fig. 62 *d* also gives a straight line, provided the logarithm of the molecular weights is plotted against relative migration distance.

This example shows that when conditions for electrophoresis are correctly adjusted, the SDS microgradient gel system can be used for the determination of molecular weights. It is advantageous to use for calibration only a narrow range of molecular weights so that the molecular weights of the fractionated proteins may be determined. The slope of the polyacrylamide gradient has also to be adapted to the molecular weight of proteins to be fractionated. The molecular weight determination is carried out after electrophoresis, staining with amido black, and analysis of the respective microdensitometer curve. A perpendicular is dropped from each peak maximum of the protein bands to the baseline, and any point of the curve, e.g. the gel top, can be used to measure the relative migration distance. The migration distance is plotted against the logarithm of known molecular weights of the reference proteins and the resulting regression line calculated; the molecular weight of the protein under investigation can then be determined relative to the curve.

Observations comparable to those described above for SDS-charged albumin have been made for other proteins decomposed into subunits by SDS treatment. It is therefore necessary to check the type of electrophoresis system used both to determine the molecular weight of subunits by SDS gel electrophoresis, or for analytical SDS gel fractionation. The continuous SDS gel electrophoresis system is usually preferable, since fewer problems of interpretation arise, but the continuous electrophoresis system, even without SDS, has the disadvantage of requiring appreciably longer electrophoresis times and the bands are not as sharp, even in gradient gels.

Autoradiography of Microgels

The autoradiography of microgels is of particular interest for recording, for example, incorporation in or binding of radioactive substances to proteins. Placing the round, moist gels on a sensitive X-ray film is unsatisfactory because of the small area of contact and because the gels dry out and shrink unevenly during the exposure period. Both draw-backs are eliminated by freeze-drying the microgels before autoradiography. The gels can be initially stained with amido black and evaluated densitometrically. They are then transferred to a Petri dish containing about 2 mm of water, and the whole is frozen; then the Petri dish is placed on a second, precooled plate and the contents are lyophilized. The embedding in ice ensures that the gels retain their external form on freeze-drying and are therefore unable to shrink. If further stabilization of the gels during freeze-drying is desired, they can be embedded in a 2 mm layer of 0.5–1% agarose from which they are easily separated after drying. A more rapid method of drying is possible by holding the gels for a few minutes vertically in absolute ethanol with a pair of soft tweezers. For complete dehydration they have to stand approximately 30 min in abs. ethanol. During this procedure the gels shrink more or less evenly so that autoradiograms may be prepared.

If the gels are not already flat after drying, they may be pressed between two glass plates. They are then placed on a 3×3 cm glass plate and fixed to the plate at each end with a small drop of celloidin (1 % in ether/ethanol 50 : 50). Finally the sensitive film from the radiation protection badges (see Chapter 2) and a second glass plate are laid on the top in the darkroom, and the two plates fixed in position with adhesive tape. After exposure for a suitable length of time, the film is developed as described in chapter 2. For the detection of very weak radio-activity, e.g. ^3H radioactivity in faint protein bands, it may be advisable to use in stead of the X-ray film the striping film technique (see Chapter 11). When dried gels are placed in 7.5 % acetic acid, they swell up and assume their original shape, and the previously stained bands appear unchanged. The gels could, therefore, be cut into slices and further evaluated by liquid scintillation counting. For the example in Fig. 63 a human serum albumin preparation containing polymers as well as monomer was treated with ^{14}C-dansyl chloride (specific activity 98 Ci/mole) and fractionated in a gradient gel as described previously. After staining with amido black, the gels were lyophilized and exposed to the X-ray film for 8 hrs. In addition to the intense blackening of the albumin monomer, 3 further bands of polymeric forms are clearly demonstrable.

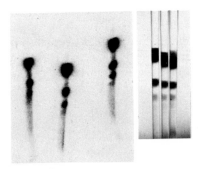

Fig. 63. On the left is the autoradiogram obtained from three freeze-dried gels after fractionation and staining with amido black of a human albumin solution (right side) containing the monomer and three polymeric forms. The albumin solution was reacted with ^{14}C-dansyl chloride prior to electrophoresis. Exposure time 8 hours, magnification $\times 3$

Sources of Error in Gradient Gel Electrophoresis

The stock solutions, particularly the concentrated solution C, must be warmed to room temperature before use. If the solutions are too cold, acrylamide can crystallize out in the capillary at the lower, concentrated, end of the gel. If solution C, which crystallizes easily in the cold, is not warmed carefully and shaken well, under certain conditions it becomes heterogeneous, and gels with variable gradients are formed.

If gels are stored for longer than one week, and do not contain Triton, they can become partially detached from the capillary wall. On protein fractionation some of the proteins leaks into the space between the gel and the capillary wall, and on staining with amido black appears as streaks at the edge.

If too little solution D is used in relation to the solution A/C, complete polymerisation of the gel results, and gradients are then no longer well-defined. Since no aqueous phase remains above the head of the gel, a gel tip is formed which is unusable for fractionations. At bisacrylamide concentrations under 1% of the acrylamide concentration, gels are formed which are very sticky and consequently are easily contaminated and are unsuitable for densitometric estimations. With bisacrylamide concentrations over 3% the resulting gels are very rigid.

If the capillaries are not carefully cleaned as described, bands due to contaminating proteins may appear when fractionation is performed. A similar effect is also be observed with siliconised capillaries; siliconising the capillaries is not recommended, since, even if they contain no Triton, the gradient gels slip out of the capillaries during electrophoresis. In addition, bands are produced which are convex to the direction of migration.

If polymerisation is too fast (due to too high a concentration of TEMED or ammonium peroxydisulphate), heterogeneous gradient gels are formed. On fractionation in such gels, the bands are arrow-shaped, the head of the arrow pointing away from the direction of migration. Bands which are arrow-shaped in the direction of migration are formed if the fractionation is carried out with too high a voltage or current; the resultant local heating in the centre of the gel leads to an increase in the mobility of the proteins which are migrating there.

If polymerisation occurs with the capillary leaning to one side instead of in an upright position, diagonal bands result on fractionation. Deformed or tailing bands are a sign of overloading; in this case the very soft head of the gradient gel frequently breaks down also.

A milky turbidity in the gel is either due to too rapid a polymerisation, too high a concentration of bisacrylamide, or to traces of acetone in the capillaries.

If incompletely filled capillaries are placed in solutions of sucrose to seal them, the capillaries fill up completely. In this case there is no longer any space for applying the sample; alterations in the gradient can also occur.

Air bubbles in the gel are due to incomplete sealing of the capillaries; this can be overcome by pushing the gel capillaries into a pre-cooled plasticine layer, or by placing the end in a 50% sucrose solution.

Concluding Remarks

It is not surprising that the first applications of polyacrylamide electrophoresis on the microscale (PUN and LOMBROZO, 1964) were in the field of neurochemistry, where it is often necessary to undertake assays with only small quantities of tissue. As a consequence, developments and refinements of this micromethod have frequently come from neurochemical laboratories. In recent years it has also been found that micro-electrophoresis on polyacrylamide gels is very suitable for many different problems of fractionation. This method has the advantage of requiring a very small quantity of starting material, and also saves a considerable amount of time. Furthermore, the cost of the basic equipment is small, the microdensitometer being the only expensive item. When the initial difficulties (which are often due more to an over-critical standardisation of all micromethods rather than to difficulties in the specific technical procedure), have been overcome, micro-electrophoresis can be used with advantage even when the quantity of material is sufficient for polyacrylamide gel electrophoresis on the normal scale.

The examples described serve to show that practically all fractionations which are performed on polyacrylamide gels on the macro scale can also be carried out with micro-electrophoresis. For some applications, e.g. for investigations of enzyme kinetics, the micromethod has a distinct advantage, (and indeed may be the only possible method of analysis) owing to the short distances for diffusion in the microgels.

As the examples cited demonstrate, the buffer systems, etc. which are used for the macro scale cannot always be transferred without alteration to the micro system. As a rule, however, only very small changes in the standard macro procedures are necessary. An experienced experimenter conversant with the technique described should find it easy to adapt micro-electrophoresis on polyacrylamide gels to any specific problem of separation.

Literature

AGRAWAL, H.C., BURTON, R.M., FISHMAN, M.A., MITCHELL, R.F., PRENSKY, A.L.: Partial characterization of a new myelin protein component. J. Neurochem. **19**, 2083–2089 (1972).

ALLEN, R.C., MOORE, D.J., DILWORTH, R.H.: New rapid electrophoresis procedure employing pulsed power in gradient gels at a continuous pH: The effect of various discontinuous buffer systems on esterase zymograms. J. Histochem. Cytochem. **17**, 189–190 (1969).

ALTHAUS, H.H., BRIEL, G., DAMES, W., NEUHOFF, V.: Zelluläre und molekulare Grundlagen der nervösen Erregungsspeicherung. 2. Neurochemische Mikroanalysen des Rückenmarks der Katze nach posttetanischer Potenzierung monosynaptischer Reflexe. In: Sonderforschungsbereich 33, Nervensystem und biologische Information, Göttingen 1969–1972, S. 107–121.

ANSORG, R., DAMES, W., NEUHOFF, V.: Micro disc electrophorese von Hirnproteinen. II. Untersuchung verschiedener Extraktionsverfahren. Arzneimittel-Forsch. (Drug. Res.) **21**, 699–710 (1971).

ANSORG, R., NEUHOFF, V.: Micro disc electrophoresis of brain proteins. III. Heterogenity of the nervous specific protein S-100. Int. J. Neurosci. **2**, 151–160 (1971).

APELLA, E., MARKERT, C.L.: Dissociation of lactate dehydrogenase into subunits with guanidine hydrochloride. Biochem. biophys. Res. Commun. **6**, 171–176 (1961).

BISPINK, G., DAMES, W.: Personal commutation.

BONAVITA, V., GUARNERI, R.: Lactate-dehydrogenase isoenzymes in nervous tissue: III. Regional distribution on ox brain. J. Neurochem. **10**, 755–764 (1963).

BREWER, G.J., SING, C.F.: An introduction to isoenzyme techniques. New York and London: Academic Press 1970.

BRINSTER, R.L.: A method for *in vitro* cultivation of mouse ova from two-cell to blastocyst. Exp. Cell Res. **32**, 205–208 (1963).

BURGESS, R.R.: Separation and characterization of the subunits of ribonucleic acid polymerase. J. biol. Chem. **244**, 6168–6178 (1969).

Burgess, R. R.: RNA-Polymerase. Ann. Rev. Biochem. **40**, 711–740 (1971).

Burgess, R. R., Travers, A.: *Escherichia coli* RNA polymerase: purification, subunit structure, and factor requirements. Fed. Proc. **29**, 1164–1169 (1970).

Cahn, R. D., Kaplan, N. O., Levene, L., Zwilling, E.: Nature and development of lactic dehydrogenase. Science **136**, 962–969 (1962).

Catsimpoolas, N.: Micro isoelectric focusing in polyacrylamide gel columns. Analyt. Biochem. **26**, 480–482 (1968).

Choules, G. L., Zimm, B. H.: An acrylamide gel soluble in scintillation fluid. Its application to electrophoresis at neutral and low pH. Analyt. Biochem. **13**, 336–344 (1965).

Conway-Jacobs, A., Lewin, L. M.: Isoelectric focusing in acrylamide gels: Use of amphoteric dyes as internal markers for determination of isoelectric points. Analyt. Biochem. **43**, 394–400 (1971).

Cremer, Th., Dames, W., Neuhoff, V.: Micro-disc electrophoresis and quantitative assay of glucose-6-phosphate dehydrogenase at the cellular level. Hoppe-Seylers Z. physiol. Chem. **353**, 1317–1329 (1972).

Dale, G., Latner, A.: Isoelectric focusing in polyacrylamide gels. Lancet **1968 I**, 847–848.

Dahlberg, A. E., Dingman, C. W., Peacock, A. C.: Electrophoretic characterization of bacterial polyribosomes in agarose acrylamide composite gels. J. molec. Biol. **41**, 139–147 (1969).

Dames, W., Maurer, H. R., Neuhoff, V.: In preparation.

Davies, B. J.: Disc-electrophoresis. II. Method and application to human serum proteins. Ann. N. Y. Acad. Sci. **121**, 404–427 (1964).

Doerr, P., Chrambach, A.: Anti-estradiol antibodies: isoelectric focusing in polyacrylamide gel. Analyt. Biochem. **42**, 96–107 (1971).

Egyhazi, E., Daneholt, B., Edström, J.-E., Lambert, B., Ringborg, U.: Low molecular weight RNA in cell components of *chironomus tentans* salivary glands. J. molec. Biol. **44**, 517–532 (1969).

Gainer, H.: Isoelectric focusing of proteins at 10^{-10} to 10^{-9} gram level. Analyt. Biochem. (in press).

Gainer, H.: Micro disc electrophoresis in sodium dodecyl sulphate: An application to the study of protein synthesis in individual, identified neurons. Analyt. Biochem. **44**, 589–605 (1971).

Glaser, L., Brown, D. H.: Purification and properties of D-glucose-6-phosphate dehydrogenase. J. biol. Chem. **216**, 67–69 (1955).

Griffith, A., LaVelle, A.: Developmental protein changes in normal and chromatolytic facial nerve nuclear regions. Exp. Neurol. **33**, 360–371 (1971).

Grossbach, U.: Acrylamide gel electrophoresis in capillary columns. Biochim. biophys. Acta (Amst.) **107**, 180–182 (1965).

Grossbach, U.: Chromosomen-Aktivität und biochemische Zelldifferenzierung in den Speicheldrüsen von *Camptochironomus*. Chromosoma (Berl.) **28**, 136–187 (1969).

Grossbach, U.: Chromomeren-Aktivität und zellspezifische Proteine bei *Camptochironomus*. Untersuchungen zur Zelldifferenzierung in den Speicheldrüsen von *C. tentans* und *C. pallidivittatus*. Habilitationsschrift, Universität Stuttgart-Hohenheim, 1971.

Grossbach, U.: Chromosomen-Struktur und Zell-Funktion. Mitt. Max-Planck-Ges., H. 2. 93–108 (1971).

Grossbach, U.: Micro electrophoresis and micro isoelectric focusing in capillary gels. In: Small conference on electrophoresis and isoelectric focusing in polyacrylamide gel: standardization, biochemical and clinical needs, new methods. October 1972 at Tübingen, Germany.

Grossbach, U., Weinstein, I. B.: Separation of ribonucleic acids by polyacrylamide gel electrophoresis. Analyt. Biochem. **22**, 311–320 (1968).

Grzeschik, K. H., Allderdice, P. W., Grzeschik, A., Opitz, J. M., Müller, O. J., Siniscalco, M.: Cytological mapping of human X-linked genes by use of somatic cell hybrids involving an X-autosome translocation. Proc. nat. Acad. Sci. (Wash.) **69**, 69–73 (1972).

Haglund, H.: Isoelectric focusing in pH gradients. A technique for fractionation and characterization of ampholytes. In: Methods of biochemical analysis, ed. by D. Glick, vol. 19, p. 1–104. New York, London, Sydney, Toronto: Intersci. Publ. J. Wiley & Sons, Inc. 1971.

Harris, H., Watkins, J.: Hybrid cells from mouse and man: artificial heterokaryons of mammalian cells from different species. Nature (Lond.) **205**, 640–646 (1965).

Hydén, H., Bjurstam, K., McEwen, B.: Protein separation at the cellular level by micro-disc electrophoresis. Analyt. Biochem. **17**, 1–15 (1966).

Hydén, H., Lange, P. W.: Micro-electrophoretic determination of protein and protein synthesis in the 10^{-9} to 10^{-7} gram range. J. Chromatog. **35**, 336–351 (1968).

HYDÉN, H., LANGE, P. W.: Protein changes in different brain areas as a function of intermittent training. Proc. nat. Acad. Sci (Wash.) **69**, 1980–1984 (1972).

HYDÉN, H., LANGE, P. W.: Protein synthesis in hippocampal nerve cells during re-reversal of handedness in rats. Brain Res. **45**, 314–317 (1972).

KATZMAN, R. L.: The inadequacy of sodium dodecyl sulfate as a dissociative agent for brain proteins and glycoproteins. Biochim. biophys. Acta (Amst.) **266**, 269–272 (1972).

KOLIN, A.: Separation and concentration of proteins in a pH field combined with an electric field. J. Chem. Phys. **22**, 1628–1629 (1954).

KONINGS, R. N. H., BLOEMENDAL, H.: Synthesis of lens protein in vitro. 3. Ribonucleic acid with template activity isolated from calf lens tissue. Europ. J. Biochem. **7**, 165–173 (1969).

KOPPERSCHLÄGER, G., DIEZEL, W., BIERWAGEN, B., HOFMANN, E.: Molekulargewichtsbestimmungen durch Polyacrylamid Gel-Elektrophorese unter Verwendung eines linearen Gelgradienten. FEBS Letters **5**, 221–224 (1969).

KOWALEWSKI, S. L.: Die Isoenzyme der Lactatdehydrogenase. In: Biochemie und Klinik, Monographien in zwangloser Folge, ed. by G. WEITZEL und N. ZÖLLNER. Stuttgart: G. Thieme 1972.

LOENING, U. E.: The determination of the molecular weight of ribonucleic acid by polyacrylamide gel electrophoresis. Biochem. J. **113**, 131–138 (1969).

LÖHR, G. W., WALLER, H. D.: Glucose-6-phosphate dehydrogenase. In: Methoden der enzymatischen Analyse, ed. H. U. BERGMEYER, 2nd ed., p. 599–606. Weinheim: Verlag Chemie 1970.

LOUIS, B. G., FITT, P. S.: Isolation and properties of highly purified *Halobacterium cutirubrum* deoxyribonucleic acid-dependent ribonucleic acid polymerase. Biochem. J. **127**, 69–80 (1972).

LOWRY, O. H., PASSONNEAU, J. V.: A flexible system of enzymatic analysis. New York and London: Academic Press 1972.

LOWRY, O. H., PASSONNEAU, J. V., SCHULZ, D. W., ROCK, M. K.: The measurement of pyridine nucleotides by enzymatic cycling. J. biol. Chem. **236**, 2746–2755 (1961).

MATSUOKA, Y., MOORE, G. E., YAGI, Y., PRESSMANN, D.: Production of free light chains of immunoglobulin by a hematopoetic cell line derived from a patient with multiple myeloma (32327). Proc. Soc. exp. Biol. (N. Y.) **125**, 1246–1258 (1967).

MAURER, H. R.: Disc electrophoresis and related techniques of polyacrylamide gel electrophoresis. Berlin-New York: Walter de Gruyter 1971.

MAURER, H. R., ALLEN, R. C.: Useful buffer and gel systems for polyacrylamide gel electrophoresis. Z. klin. Chem. **10**, 220–225 (1972).

McEWEN, B., HYDÉN, H.: Study of specific brain proteins on the semi-micro scale. J. Neurochem. **13**, 823–833 (1966).

McPHIE, P., HOUNSELL, J., GRATZER, W. B.: The specific cleavage of yeast ribosomol RNA with nuclease. Biochemistry **5**, 988–993 (1966).

MILLER, O. J., COOK, P. R., MEERA KHAN, P., SHIN, S., SINISCALCO, M.: Mitotic separation of two human X-linked genes in man-mouse somatic cell hybrids. Proc. nat. Acad. Sci. (Wash.) **68**, 116–120 (1971).

NAUMOVA, L. P.: Technika Mikroelektrophoresa na Smeshannych Blyakrylamid Agarosnych Geliach. Shkola Seminar Mikrometody Analysa Nukleinowych Kislot. Inst. Org. Cem. Novosibirsk/UdSSR 17–23 (1971).

NEUHOFF, V.: Micro-Disc-Electrophorese von Hirnproteinen. Arneimittel-Forsch. **18**, 35–39 (1968).

NEUHOFF, V., LEZIUS, A.: Nachweis der Substruktur von DNA-Polymerasen, der enzymatisch aktiven Proteinkomponente und ihrer Enzym-Substrat-Komplexe mit der Micro-Disc-Electrophorese. Hoppe-Seylers Z. physiol. Chem. **348**, 1239 (1967).

NEUHOFF, V., LEZIUS, A.: Nachweis und Charakterisierung von DNS Polymerasen durch Micro-Disc-Electrophorese. Z. Naturforsch. **23**b, 812–819 (1968).

NEUHOFF, V., MÜHLBERG, B., MEIER, J.: Strom- und spannungskonstantes Netzgerät für die Micro-Disc-Electrophorese. Arzneimittel-Forsch. **17**, 649–651 (1967).

NEUHOFF, V., SCHILL, W.-B.: Kombinierte Mikro-Disk-Elektrophorese und Mikro-Immunpräzipitation von Proteinen. Hoppe-Seylers Z. physiol. Chem. **349**, 795–800 (1968).

NEUHOFF, V., SCHILL, W.-B., JACHERTS, D.: Nachweis einer RNA-abhängigen RNA-Replicase aus immunologisch kompetenten Zellen durch Mikro-Disk-Elektrophorese. Hoppe-Seylers Z. physiol. Chem. **351**, 157–162 (1970).

NEUHOFF, V., SCHILL, W.-B., STERNBACH, H.: Mikro-Disk-elektrophoretische Analyse reiner DNA-abhängiger RNA-Polymerase aus *Escherichia coli*. I. Struktur und Matrizen-abhängige Funktion. Hoppe-Seylers Z. physiol. Chem. **349**, 1126–1136 (1968).

NEUHOFF, V., SCHILL, W.-B., STERNBACH, H.: Mikro-Disk-elektrophoretische Analyse reiner DNA-abhängiger RNA-Polymerase aus *E. coli*. II. Vergleichende Analyse verschiedener Enzympräparate. Arzneimittel-Forsch. (Drug Res.) **19**, 336–339 (1969 a).

NEUHOFF, V., SCHILL, W.-B., STERNBACH, H.: Mikro-Disk-elektrophoretische Analyse reiner DNA-abhängiger RNA-Polymerase aus *Escherichia coli*. IV. Isolierung und Charakterisierung elektrophoretisch getrennter Enzymfraktionen. Hoppe-Seylers Z. physiol. Chem. **350**, 767–774 (1969 b).

NEUHOFF, V., SCHILL, W.-B., STERNBACH, H.: Microanalysis of pure deoxyribonucleic acid-dependent ribonucleic acid polymerase from *Escherichia coli*. Biochem. J. **117**, 623–631 (1970).

NOVOTNY, G. E. K.: Untersuchung der Proteinbeschaffenheit funktioneller Systeme des Zentralnervensystems und deren Veränderung durch experimentelle Eingriffe. In: Sonderforschungsbereich 33, Nervensystem und biologische Information, Göttingen 1969–1972, p. 247–280.

OELSCHLEGEL, F. J., Jr., STAHLMANN, M. A.: Cyanide sensitive tetrazolium oxidase and its role in dehydrogenase staining. Analyt. Biochem. **42**, 338–341 (1971).

ORNSTEIN, L.: Disc-electrophoresis. I. Background and theory. Ann. N. Y. Acad. Sci. **121**, 321–349 (1964).

PASANTES-MORALES, H., KLETHI, J., URBAN, P. E., MANDEL, P.: Changes in the lactate and malate dehydrogenase isoenzyme patterns of chicken embryo brain and retina. J. Neurochem. **19**, 1183–1188 (1972).

PEACOCK, A., DINGMAN, C. W.: Analytical studies on nuclear ribonucleic acid using polyacrylamide gel electrophoresis. Biochemistry **7**, 659–668 (1968).

PEAKCOCK, A. C., DINGMAN, C. W.: Resolution of multiple ribonucleic acid species by polyacrylamide gel electrophoresis. Biochemistry **6**, 1818–1827 (1967).

PLAGEMANN, P. G. W., GREGORY, K. F., WROBLEWSKI, F.: The electrophoretically distinct forms of mammalian lactic dehydrogenase: I. Distribution of lactic dehydrogenases in rabbit and human tissues. J. biol. Chem. **235**, 2282–2287 (1960).

POTTER, M., KUFF, E. L.: Disorders in the differentiation of protein secretion in neoplastic plasma cells. J. molec. Biol. **9**, 537–544 (1964).

PUN, J. Y., LOMBROZO, K.: Microelectrophoresis of brain and pineal protein in polyacrylamide gel. Analyt. Biochem. **9**, 9–20 (1964).

QUENTIN, C.-D., NEUHOFF, V.: Micro-isoelectric focusing for the detection of LDH isoenzymes in different brain regions of rabbit. Int. J. Neurosci. **4**, 17–24 (1972).

RAYMOND, S., WEINTRAUB, L.: Acrylamide gel as a supporting medium for zone electrophoresis. Science **130**, 711 (1959).

REIFENRATH, R., ELLNEL, J.: In preparation.

REYNOLDS, J. A., TANFORD, CH.: The gross conformation of protein-sodium dodecyl sulfate complexes. J. biol. Chem. **245**, 5161–5165 (1970).

RILEY, R. F., COLEMAN, M. K.: Isoelectric fractionation of proteins on a microsclae in polyacrylamide and agarose matrices. J. Lab. clin. Med. **72**, 714–720 (1968).

RINGBORG, U., EGYHAZY, E., DANEHOLT, B., LAMBERT, B.: Agarose acrylamide composite gels for microfractionation of RNA. Nature (Lond.) **220**, 1037–1039 (1968).

ROSE, I. A.: The use of kinetic isotope effects in the study of metabolic control. I. Degradation of glucose-1-D by the hexose monophosphate pathway. J. biol. Chem. **236**, 603–609 (1961).

RÜCHEL, R., MESECKE, S., WOLFRUM, D. I., NEUHOFF, V.: In preparation (1973).

RUDDLE, F. H., CHAPMAN, V. M., RICCIUTI, F., MURNANE, M., KLEBE, R., MEERA KHAN, P.: Linkage relationships of seventeen human gene loci as determined by man-mouse somatic cell hybride. Nature (Lond.) New Biol. **232**, 69–73 (1971).

SIEPMANN, R., STEGEMANN, H.: Enzym-Elektrophorese in Einschluß-Polymerisaten des Acrylamids. A: Amylasen, Phosphorylasen. Z. Naturforsch. **22 b**, 949–955 (1967).

SMEDS, S., BKÖRKMAN, U.: Micro-scale protein separation by electrophoresis in continuous polyacrylamide concentration gradients. J. Chromatogr. **71**, 499–505 (1972).

STEGEMANN, H.: Enzym-Elektrophorese in Einschluß-Polymerisation des Acrylamids. B. Polygalakturonasen (Pektinasen). Hoppe-Seylers Z. physiol. Chem. **348**, 951–952 (1967).

SVENSSON, H.: Isoelectric fractionation, analysis, and characterization of ampholytes in natural pH gradients: I. The differential equation of state of solute concentrations at a steady state and its solution for simple cases. Acta chem. scand. **15**, 325–341 (1961).

VESTERBERG, O., SVENSSON, H.: Isoelectric fractionation, analysis, and characterization of ampholytes in natural pH gradients: IV. Further studies on the resolving power in connection with separation of myoglobins. Acta chem. scand. **20**, 820–834 (1966).

WAEHNELDT, T V., NEUHOFF, V.: Membrane proteins of the nervous system. Demonstration of different protein profiles in whole brain and its subcellular particles. Naturwissenschaften **59**, 232–239 (1972).

WIELAND, T., PFLEIDERER, G., HAUPT, J., WÖRNER, W.: Über die Verschiedenheit der Milchsäure-dehydrogenasen: IV. Quantitative Ermittlung einiger Enzymverteilungsmuster. Biochem. Z. **332**, 1–10 (1969).

WILKINSON, J.H.: Isoenzymes, 2nd ed. London: Chapman and Hill Ltd. 1970.

WOLFRUM, D.I., RÜCHEL, R., MESECKE, S., NEUHOFF, V.: In preparation (1973).

WRIGLEY, C.W.: Analytical fractionation of plant and animal proteins by gel electrofocusing. J. Chromatog. **36**, 362–365 (1968).

YOSHIDA, A.: Glucose-6-phosphate dehydrogenase of human erythrocytes. I. Purification and characterization of normal (B) enzyme. J. biol. Chem. **241**, 4966–4976 (1966).

ZACHARIUS, R.CH., ZELL, T.E.: Glycoprotein staining following electrophoresis on acrylamide gel. Analyt. Biochem. **30**, 148–152 (1969).

ZILLIG, W., FUCHS, E., MILETTE, R.L.: DNA-dependent RNA polymerase (E.C.2.7.7.6.). In: CANTONI, G.L., and DAVIES, D.R., Procedure in nucleic acid res., p. 323–339. New York: Harper and Row 1966.

Chapter 2

Micro-Determination of Amino Acids
and Related Compounds with Dansyl Chloride

Dansyl chloride (Dans-Cl, 1-dimethylamino-naphthalene-5-sulfonyl chloride) was used by WEBER in 1952 for the first time for the preparation of fluorescent conjugates of albumin. It has subsequently found almost as wide an application as FISCHER'S naphthalene sulfonyl-chloride or SANGER'S 2,4-dinitroflurobenzene (reviews, see GRAY, 1967a; SEILER, 1970; SEILER and WIECHMANN, 1970). Its usefulness is due to the fact that its reaction products with amino acids, amines, peptides, proteins, phenols, imidazoles, and sulphydryl groups have an intense yellow to yellow-orange fluorescence and can be separated easily with suitable chromatographic systems (SEILER, 1970). WOODS and WANG (1967) first described the fractionation of dansylated amino acids on polyamide sheets (review see WANG and WEINSTEIN, 1972), and GRAY and HARTLEY (1969) introduced this method of separation for the determination of end-groups and in sequence analysis of proteins and peptides. It was HARTLEY who, during an EMBO summer-school in Elmau in 1969, suggested adapting this technique to the micro-scale. This was immediately successful, the normal 15×15 cm polyamide sheets were simply replaced by 3×3 cm ones, and the application of the dansylated sample was performed with a very fine capillary under the stereomicroscope. On a 15×15 cm polyamide layer about 10^{-9} moles of each dansylated amino acid is detectable; using 3×3 cm polyamide layers as little as 10^{-12} moles can be detected. However, it subsequently became obvious that, in order to obtain consistent data, even when only semiquantitative results are required, a number of factors have to be taken into account. This is particularly important when working on the micro scale.

Dansylation Reaction

The reaction of the amino acids with dansyl chloride depends upon a variety of conditions, e.g. concentration of dans-Cl, time of reaction, temperature, and pH. NEADLE and POLLIT (1965), and SEILER (1970), have shown that the carboxyl group of the N-dansyl amino acid formed can react further with excess dansyl chloride to give dansyl amine with the degradation of the dansyl amino acid (compare Scheme 1). This is confirmed by the data shown in Table 1, which also demonstrates that when a single amino acid (10^{-4} M each) is treated with dans-Cl, the yield of dansyl amino acid depends on the concentration of dansyl-Cl applied, and that 100% reaction of individual amino acids is not obtained in

Scheme 1. Reaction of dansyl chloride with amino acids and dansyl amino acids. (From SEILER, 1970)

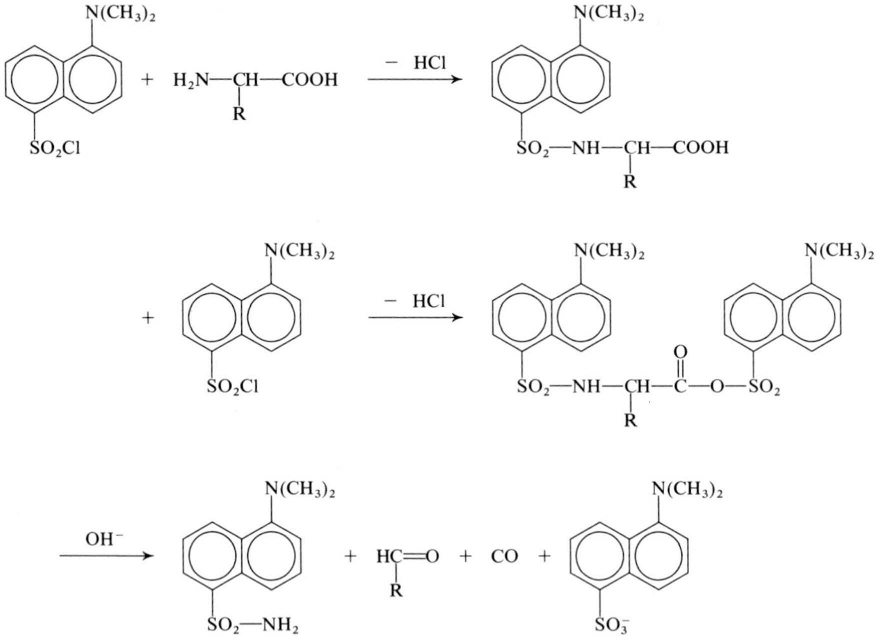

practice. The yield of dansyl amino acid is consistent for individual amino acids, and so is the quantity of dansyl amine which is formed simultaneously. However, taking into account that the dansyl amine is a product of the secondary reaction between dansyl amino acid and excess dans-Cl, the amount of amino acid which has reacted with dans-Cl ($=$ dansyl amino acid + dansyl amine) approaches 100%. Proline reacts most efficiently (93–95% reaction at a 60–100 fold excess of dansyl chloride). Glycine, which could react completely in a mixture (compare Table 2 and 3), undergoes only 76% reaction with a 30–60 fold excess of dans-Cl (BRIEL, 1972; BRIEL and NEUHOFF, 1972).

If a mixture of 7 amino acids (tryptophan, lysine, phenylalanine, histidine, proline, glycine and isoleucine, 10^{-4} M each) is allowed to react with different concentrations of dans-Cl (ratio of dans-Cl to reactive groups 2.8:1, 5.9:1, 11.7:1, 23.3:1 and 35.7:1) the best yield of dansyl amino acid is obtained with a 5.9 fold excess per reactive group (Table 2). The individual amino acids behave in different ways; proline and glycine react best at different concentrations of dansyl chloride (compare Figs. 13 and 15).

Again, differing reactions are observed for the individual amino acids when a mixture of 24 amino acids is treated with dansyl chloride. A constant amount of ^{14}C-dansyl chloride (4 μl, 10.63×10^{-3} M) was incubated with the following concentrations of each amino acid: 4×10^{-4}, 10^{-4}, 5×10^{-5}, 10^{-5} M. The total reaction volume was kept constant (2 μl amino acid sample + 2 μl acetone + 4 μl ^{14}C-dansyl chloride). The ratios of dansyl chloride to the total amino acid content in this pool was 2:1, 9:1, 18:1, and 89:1. Table 3 shows the yields of

Table 1. Recovery, as % of the theoretical value ($=100\%$), on reaction of different concentrations of ^{14}C-dansyl chloride (11.0, 5.0, 3.02, 1.51, 1.18, 0.92, 0.68, 0.11×10^{-3} M) with *single amino acids* (concentration of each 10^{-4} M). Reaction mixture and incubation conditions as in Table 3. Percentage of dansylamine formed is given in brackets

$\dfrac{[^{14}C]Dns\text{-}Cl}{\text{amino acid}}$	220	100	60	30	24	18	14	2.2
Dns-amino acid (and Dns-NH$_2$)				Dansylation [%]				
Dns-Trp	—	52.9	38.1	55.4	49.5	—	—	—
(Dns-NH$_2$)	—	(24.2)	(19.7)	(13.5)	(3.0)	—	—	—
Dns$_2$-Lys	—	12.2	40.7	48.8	37.3	—	—	—
N^α-Dns-Lys	—	4.0	3.1	2.9	5.2	—	—	—
(Dns-NH$_2$)	—	(21.5)	(25.5)	(15.2)	(7.3)	—	—	—
Dns-Phe	—	27.2	63.1	60.6	57.6	—	—	—
(Dns-NH$_2$)	—	(41.9)	(21.0)	(10.9)	(4.8)	—	—	—
Dns$_2$-His	—	11.8	16.7	5.4	1.4	—	—	—
N^α-Dns-His	—	20.7	18.0	20.8	27.6	—	—	—
(Dns-NH$_2$)	—	(26.8)	(27.2)	(14.5)	(3.1)	—	—	—
Dns-Ile	32.5	52.2	64.0	58.1	56.4	—	41.1	7.2
(Dns-NH$_2$)	(45.6)	(50.0)	(22.2)	(14.2)	(2.8)	—	(2.6)	(0.0)
Dns-Pro	77.7	95.3	93.4	—	—	81.3	—	43.9
(Dns-NH$_2$)	(16.5)	(17.8)	(4.8)	—	—	(1.5)	—	(0.0)
Dns-Gly	—	60.9	76.5	76.4	60.8	—	—	8.8
(Dns-NH$_2$)	—	(20.3)	(19.2)	(13.7)	(4.3)	—	—	(0.0)

dansylated amino acids as percentages of the possible (100%) yields. The yield varies for the different amino acids, but the reaction is optimal at a ratio of 2:1 for all amino acids except for 5-hydroxyindole, which reacts optimally at a ratio of 89:1. For most of the amino acids the degree of dansylation decreases as the amount of dansyl chloride increases (2:1) except for proline, glycine, and alanine, which react equally well at 8 or 16 fold excess.

The results suggest that optimal concentration of dansyl chloride (total incorporation of ^{14}C-dansyl chloride into the amino acid mixture) should first be determined before a tissue sample is analyzed. The necessity for this is shown by the sample in Table 4, in which the dansylation of free amino acids in a biological material was carried out at various concentrations of dansyl chloride. Cat's spinal cord (0.9 mg) was homogenized in 20 µl 0.05 M NaHCO$_3$ and adjusted to pH 10.2 with NaOH so that the clear supernatant obtained after centrifugation had a pH of 10.0. Dansylation of 4 µl portions (compare Scheme 2) was carried out as described later. The 4 µl portions added to the samples had the following concentrations of dansyl chloride: 9.97×10^{-3} M (stock solution), 4.99×10^{-3} M, 3.32×10^{-3} M, 2.45×10^{-3} M, and 1.66×10^{-3} M. After incubation at 37° C for 30 min, two portions from each sample were applied to a 3×3 cm micro-polyamide layer and subjected to chromatography in two

Table 2. Recovery, as % of the theoretical value ($=100\%$), on reaction of different concentrations of ^{14}C-dansyl chloride (16.04, 10.5, 5.25, 2.63, 1.25×10^{-3} M) with a *mixture of 7 amino acids* (concentration of each amino acid $= 10^{-4}$ M). Reaction mixture and incubation conditions as in Table 3

$\dfrac{[^{14}\text{C}]\text{Dns-Cl}}{\text{each amino acid}}$	321	210	105	53	25
$\dfrac{[^{14}\text{C}]\text{Dns-Cl}}{\text{no. of reactive groups}}$	35.7	23.3	11.7	5.9	2.8
			Dansylation [%]		
Dns-Trp	55.5	62.8	76.2	92.8	77.8
Dns$_2$-Lys	50.0	62.6	79.2	91.3	47.4
Dns-Phe	66.7	78.2	94.7	101.6	65.7
Dns$_2$-His	42.2	53.4	59.2	48.2	10.5
Dns-Ile	46.5	62.7	79.3	98.7	74.2
Dns-Pro	88.7	93.1	100.5	108.0	101.1
Dns-Gly	88.6	86.6	106.2	114.0	88.1
N^α-Dns-His + N^α-Dns-Lys	37.8	24.9	14.4	20.1	46.3
Dns-NH$_2$	20.3	11.9	8.0	3.7	1.1

Table 3. Recovery, as % of the theoretical value ($=100\%$), on reaction of different concentrations of amino acids with a constant concentration of ^{14}C dansyl chloride (10.63×10^{-3} M). Reaction mixture: 2 µl amino acid solution + 2 µl acetone + 4 µl ^{14}C-dansyl chloride. Incubation 30 min at 37° C

Dns-Amino acid [a]	Dansylation [%] at amino acid concentrations of			
	4×10^{-4} M	10^{-4} M	5×10^{-5} M	10^{-5} M
Dns-Trp	96.6	74.6	61.1	49.6
Dns$_2$-Lys	95.1	71.8	52.4	30.6
Dns-Phe	105.2	81.5	64.9	38.3
Dns$_2$-His	85.8	61.5	51.2	28.6
Dns-Ile	92.1	69.0	56.7	28.0
Dns-Pro	118.5	107.6	101.6	—
Dns-Gly	116.5	110.4	103.7	—
Dns-Glu	98.6	83.5	66.7	—
Dns-Asp	113.1	93.5	85.2	—
Dns-γ-Abu	102.3	90.7	68.2	—
Dns-Met	38.4	28.3	22.2	—
Dns-Taurine	—	111.7	118.7	—
Dns$_2$-Trp (5 OH)	52.0	64.0	67.4	49.3
Dns-Trp (5 OH)	26.1	12.4	12.6	—
Dns-Ala	102.7	111.2	96.8	—
Dns-Leu	88.6	73.3	55.1	37.2
Dns$_2$-Tyr	87.2	76.3	58.7	40.8
Dns-5-Hydroxyindole	57.3	79.9	96.3	102.6
Dns$_2$-Orn	92.0	75.3	57.5	30.7
Dns-Val	97.8	81.6	67.9	63.2
Dns-Arg + N^α-Dns-His + N^α-Dns-Lys	88.7	72.4	97.9	—
$\dfrac{[^{14}\text{C}]\text{Dns-Cl}}{\text{pool of amino acids}}$	2	9	18	89
$\dfrac{[^{14}\text{C}]\text{Dns-Cl}}{\text{each amino acid}}$	53	212	425	2 125

[a] Dns-Ser, Dns-Thr, Dns-Asn and Dns-Gln were not estimated because they are not well separated.

Table 4. Dansylation of free amino acids from cat's spinal cord (ventral horn region). Portions of the clear supernatant were treated with different concentrations of ^{14}C-dansyl chloride. From each sample two chromatograms on 3×3 cm micropolyamide layers were prepared as described in the text. For the calculation of the percentage of each amino acid in the sample, the mean value of the total dpm was used. For details see text. The unknown ^{14}C-dansyl compounds are also recorded in the table, but not calculated in the total percentage

Dns-Amino acid	Concentration of [^{14}C]Dns-Cl									
	9.97×10^{-3} M		4.99×10^{-3} M		3.32×10^{-3} M		2.45×10^{-3} M		1.66×10^{-3} M	
	[dpm]	[%]	[dpm]	[%]	[dpm]	[%]	[dpm]	[%]	[dpm]	[%]
Dns-Cys	372	2.0	430	1.95	445	2.65	291	1.9	137	1.1
Dns-Cystine	277		350		463		314		113	
Dns-Taurine	667	4.1	566	2.95	401	2.5	345	2.4	211	2.0
	656		591		457		413		249	
Dns-Trp	54	0.3	38	0.2	30	0.2	32	0.2	26	0.25
	—		44		36		30		29	
Dns$_2$-Orn	155	0.7	184	0.85	206	1.25	207	1.35	159	1.35
	81		148		219		222		149	
Dns$_2$-Lys	220	1.45	351	1.85	407	2.3	352	2.15	241	1.9
	244		375		376		336		205	
Dns-Phe	58	0.3	59	0.35	59	0.3	54	0.35	42	0.35
	47		68		53		57		36	
Dns$_2$-His	71	0.4	100	0.45	90	0.5	67	0.4	43	0.35
	64		71		90		65		32	
Dns-Leu	86	0.55	107	0.55	98	0.6	110	0.7	111	0.85
	85		112		94		107		90	
Dns-Ile	84	0.4	62	0.4	94	0.4	63	0.4	60	0.5
	48		86		51		65		57	
Dns$_2$-Tyr	—	—	47	0.25	36	0.2	56	0.4	—	—
	—		59		—		—		—	
Dns-5-Hydroxyindole	1144	6.75	1581	7.3	1378	7.7	909	5.8	713	5.8
	1034		1314		1267		946		630	
Dns-Pro	66	0.4	98	0.45	77	0.4	71	0.4	81	0.65
	63		69		66		56		70	

Table 4 (continued)

Dns-Amino acid	9.97×10^{-3} M [dpm]	[%]	4.99×10^{-3} M [dpm]	[%]	3.32×10^{-3} M [dpm]	[%]	2.45×10^{-3} M [dpm]	[%]	1.66×10^{-3} M [dpm]	[%]
Dns-Val	122 / 84	0.65	151 / 133	0.7	148 / 137	0.8	159 / 141	0.95	161 / 120	1.25
Dns-γ-Abu	516 / 375	2.75	594 / 530	2.85	573 / 483	3.1	543 / 485	3.2	433 / 395	3.6
N-Dns-Ethanolamine	685 / 650	4.15	838 / 783	4.1	746 / 765	4.4	634 / 594	3.85	468 / 432	3.9
Dns-Ala	659 / 606	3.9	904 / 844	4.45	772 / 779	4.55	747 / 810	4.9	572 / 585	5.05
Dns-Gly	3404 / 3362	21.0	4408 / 4159	21.6	4123 / 4096	24.05	4543 / 4661	28.85	4142 / 3943	35.05
Dns-Glu	3867 / 2833	20.6	4274 / 3812	20.35	3269 / 3085	18.55	2786 / 2784	17.45	1713 / 1649	14.55
Dns-Asp	2382 / 2343	14.7	2603 / 2444	12.75	1839 / 1924	10.4	1612 / 1650	10.2	948 / 916	8.05
Dns-Gln +Dns-Thr +Dns-Asn	1769 / 1437	9.9	2060 / 1705	9.45	1826 / 1604	10.0	1335 / 1257	8.15	733 / 762	6.5
Dns-Ser	606 / 476	3.3	1448[a] / 575	3.1	690 / 458	3.3	502 / 591	3.45	247 / 313	2.45
Dns-Arg +Nα-Dns-Lys +Nα-Dns-His	174 / 358	1.7	299 / 302	1.5	313 / 225	1.6	490 / 423	2.85	560 / 505	4.65
Summe	17163 / 15144	100.0	21202 / 18572	100.0	17617 / 16718	100.0	15906 / 16007	100.0	11799 / 11278	100.0
7 unknown compounds	2000		2126		2011		1100		520	

[a] Contamination with Dns-OH.

dimensions as described later. Table 4 gives the values of the ^{14}C-dansyl amino acids in dpm obtained from each of the two chromatograms, and the percentage of each amino acid, calculated from the mean value of the sum of all measured counts belonging to dansyl amino acids. Since the highest total of counts reflects the greatest efficiency of reaction between ^{14}C-dansyl chloride and the amino acids in the mixture, it can be seen that in this tissue sample, the optimal reaction conditions are obtained with a 1:2 diluted stock solution of dansyl chloride. It can be seen that the results from each chromatogram are similar, so the application of dansyl compounds with calibrated micropipettes, the separation, and the counting of radioactivity associated with each dansyl spot is reproducible.

Furthermore, Table 4 shows clearly that for any determination it is extremely important to carry out the dansyl reaction under optimal conditions. For example, glycine is one of the very few amino acids which react to almost 100% extent even when the dansyl chloride is present in the reaction mixture in low concentration. This is demonstrated by the four experiments where the stock solution of dansyl chloride is diluted 1:2, 1:3, 1:4, and 1:6, while the dpm values for dansyl glycine remain relatively constant. However, if the percentage of glycine in the mixture is calculated, the proportion changes from 22% to 35%. Amino acids which do not behave so consistently as glycine give variable dpm values and percentages in mixtures. The radioactivity associated with 7 unknown dansyl compounds from spinal cord is also recorded in Table 4; it can be seen that the yield also varies with the dansyl concentration in the reaction mixture.

That the optimal reaction conditions have to be determined separately for each tissue sample is demonstrated by the results of the determinations of free amino acids in rabbit ova (Table 5). In this case the best recovery of ^{14}C-dansyl amino acids is obtained with a 1:6 dansyl stock solution (PETZOLDT et al., 1973). If the optimal concentrations of dansyl chloride are known, the percentage of the individual amino acids can be estimated by using calibration curves (BRIEL,

Table 5. Determination of free amino acids in rabbit ova. 40 rabbit ova at different stages of cleavage (prepared by Dr. U. PETZOLDT, Freiburg) were homogenized in 5 µl 0.05 M NaHCO$_3$, pH 10.0, and the clear supernatant treated with different concentrations of ^{14}C dansyl chloride. The amino acids were counted after separation on 3 × 3 cm polyamide layers. It can be seen that the highest yield is given by a 1:6 diluted stock solution of ^{14}C-dansyl chloride. The percent values were not calculated since not all amino acids on the chromatogram were removed and counted. (Tables 1–5 from BRIEL and NEUHOFF, 1972)

Concentration of [^{14}C]Dns-Cl Dilution of stock solution	4.99×10^{-3} M 1:2 [dpm]	3.32×10^{-3} M 1:3 [dpm]	1.66×10^{-3} M 1:6 [dpm]
Dns-Amino acid:			
Dns$_2$-Lys	42.6	38.9	50.5
Dns-Leu	40.7	71.9	89.7
Dns-Phe	26.4	31.7	30.9
Dns-Pro	48.8	74.2	98.2
Dns-Val	44.0	66.6	70.5
Dns-Gly	458.3	1015.2	1170.0

1972). Figs. 1–16 illustrate such curves for leucine, lysine, isoleucine, histidine, tyrosine, tryptophan, valine, phenylalanine, ornithine, glutamic acid, γ-amino butyric acid, aspartic acid, glycine, alanine, proline and 5-hydroxyindole. The ordinates of the calibration curves give the dpm/μl of the reaction mixture, and the abscissa gives the known concentration of the amino acids in the pool. The calibration value for a 100 % recovery is shown by the thick line in the calibration curves. The standard curves were calculated according to the method of linear regression: $\log y = \log b + a \cdot \log x$. The calculated values of the regression lines, the standard errors of the coefficients, the standard observation error σ (dotted line), and the correlation coefficient $r_{\lg x \lg y}$ for the standard curves are given in the captions to the figures.

On the basis of the specific activity of the ^{14}C-dansyl chloride (49 mCi/mMol) used, it can be calculated that 108 dpm is equivalent to 1 pico mole of ^{14}C-dansyl chloride. Since one mole of dansyl chloride reacts with one mole of a reactive

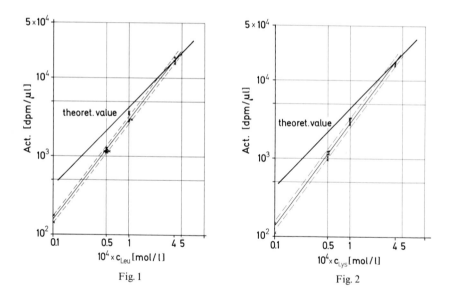

Fig. 1   Fig. 2

Fig. 1. Activity of leucine at different concentrations when reacted with a constant amount of ^{14}C-dans-Cl. The results (thin line) show that the experimental values differ from the calculated ones (thick line). It is clear that dansylation at lower concentrations (10^{-5} M) results in the greatest deviation from the theoretical values.

Calculations: $\lg y = (8.565 \pm 0.094) + (1.273 \pm 0.023) \cdot \lg x$
$\sigma = \pm 0.042, \quad r = 0.998$

σ is the standard error of the dependent variable. If the observations are distributed normally, nearly one third of the total are outside the scatter range (dotted line) $\pm \sigma$

Fig. 2. Reaction curve of lysine (dans$_2$-lysine)

Calculations: $\lg y = (8.719 \pm 0.109) + (1.317 \pm 0.026) \cdot \lg x$
$\sigma = \pm 0.049, \quad r = 0.997$

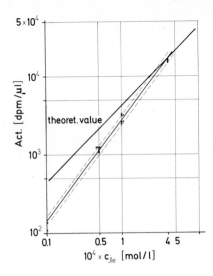

Fig. 3. Reaction curve of isoleucine

Calculations:
$$\lg y = (8.697 \pm 0.108) + (1.311 \pm 0.260) \cdot \lg x$$
$$\sigma = \pm 0.048, \quad r = 0.997$$

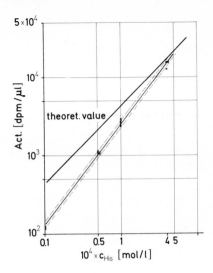

Fig. 4. Reaction curve of histidine
(dans$_2$-histidine)

Calculations:
$$\lg y = (8.573 \pm 0.075) + (1.290 \pm 0.018) \cdot \lg x$$
$$\sigma = \pm 0.033, \quad r = 0.999$$

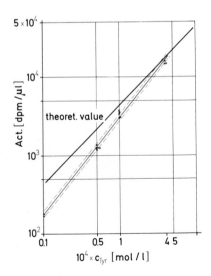

Fig. 5. Reaction curve of tyrosine (dans$_2$-tyrosine)

Calculations:
$$\lg y = (8.337 \pm 0.079) + (1.215 \pm 0.019) \cdot \lg x$$
$$\sigma = \pm 0.035, \quad r = 0.997$$

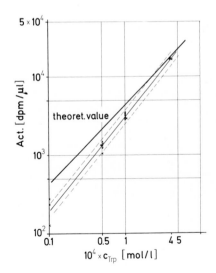

Fig. 6. Reaction curve of tryptophan

Calculations:
$$\lg y = (8.301 \pm 0.155) + (1.202 \pm 0.037) \cdot \lg x$$
$$\sigma = \pm 0.075, \quad r = 0.992$$

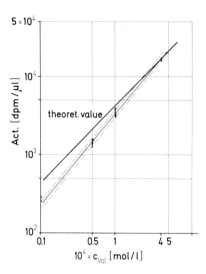

Fig. 7. Reaction curve of valine

Calculations:
$$\lg y = (8.067 \pm 0.089) + (1.134 \pm 0.021) \cdot \lg x$$
$$\sigma = \pm 0.039, \quad r = 0.997$$

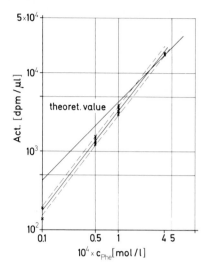

Fig. 8. Reaction curve of phenylalanine

Calculations:
$$\lg y = (8.622 \pm 0.108) + (1.276 \pm 0.026) \cdot \lg x$$
$$\sigma = \pm 0.048, \quad r = 0.997$$

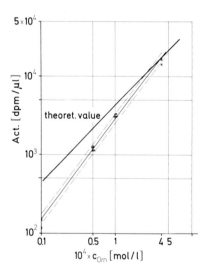

Fig. 9. Reaction curve of ornithine
(dans$_2$-ornithine)

Calculations:
$$\lg y = (8.657 \pm 0.111) + (1.297 \pm 0.027) \cdot \lg x$$
$$\sigma = \pm 0.049, \quad r = 0.997$$

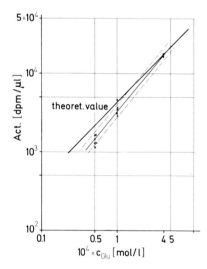

Fig. 10. Reaction curve of glutamic acid

Calculations:
$$\lg y = (8.292 \pm 0.206) + (1.192 \pm 0.051) \cdot \lg x$$
$$\sigma = \pm 0.066, \quad r = 0.989$$

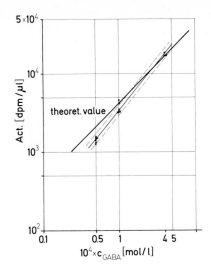

Fig. 11. Reaction curve of γ-aminobutyric acid
(GABA)

Calculations:
$$\lg y = (8.330 \pm 0.173) + (1.195 \pm 0.043) \cdot \lg x$$
$$\sigma = \pm 0.055, \quad r = 0.992$$

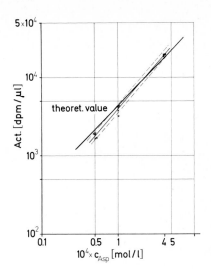

Fig. 12. Reaction curve of aspartic acid

Calculations:
$$\lg y = (8.151 \pm 0.123) + (1.137 \pm 0.031) \cdot \lg x$$
$$\sigma = \pm 0.038, \quad r = 0.996$$

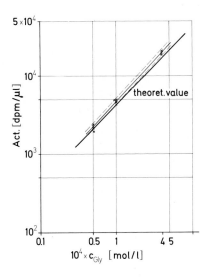

Fig. 13. Reaction curve of glycine

Calculations:
$$\lg y = (7.890 \pm 0.094) + (1.056 \pm 0.024) \cdot \lg x$$
$$\sigma = \pm 0.028, \quad r = 0.998$$

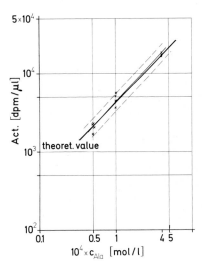

Fig. 14. Reaction curve of alanine

Calculations:
$$\lg y = (7.781 \pm 0.322) + (1.036 \pm 0.081) \cdot \lg x$$
$$\sigma = \pm 0.969, \quad r = 0.973$$

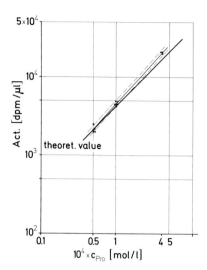

Fig. 15. Reaction curve of proline

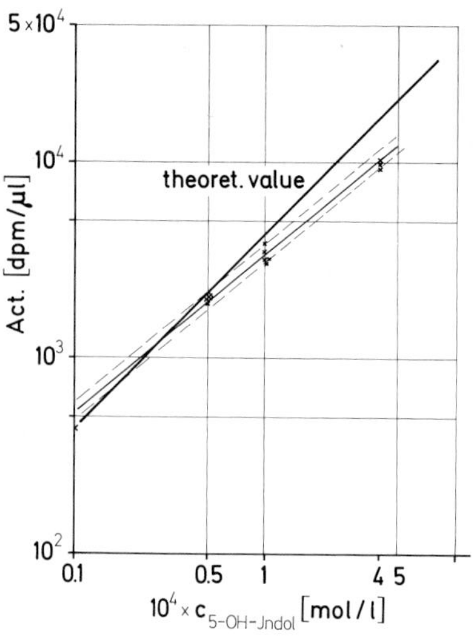

Fig. 16. Reaction curve of 5-hydroxyindole

Calculations:

$\lg y = (7.963 \pm 0.092) + (1.075 \pm 0.231) \cdot \lg x$

$\sigma = \pm 0.029, \quad r = 0.997$

Calculations:

$\lg y = (6.749 \pm 0.104) + (0.805 \pm 0.025) \cdot \lg x$

$\sigma = \pm 0.040, \quad r = 0.994$

(Figs. 1–16 from BRIEL, 1972)

group under optimal conditions, the amount of each amino acid on one chromatogram can be calculated in terms of pico mole or ng. If the recovery for each amino acid under optimal reaction conditions is known, the calibration curves can be used to obtain correction factors from which a semi-quantitative estimation of the content of free amino acids in a sample is possible. However, in a tissue extract the number of unknown compounds which can react with dansyl chloride has to be taken into account. Some tissues, for example cat's spinal cord, have only 7 unidentified dansyl products, while tissues from the snail *Helix pomatia* have up to 27 unknown products (OSBORNE, BRIEL, NEUHOFF, 1971). For the quantitative estimation of amino acids a number of factors have to be considered; one of these is the amount of unknown compounds in the mixture, since they too react with dansyl chloride.

Dansylation of a mixture of different amino acids consists of a group of coupled reactions. It seems that the yield from a mixture of dansylated amino acids is higher than that from a single amino acid (see Tables 1–3). Qualitatively one may explain the phenomenon on the basis that the dansylation of a single amino acid (AA_1-NH_2) involves at least three reactions:

1) $AA_1\text{-}NH_2 + \text{dans-Cl} \rightarrow \text{dans-NH-}AA_1 + HCl$

2) $H_2O + \text{dans-Cl} \rightarrow \text{dans-OH} \quad + HCl$

3) $\text{dans-NH-}AA_1 + \text{dans-Cl} \rightarrow \text{dans-}NH_2 \quad + \text{other products}$

The equilibrium of reaction 1 can be displaced towards the right by carrying out the reaction at high pH. However, since high pH also favours the competing reaction 2, an optimum pH must be found. Reaction 2 has a rate which depends on the concentration of dansyl chloride in the reaction mixture, and is independent of the concentration or type of amino acid present. Considering only reactions 1 and 2, a 100% yield of the dansylated amino acid in aqueous solution is only possible for a suitable ratio of the rate constants. Reaction 3 is consecutive to reaction 1. The decomposition of the dansylated amino acid in reaction 3 leads to products other than amino acids; therefore a 100% yield of the desired product cannot be obtained. The total yield of reaction 1 is the sum of the yields of dansylated amino acid and dansyl amine. One can, however, minimize the extent of the unwanted side reactions by varying the reaction conditions. Reaction 3 depends on the presence of substituted dansyl amino groups, and is, to a first approximation, independent of the nature of the amino acid involved. It also proceeds at a much slower rate than reaction 1. Thus, the yield of the desired product can be increased by adding large amounts of a second dansylated amino acid to favour the competition reaction. When a mixture of different amino acids has to be dansylated, one must first consider that the pH optimum for the coupled reaction of several amino acids differs from that of a single amino acid. However, the experimentally determined pH optimum is often found to be close to pH 10 because basic amino acids, which could lead to a lowering of the optimum pH, are usually present in small amounts. The rate of reaction 1 for the different amino acids depends on the ratio of dansyl chloride to the individual amino acids, and not to the total concentration of amino acids. The rate of reaction 3 depends on the ratio of dansyl chloride to all dansyl amino acids formed in reaction 1. Therefore, if other amino acids are present, reaction 3 has a smaller influence on the yield of an individual dansylated amino acid. The correct explanation for the complex reaction conditions observed between amino acids and dansyl chloride, however, will be possible only after studying the kinetics of these reactions.

Preparation of the Stock Solution of Dansyl Chloride

Dansyl chloride is obtainable from various firms. It is advisable to store the preparations dry in order to inhibit the formation of dans-OH. The substance is sparingly soluble in water, and is therefore dissolved in acetone (3 mg/ml, analytical grade) for the reaction. ^{14}C-dansyl chloride solution in acetone (2.7 mg/ml, spec. activity 49 mCi/mol or 98 mCi/mol) is obtainable from CEA, Gif-sur-Yvette, France, or from Schwarz/Man, Orangeburg, New York; ^{3}H-dansyl-Cl solutions in benzene (1 000 mCi/mmol) can be obtained from Amersham, GB, or Amersham/Buchler, Braunschweig, Germany). The highest practicable specific activity of ^{14}C-dans-Cl should be used. Since the results of dansylation depend to a large extent on the concentration of dans-Cl, evaporation of acetone while handling the stock solution of dans-Cl should be avoided. After opening the vial and washing it out well with acetone, the solution is allowed to evaporate completely, and the residue is dissolved in 200 µl acetone. 1 µl of dans-Cl solution is removed immediately, and the concentration is determined spectrophotometrically, using an extinction coefficient of 3.67×10^{3}

at 369 nm (GRAY, 1964). To do this, 1 µl of solution is transferred to a micro-cuvette (best carried out with a Drummond 1 µl Microcap) which has been charged previously with 100 µl absolute ethanol. The cuvette is sealed carefully with Parafilm, and the contents are mixed well by shaking. 100 µl absolute ethanol and 1 µl acetone are need in the blank cuvette. Since the extinction coefficient of dansyl chloride, and the length of the light path (1 cm) are known, the concentration can be calculated using the formula

$$OD = e \cdot c \cdot d$$

(e = extinction coefficient, c = concentration (moles/L), d = light path). The concentration of this $\left[c = \dfrac{OD}{e \cdot d} \right]$ 1 µl of the dansyl chloride-acetone solution (2.7 mg/ml) has an optical density of 0.36 when diluted one hundred-fold. The concentration of the original solution is calculated as:

$$c = \frac{0.36 \cdot 100}{3.67 \cdot 10^3} = 0.0098 \text{ Mol/l} = 9.8 \text{ µmol/ml}.$$

Commercial preparations of ^{14}C-dansyl chloride usually contain 0.1 mCi in 200 µl acetone, so the concentration can be calculated from the specific activity:

$$\frac{0.5 \text{ mCi}}{x \text{ mMol}} = \frac{49 \text{ mCi}}{1 \text{ mMol}} = 0.01 \text{ mMol/ml} = 10 \text{ µMol/ml}.$$

The results are thus in good agreement with the values determined by photometry. For a rapid, and equally exact, determination of concentration, the radioactivity of a measured volume may be ascertained and used as an alternative to the optical density, so that the concentration of the sample can be calculated. For this, 1 µl of the 200 µl stock solution is used. The 1 µl Drummond microcaps fill by capillary attraction when dipped into the solution. The filled capillary is transferred to a counting vial previously filled with scintillation solution, and broken up in the vial using a glass rod. Thorough shaking ensures an even distribution of the ^{14}C-dansyl acetone solution in the scintillator toluene. The total activity of the 200 µl of dansyl chloride stock solution is 0.1 mCi, therefore 1 µl contains 0.5 µCi. 1 Curie = 3.7×10^{10} disintegrations sec^{-1} = 2.22×10^{12} dpm. 0.5 µCi therefore gives 1.11×10^6 dpm. Since the specific activity (= Ci/mole) is 49 mCi = 108.7×10^9 dpm, the actual concentration of the solution can be ascertained easily by simple proportional calculation, taking into consideration the counting yield of the apparatus.

200 µl of the stock solution are then transferred to 3 pyrex glass tubes (length 5 cm, inner diameter 1.5 mm, wall thickness 0.5 mm), and closed with Dynal[1] (polyoxymethylene) stoppers, and stored at $-20°$ C until required. To minimize the evaporation of acetone while portions are removed, the tube in use must always be kept in ice; this can be accomplished by using the device shown in Fig. 17. Using this method, the concentration of dans-Cl remains constant, even after repeated transfers of portions, as has been proved by measuring their radioactivity.

---

[1] Dynal, Sustarin, Delrin; Mannesmann, Plastic GmbH, Lahnstein, Germany.

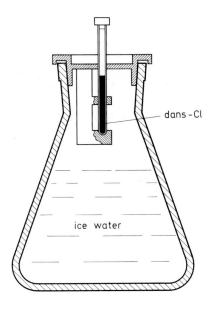

Fig. 17. Ice container for keeping the ^{14}C-dans-Cl solution cool in a pyrex tube. (From BRIEL and NEUHOFF, 1972)

Dependence of the Dansyl Reaction on pH

For the determination of N-terminal groups in peptides, using dansyl chloride, GRAY and HARTLEY (1963), and HARTLEY (1970), used 0.1 M NaHCO$_3$ as buffer to neutralize the HCl produced. SEILER and WIECHMANN (1966) found that primary and secondary amines react quickly and quantitatively with dansyl chloride in buffered acetone at pH 8.0 at room temperature. According to their investigations, a pH of 10 is optimal for the reaction with phenols and diphenols, since these compounds react only slowly and incompletely at lower pH. GROS and LABOUESSE (1969) carried out the reaction in the presence of NaHCO$_3$ at pH 8.3, and ZANETTA et al. (1970) and SPIVAK et al. (1971) used 0.2 M sodium phosphate buffer, pH 8.85, or alternatively 0.2 M KHCO$_3$. The "buffer" employed for the reaction of dansyl chloride with amino acids in micro scale is 0.05 M NaHCO$_3$, pH 10.0.

To determine the optimal pH, a mixture of 20 amino acids (lysine, histidine, ornithine, tyrosine, tryptophan, methionine, asparagine, glutamine, glutamic acid, aspartic acid, alanine, proline, leucine, glycine, phenylalanine, threonine, iso-leucine, valine, serine and taurine) is dissolved in 0.05 M NaHCO$_3$, pH 10.0 to give a concentration of 5×10^{-4} M of each. Portions are adjusted to pH 8.5, 9.0, 9.5, 10.0, 10.5, 11.0 and 11.5 using a glass electrode. 2 µl of each sample is mixed with 2 µl of acetone and 4 µl of ^{14}C-dansyl chloride (10.63 µmol/ml, undiluted stock solution) and incubated at 37° C for 30 min; the dansylated amino acids are separated by micro-chromatography. The spots containing the dansyl amino acids glycine, proline, valine, leucine, phenylalanine, tryptophan, glutamic acid,

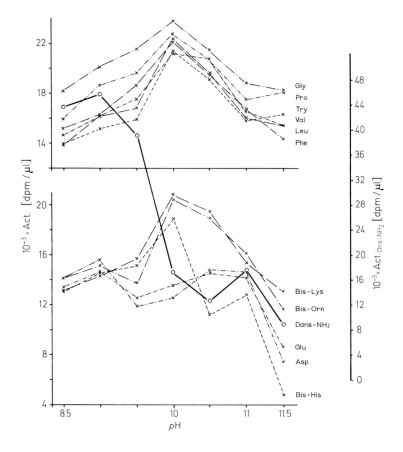

Fig. 18. Influence of pH during the dansylation of amino acids on their recovery. (From BRIEL and NEUHOFF, 1972)

aspartic acid, lysine, ornithine and histidine, and dansyl amine are recovered individually, suspended in scintillation liquid and counted in a Packard Tri-carb liquid scintillation spectrometer Model 3380. Fig. 18 shows that most of the amino acids tested, including the bifunctional amino acids lysine, ornithine, and histidine, are optimally dansylated at pH 10.0. The decrease of bis(dansyl)-histidine at pH 10.5 reflects the formation of N^1-(or N^3-)dansyl histidine (NEU-HOFF, BRIEL, MAELICKE, 1971) at high pH values.

From these results it can be concluded that the dansylation of free amino acids in tissues must be carried out at pH 10.0. If the volume of the homogenate is too small for its pH to be determined exactly, the pH of the homogenization buffer should be estimated from previous experiments, so that a pH 10.0 can be achieved by the homogenization of a defined amount of tissue per ml $NaHCO_3$ solution. The pH must be measured *before* the addition of acetone since this results in an apparent increase of pH.

Dependence of the Dansyl Reaction on the Incubation Time

Various times and temperatures of incubation have been reported for the reaction between dansyl chloride and the component to be dansylated. NEADLE and POLLITT (1965) allowed the reaction to occur at room temperature, without specifying a time, while 16 hrs. at room temperature was employed by SEILER and WIECHMANN (1966). For the end-group determination of proteins and peptides with dansyl chloride, GRAY and HARTLEY (1963), and GRAY (1967b), allowed the incubation to proceed for 3 hrs. at room temperature or for 1 hr. at 37° C; CROWSHAW, JESSUP and RAMWELL (1967) also allowed dansylation to proceed for 3 hrs. at room temperature. GROS and LABOUESSE (1969) carried out the reaction at 20° C, and found that depending on the reactive group, 95% of the dansylation occured within the first 30 min. ZANETTA et al. (1970) confirmed these findings, but recommended a time of two hours at room temperature, since the mono-aminodicarboxylic acids are not completely dansylated in less than 30 min. SPIVAK et al. (1971) used 40 min at room temperature. NEUHOFF et al. (1969) initially allowed an incubation time of 12–15 hrs. at room temperature, or 2–3 hrs. at 37° C, but on the basis of the foregoing data have since incubated for 30 min at 37° C (NEUHOFF, BRIEL, MAELICKE, 1971; OSBORNE, BRIEL, NEU- HOFF, 1971; BRIEL, NEUHOFF, OSBORNE, 1971; BRIEL et al., 1972).

To determine the best incubation time, a mixture of 24 amino acids (leucine, alanine, phenylalanine, tyrosine, serine, threonine, methionine, isoleucine, glutamic acid, γ-aminobutyric acid, taurine, aspartic acid, arginine, glycine,

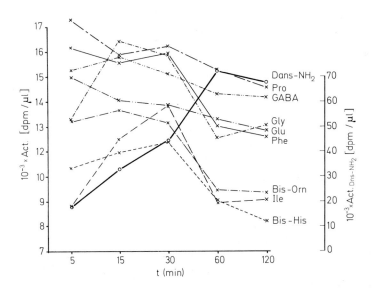

Fig. 19. Influence of incubation time on the formation of dansyl amino acids. (From BRIEL and NEUHOFF, 1972)

valine, proline, ornithine, lysine, histidine, asparagine, glutamine: 4×10^{-4} M per acid) was incubated with ^{14}C-dansyl chloride as described above. Portions were removed after 5, 15, 30, 60 and 120 min and subjected to chromatography. The spots containing the dansylated amino acids ornithine, histidine, γ-amino-butyric acid, phenylalanine, glutamic acid, isoleucine, glycine, and dansyl amine, were recovered and measured in a scintillation counter. Fig. 19 shows that the reaction reached a maximum between 5 and 30 min. After 30 min the dansyl amino acids decrease rapidly; the formation of dansyl amine seems to be suitably low after 5 min. The rate of formation of dans-NH_2 from dansyl amino acids differs for different amino acids (compare Table 1–3); increases of dansylation in the cases of histidine, isoleucine, and glycine are observed within 5–30 min, whereas dansylated phenylalanine, γ-aminobutyric acid, glutamic acid, ornithine, and proline remain more or less constant when incubated for 15 or 30 min. A reaction time of 30 min seems therefore to be suitable for the dansylation of free amino acids in biological material.

Practical Procedure

Dansyl Reaction in Micro Scale

2 µl of the amino acid solution in 0.05 M $NaHCO_3$, pH 10.0, is taken up with a 2 µl Drummond microcap, by dipping the microcap into the test solution and allowing it to fill by capillary attraction. A suitable glass capillary is then used to transfer the samples to individual reaction tubes having an inner diameter of 1.5 mm, a wall thickness of 0.5 mm, and a length of 1.5 cm. They are made from soft glass and are carefully heat sealed at one end. The 2 µl samples are placed at the bottom of the reaction tube by placing the end of the capillary directly against the bottom of the tube. It must be checked first, however, that the liquid in the capillary reaches right to the top, and that no tiny columns of air are left in the tip; their removal can only be effected by strong, and thus risky, blowing. Care must be taken that not even the tiniest drops remain on the glass wall. To facilitate this the reaction tubes are kept short (1.5 cm). In order to obtain the same conditions for the analysis of pure amino acid solutions as for amino acids from biological material, 2 µl acetone are added, so that the proportion of water in the aqueous phase of the reaction mixture is reduced. Consequently, the reaction of dansyl chloride with water to form dans-OH is diminished, resulting in an improvement in the final microchromatography. Excess dans-OH can impair the separation of the dansyl amino acids. According to SEILER and WIECHMANN (1966), the water content in the reaction mixture should only be 10–30%; in our experimental mixtures it is 25%. Care must be taken that no acetone remains on the walls of the tube; it is best to add the acetone directly into the 2 µl solution already in the tube. Then 4 µl of a suitably concentrated solution of dansyl chloride are added without letting the solution touch the glass wall. It is important to use the correct capillary, i.e. one with an inner diameter fine enough to allow good control, with a closed silicon tube. The pulled-out tip of the capillary pipette should be no longer than 2 cm, to ensure

safe transfer of the small volumes into the 1.5 cm long reaction tube. The diameter must be so small that acetone does not run out easily on its own, but must not be so narrow that pressure by blowing fairly hard has to be applied to expell the acetone completely. If pressure has to be applied to empty the capillary, the reaction mixture is often shaken up so that it clings to the upper glass wall. There the acetone evaporates rapidly, and some of the reaction mixture often dries onto the glass wall and can only be brought into the solution with difficulty. In addition, the concentration of the dansyl chloride can easily be altered in this way, thus influencing the rate of reaction (see above). The reaction tube is closed carefully with Parafilm and mixed thoroughly on a Whirlmix; this is absolutely necessary in order to complete the reaction. The lower end of the sealed reaction tube is held repeatedly against the outer edge of the vibrator for a few moments. It is advisable to practice this beforehand with a test sample, since the reaction mixture must not be sprayed against the Parafilm seal during mixing. If on shaking, a small air bubble forms in the middle of the reaction mixture it must be removed by a downward jerk, as with a clinical thermometer, so as to reduce the danger of evaporation during incubation.

The mixture is then incubated for 30 min at 37° C and dried under vacuum, a process needing only 3–5 min. For this the Parafilm seal is not perforated, but removed. The application of the vacuum should be a slow process, so that the reaction mixture is not ejected from the short tube. The dry residue is taken up in a known volume (2–5 µl, depending on the concentration of the dansyl product in the reaction) of acetone/acetic acid (3 : 2 v/v (WOODS and WANG,

Scheme 2. Summary of extraction of free amino acids form tissues, and their subsequent dansylation

1 mg Tissue/20 µl buffer (0.05 M NaHCO$_3$, pH 10.2)
↓
Homogenization (2 min)
↓
30 min at 15000 rpm in capillary centrifuge
↓
Equal volume of acetone to the supernatant
↓
Keep 60 min at −20° C
↓
30 min at 15000 rpm in capillary centrifuge
↓
4 µl portion + 4 µl dansyl chloride solution (stock solution, diluted with acetone)
↓
Incubation 30 min at 37° C
↓
Dry *in vacuo*
↓
+ 5 µl acetone/acetic acid 3 : 2 (v/v)
↓
0.2 µl to 0.5 µl portions per chromatogram

1967)), the tube is closed with Parafilm, shaken on the Whirlmix as described, and an aliquot (0.2–0.5 µl) is applied to a 3×3 cm or 5×5 cm polyamide sheet using a very fine capillary pipette previously calibrated for microchromatography. If several chromatograms have to be prepared from one mixture, for multiple estimations, the mixture must be kept cooled so that the volume remains as constant as possible. A small piece of plasticine, in which the reaction tube may be placed conveniently and safely, is best suited for this. The plasticine with the tube can then be kept cool in a covered cold-box filled with ice (polystyrene box or suitable packing material).

Scheme 2 shows the analysis of amino acids from biological material. It differs from the method for solutions of pure amino acids only in that an equal volume of acetone is added to the clear supernatant obtained after homogenization and centrifugation of the material, and that this mixture is maintained for a further 60 min at $-20°$ C to precipitate any proteins. After another centrifugation, 4 µl of solution is removed and treated with 4 µl dansyl chloride as described.

Micro-Chromatography

SEILER (1970) has reported numerous systems for the fractionation of dansyl products on layers of silica gel on the macro scale, but these are not suitable for microchromatography. WOODS and WANG (1967) first separated dansyl amino acids by thin layer chromatography using 15×15 cm polyamide sheets[2] (review see WANG and WEINSTEIN, 1972). These sheets were then reduced to 3×3 cm, and dansyl amino acids were detected in the picomole range after two-dimensional chromatography (NEUHOFF et al., 1969). Micro-polyamide sheets[3] proved (NEUHOFF, BRIEL, MAELICKE, 1971) to be more suitable than the polyamide layer for microchromatography, since micro-polyamide layers are more homogeneous and give a sharper resolution of the dansyl derivatives. A mixture of 30–40 dansyl derivatives can be separated on 3×3 cm micro-polyamide layers after two dimensional chromatography (e.g. Figs. 22, 40, 42, 43, 46).

The polyamide sheets suitable for microchromatography, delivered in size 15×15 cm, are cut to either 3×3 cm or 5×5 cm. Great care must be taken not to contaminate the sheets with fingerprints, since the developing solvents do not flow properly over such regions, resulting in bad chromatograms. The sheets, coated on both sides with polyamide (layer thickness 25 µ), are packed in boxes with a transparent sheet of paper between each layer. 3×3 cm grids are drawn on a single sheet of the transparent papers placed directly over a polyamide sheet. The outline of the grid imprinted on the polyamide sheet is used as a guide for cutting, which is done with a pair of well-sharpened large scissors or a photographic cutter. The polyamide layer is bound sufficiently tightly to the plastic sheet so that very little of the layer peels off when the edges are cut. Occasionally the thin polyamide layer is raised slightly from the support, and torn. More often the polyamide layer overlaps the supporting foil slightly after

---

[2] Cheng-Ching Trading Co., Ltd., Hankow St. Taipeh, Taiwan.
[3] Schleicher and Schüll TLC Ready-Plastic Sheets F 1700 Micro-Polyamide, Dassel, Germany.

cutting and can be scraped off carefully with a spatula. If this is not done the solvent forms a bow front at these positions, resulting in bad chromatograms.

In microchromatography of dansylated amino acids on polyamide sheets, the quality of the sample applied is of critical importance. Buffers of low molarity are recommended since those of high molarity impair the chromatography significantly. If 0.05 M $NaHCO_3$ is employed, so little salt precipitates at the origin when applying 0.5 μl of the reaction mixture that the chromatographic separation of the substances is not affected. In this case the volume which can be applied can be as much as 1 μl, if only pure model mixtures are applied. If other impurities, e.g. peptides or salt from the biological material, are present, the maximum volume which can be applied is significantly lower. Using the stereo microscope, it is very easy to see if too many foreign substances are present in the sample, since the application spot appears as a micro atoll, with a plateau in the middle, surrounded by a yellow bank of dansylated product. If, due to technical difficulties, better samples cannot be prepared, it is advisable to follow the fractionation by viewing the developing chromatogram under UV light. If this is not done, and the normal conditions for chromatography are used, i.e. allowing development to proceed until the solvent front reaches the upper edge of the layer, it is possible to find that the front of the dansyl products do not reach the end of the microchromatogram. For development in the first dimension it is convenient to use the fluorescent blue dans-OH, produced by the reaction of dansyl chloride with the water in the mixture, as a reference point. Under normal conditions it moves beyond the middle of a 3×3 cm chromatogram in the first dimension. When the conditions are unfavourable, e.g. the salt concentration in the sample is too high, the chromatogram should be developed under UV control, and the dans-OH spot allowed to run at least to the middle. Under such conditions the dansyl amino acids show some tailing and are not separated clearly.

The way in which the sample is applied determines the quality of a microchromatogram. Special fine microcapillaries should be used; the orifice must be of such a size that the acetone/acetic acid solution can be released by applying slight pressure via the mouth piece. To apply the sample, the capillary tip is brought carefully into contact with the foil so that the latter is not damaged; if the origin is damaged the spots of dansyl compounds show some tailing. If the capillary tip is not in direct contact with the strongly-absorbent polyamide layer, some of the acetone solution to be applied may flow onto the outer wall of the capillary, and so be lost. The point of application should be at one corner of the sheet, 3 mm, or at most 4 mm, from the edges. It is recommended to mark this point initially with a soft lead pencil, without damaging the thin polyamide layer in the process. The marking is a help during application, which is carried out under a stereo microscope. After some practice the application point can often be recognized under the stereo microscope without marking.

The diameter of the application point should not be more than 0.5 mm if possible, the maximum permissible being 1 mm. 0.25 to 0.5 μl of the sample is taken up in the graduated microcapillary, and applied in small amounts, ca. 20–70 applications, to the polyamide layer. The smaller the applied portions, the sharper is the application point. A stream of hot air from a hair dryer is

used during application. The polyamide sheet must be fixed to the base plate
of the stereo microscope, using the springs normally used for holding microscope
slides (see Fig. 20), and taking care that the polyamide layer is not damaged in
the process. While drying a portion of the sample on the chromatogram, the
capillary containing the remaining sample for application must be kept out of
the stream of warm air so as to avoid evaporation. Also, the liquid in the capillary
tip may get into the wider part of the capillary, so making it very difficult to get
it completely back into the tip. An experienced worker needs ca. 2–3 min for
the careful application of 0.2 to 0.5 µl.

Fig. 20. Arrangement of a 3 × 3 cm polyamide layer under a stereo microscope for the application of
the reaction mixture

To graduate the capillary pipettes for the application of different volumes of
liquid, a heating wire of the de Fonbrune micro-forge is used (details see on p. 230
in chapter 6). Calibration markings with ink are not very accurate because of the
relatively broad line, and should not be used. The volume between the tip of the
capillary and the graduation is determined by using a solution of a ^{14}C compound,
e.g. ^{14}C-dans-Cl, of known activity and scintillation counts. The volume determination should be performed several times.

Chromatographic development is carried out in 50 ml beakers, in which the
floor is just covered with the appropriate solvent. Under no circumstances should
the origin dip into the solvent. Development takes about 3 min for the first
dimension (formic acid/water), after which the chromatogram must be dried

carefully with warm air. The plate can be checked for complete dryness by smelling it (being careful not to touch the foil with one's nose); when the chromatogram is dry there is no longer a smell of formic acid. The plate must then be cooled again by waving it gently in the air. If plates which are still warm are developed in the second dimension (benzene/acetic acid) a distinct tailing of dansyl substances occur. Development in the second dimension lasts about 4 min. The plate is then re-dried. As a rule development is carried out until the solvent front is about 1 mm from the upper edge of the foil. During chromatography in the first dimension the beaker can be left open, but in the second dimension it must be covered with a Petri dish. The solvent systems may be used for about a week if stored in the cold in closed vessels. The development takes longer if the developing solvents are cold when used for the chromatography. This effect can be useful under certain circumstances, especially when chromatograms are overloaded. Owing to the appreciably slower chromatography more time is available for the distribution of substances between solvent and polyamide sheet, and the chromatography is thereby improved. This effect is more important when developing the chromatogram in the first dimension, and less so in the second dimension.

Solvent systems:

First dimension for micro-polyamide foils from Schleicher and Schüll: formic acid/water 1.5 : 50 (v/v) (NEUHOFF, BRIEL, MAELICKE, 1971).

First dimension for polyamide foils from Cheng Chin Trading Co.: formic acid/water 1.5 : 100 (v/v) (WOODS and WANG, 1967).

Second dimension for both types of foil: benzene/acetic acid 9 : 1 (WOODS and WANG, 1967).

In practice, it has been found to be expedient to handle the polyamide plates with "cross-action" cover-glass forceps of the type shown in Fig. 21.

Fig. 21. Cross-action forceps for cover glasses

Fig. 22 shows an autoradiogram (preparation see p. 117) after the two-dimensional chromatography of a mixture of 23 amino acids (tyr, lys, orn, phe, his, leu, ile, pro, val, GABA, gly, glu, asp, glu-NH_2, asp-NH_2, ala, arg, ser, thr, tau, cys, 5-HT and 5-HO-indole, 4×10^{-5} m each) which were reacted with ^{14}C-dansyl chloride as described, and separated on 3×3 cm micro-polyamide. As a guide to the locations, a map giving the positions of the individual amino acids is shown alongside. Black areas on the film which are not identified are due to impurities in the preparation of ^{14}C-dansyl chloride.

A third development in the direction of the second dimension, using ethyl acetate/methanol/acetic acid 20 : 1 : 1 (v/v) (CROWSHOW, JESSUP, RAMWELL, 1967) is necessary to separate dansyl alanine from dansyl amine, dansyl glutamic acid from dansyl aspartic acid, and dansyl glutamine and dansyl threonine from dansyl asparagine and dansyl serine. If the radioactivity of the individual dansyl spots is to be measured for quantitative analysis, spots 3–18 are removed for measure-

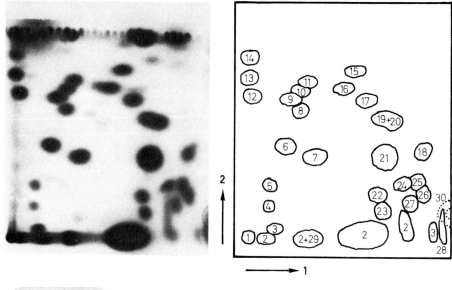

Fig. 22. Autoradiogram (original size) after micro-chromatography of a mixture of 25 ^{14}C-dansyl amino acids. The directions of chromatography are indicated by the arrows.

1st direction: water/formic acid 100 : 3.

2nd direction: benzene/acetic acid 9 : 1. A map of the chromatogram is shown to assist in the identification of individual dansyl compounds.

1 origin, *2* dans-OH, *3* dans-cystine/cysteine, *4* dans-N-serotonin, *5* dans-tryptophan, *6* dans$_2$-lysine, *7* dans$_2$-ornithine, *8* dans-phenylalanine, *9* dans$_2$-histidine, *10* dans-leucine, *11* dans-isoleucine, *12* dans$_2$-tyrosine, *13* dans$_2$-serotonin, *14* dans-5-O-indole, *15* dans-proline, *16* dans-valine, *17* dans-GABA, *18* not identified, *19* dans-alanine, *20* dans-NH$_2$, *21* dans-glycine, *22* dans-glutamic acid, *23* dans-aspartic acid, *24* dans-threonine, *25* dans-glutamine, *26* dans-asparagine, *27* dans-serine, *28* dans-(arginine, α-amino-histidine, α-lysine), *29* dans-taurine, *30* lable of the x-ray film. (From BRIEL, 1972)

ment, and the chromatogram is then developed in the third solvent system. Subsequently the dansyl alanine, glycine, glutamine, threonine, asparagine, and serine may be removed for measurement. Fig. 23 shows the autoradiogram of the same chromatogram as in Fig. 22 after chromatography in the third solvent system.

The majority of the dansyl amino acids recorded have a bright yellow fluorescence under UV light (248 nm). Bis-dansyl histidine, bis-dansyl tyrosine and dansyl-1(-3)-N-histidine (NEUHOFF, BRIEL, MAELICKE, 1971) have a markedly orange fluorescence. 1-Dimethylamino-naphthalene-5-sulfonic acid, dans-OH, has a bright blue fluorescence, and on chromatography it separates into several frac-

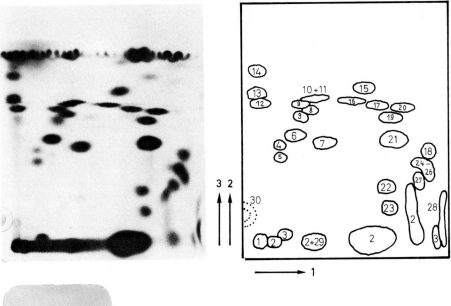

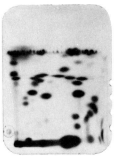

Fig. 23. Autoradiogram (original size) of the same chromatogram as in Fig. 22 after chromatography with the third solvent system (ethyl acetate/methanol/acetic acid 20:1:1 v/v) in the direction of the 2nd dimension to separate dans-alanine (spot *19*) from dans-NH$_2$ *20*, dans-glutamic acid *22* from dans-aspartic acid *23*

tions (see Fig. 22, spot 2) which have not been characterised chemically. After development in the first dimension minor fractions are found near the origin, between it and the main fractions. Dansyl taurine is also found in this region. In order to separate dansyl taurine, it is necessary to develop yet again in the direction of the second dimension, using water/formic acid (50:1.5 v/v), after development of the chromatogram in reagents 1–3 and removal of the other dansyl amino acids. BRIEL et al. (1972) combined the developing systems given in Fig. 24 to fractionate the dansyl amino acids obtained from the free amino acids from isolated pancreatic islets of mice.

0.05 M trisodium phosphate/ethanol (3:1 v/v) may be used to separate dansyl amino acids which run near the solvent front in the first dimension, and which are insufficiently separated in solvent systems 2 and 3 (ε-lysine, α-aminohistidine, and arginine) (TANG and HARTLEY, 1970). Development is in the same direction as the second chromatography. Under these conditions α-amino-dansyl histidine travels in front of ε-dansyl lysine, and dansyl arginine has the slowest mobility. Another method for the quantitative determination of histidine is to transform the mono-dansyl compound to bis-dansyl histidine by a second dansylation. If this is

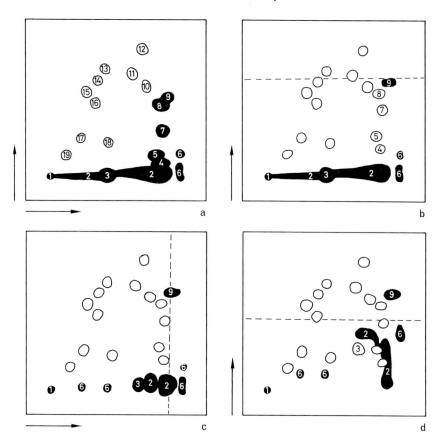

Fig. 24 a–d. Chromatography of dansylated amino acids on micro-polyamide layer plates (5 × 5 cm). The amino acids (yellowish fluorescence), as well as some spots of constantly-occurring by-products (blue fluorescence), are shown. Volume compositions of the solvents were: *a* first direction: 3 % formic acid; second direction: benzene/acetic (9:1): *b* ethyl acetate/methanol/acetic acid (20:1:1); *c* and *d* 3 % formic acid. Spots indicated (●) are not separated sufficiently, those indicated (⑩) are ready to be scraped off. Spots indicated (○) represent the holes remaining after the fluorescent spots have been scraped off. As can be seen, 10 spots were removed after the first 2 separations, = *a*. After separation *b*, run to the dashed line, another 4 spots were scraped off. Separations *c* and *d* were performed to separate dans-taurine from dans-OH. The solvents were allowed to run as far as indicated by the dashed lines. The spots are indicated as follows: *1* origin, *2* dans-OH, *3* dans-taurine, *4* dans-aspartic acid, *5* dans-glutamic acid, *6* unknown by-products, *7* dans-glycine, *8* dans-alanine, *9* dans-NH₂, *10* dans-γ-aminobutyric acid, *11* dans-valine, *12* dans-proline, *13* dans-isoleucine, *14* dans-leucine, *15* dans₂-histidine, *16* dans-phenylalanine, *17* dans₂-lysine, *18* dans₂-ornithine, and *19* dans-tryptophan.
(From BRIEL *et al.*, 1972)

performed with biological materials, the two-dimensional chromatogram shows only bis-dansyl histidine and bis-dansyl lysine, at their characteristic positions, and in the region where all three were formerly mixed together only the dansyl arginine spot is found.

Fig. 25 illustrates the chromatographic behaviour of pure substances which were reacted with unlabelled dansyl chloride as described, and separated by two dimensional chromatography on 3 × 3 cm micro-polyamide sheets.

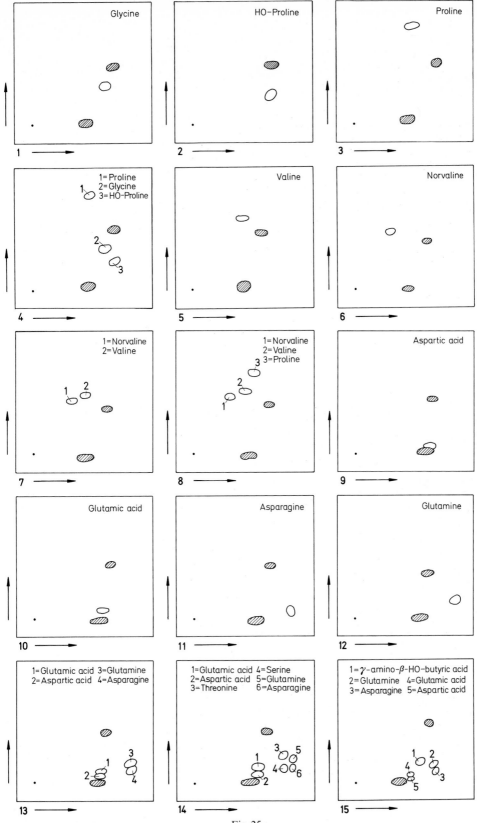

Fig. 25

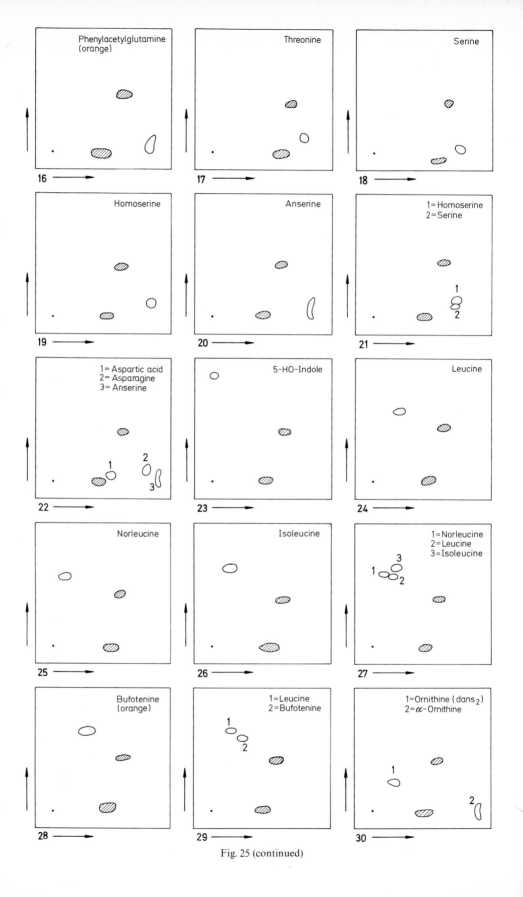

Fig. 25 (continued)

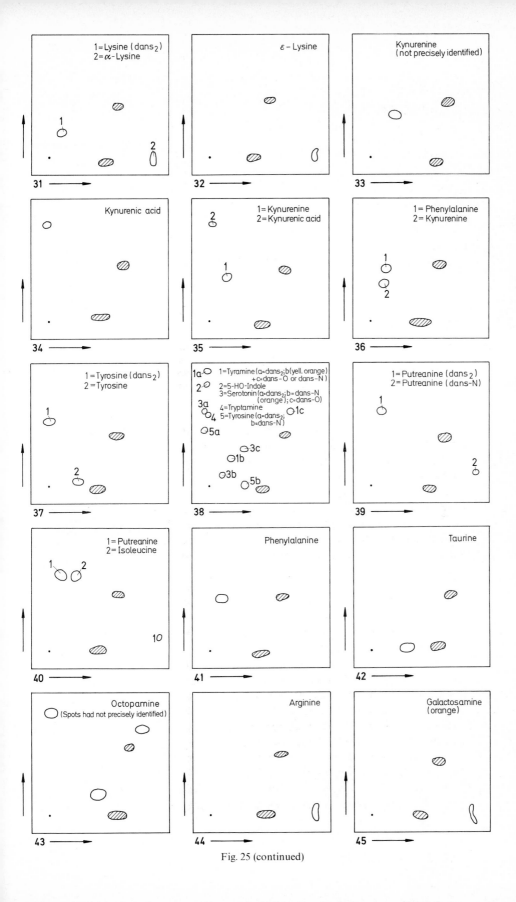

Fig. 25 (continued)

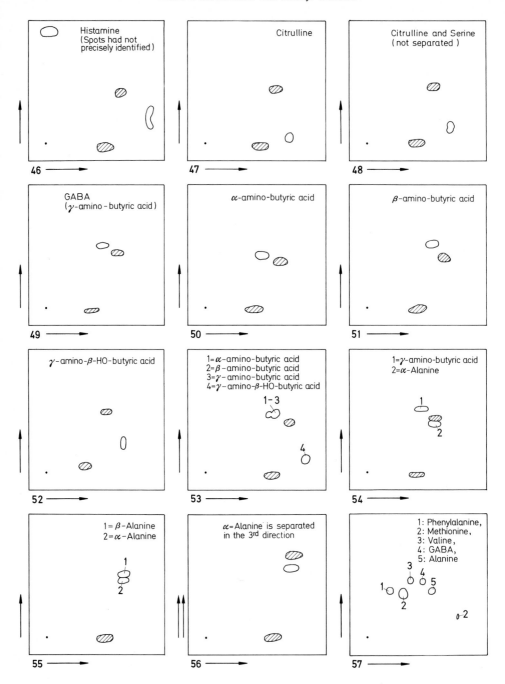

Fig. 25. Behaviour of pure compounds on reaction with dans-Cl and separation by two-dimensional chromatography on 3×3 cm micro-polyamide layers.
1st direction: water/formic acid $100 : 3$ v/v.
2nd direction: benzene/acetic acid $9 : 1$ v/v.
The two spots indicated ⊘ refer to dans-OH and dans-NH$_2$

Amino acid	Chromatogram No.	Amino acid	Chromatogram No.
α-Alanine	54, 55, 56	Kynurenine	33, 35, 36
β-Alanine	55	Leucine	24, 27, 29
α-amino-butyric acid	50, 53	α-Lysine	31
β-amino-butyric acid	51, 53	ε-Lysine	32
γ-amino-butyric acid	49, 53, 54	Norleucine	25, 27
γ-amino-β-HO-butyric acid	15, 52, 53	Norvaline	6, 7, 8
Anserine	22	Octopamine	43
Arginine	44	Ornithine	30
Asparagine	11, 13, 14, 15, 22	Phenylacetylglutamine	16
Aspartic acid	9, 13, 14, 15, 22	Phenylalanine	36, 41
Bufotenine	28, 29	Proline	3, 4, 8
Citrulline	47, 48	HO-Proline	2, 4
Galactosamine	45	Putreanine	39, 40
Glutamic acid	10, 13, 14, 15	Serine	14, 18, 21, 48
Glutamine	12, 13, 14, 15	Serotonin	38
Glycine	1, 4	Taurine	42
Histamine	46	Threonine	14, 17
Homoserine	21	Tryptamine	38
5-HO-Indole	23, 38	Tyramine	38
Isoleucine	26, 27, 40	Tyrosine	37, 38
Kynurenic acid	34, 35	Valine	5, 7, 8, 57

Evaluation of Microchromatograms

The fluorescence spectra of all aliphatic dansyl amino acids display, both for excitation and emission, wavelengths specific for the dansyl chromophore: for example dansyl alanine and dansyl γ-aminobutyric acid (see Figs. 37 and 38). The positions of the maxima depend on the solvent in which they are measured; in water they are at longer wavelengths and have lower-quantum yields than in organic solvents. Moreover, since the solubility of dansyl derivatives in organic solvents is far better than in water, the use of organic solvents (analytical grade) for recording spectra is recommended. If the corresponding standard curves are available, quantitative determination of the dansyl compounds is possible. This, however, raises the problem of quantitative elution from the carrier layer after chromatographic separation, since not all dansyl compounds can be easily and quantitatively eluted from polyamide. After two-dimensional chromatography on 3 × 3 cm sheets, the spots of the individual dansyl compounds are as a rule so small that, even after quantitative elution, the content of the dansyl compound is too small for fluorometric determination to be possible.

In contrast, if ^{14}C-dansyl chloride has been used, the quantitative estimation of substances on micro-chromatograms presents no problems. The main difficulty in this case is the quantitative evaluation of substances on chromatograms, since the rates of reaction of the individual amino acids vary as has been previously explained.

For evaluation, the dansyl spots are marked under UV light with a sharp soft lead pencil. Hard pencils are unsuitable since the polyamide layer is very easily damaged during marking, and small pieces of polyamide may be broken off and lost. The pencil markings should always be slightly larger than the diameter of the fluorescent spots. If several chromatograms are to be marked it is advisable to wear goggles with UV-filters; if this is not done, a painful irritation to the eyes can occur, often leading to conjunctivitis.

To remove the polyamide layer from the carrier foil, a special knife, shown in Fig. 26, is used, under a stereomicroscope. A slightly bent steel wire, 1 mm thick, with the end flattened on one side so that a fragment of razor blade may be soldered onto it, is inserted into a pencil-shaped plastic holder. The side away from the razor edge is then also flattened. The edge of the fragment of razor blade (preparation see p. 217) should be between 0.5 and at most 1 mm wide. Such

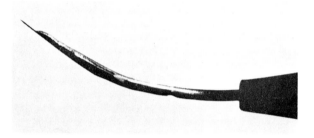

Fig. 26. Microknife for removal of polyamide layer

knives can be used for a relatively long time; if they become so blunt that difficulty is encountered in removing the foil, a new fragment of blade can be soldered on to the end. To remove the polyamide layer, the layer is first cut through down to the carrier foil by gentle pressure with the knife, and then the layer is scraped off the carrier foil in parallel, overlapping strips. In this way the small strips of polyamide keep their consistency, and can be removed along the pencil marking more easily and transferred completely to a counting vial. The fragments of polyamide often cling to the knife, and it can be difficult to get the fragments off the knife and into the vial. Inspection under the stereomicroscope is necessary to see whether all the polyamide layer within the marking has been scraped off. After some pratice, removal of the layer without loss is quite easy.

If several dansyl spots are to be removed from several chromatograms, a definite and unaltered sequence of procedure should be observed, including the micro-chromatographic separation. It is not possible to label the individual spots with, for instance, numbers on the polyamide layer, as they are much too small. The beginner is recommended to keep a good, numbered sketch of the dansyl spots near the stereomicroscope, as a guide. The quenching effect of the fragments of polyamide in 10 ml of the scintillation solution (4 g PPO and 0.1 g POPOP/1 L toluene) is negligibly small. Ten such pieces, with diameters of 1–2 mm^2, have a quenching effect of 0.7 % (NEUHOFF, 1971). If the counting efficiency of the apparatus is known, a quantitative determination can be performed easily.

Evaluation by Autoradiography

Autoradiograms from 3×3 cm chromatograms are conveniently produced with the 3×4 cm Personal-Monitoring-Film D2/D10 from Agfa-Gevaert. Each pack contains two X-ray films; the film (Structurix D10) at the front (the rear side of the pack is indicated) is about 130 times more sensitive than the film at the rear. For easier identification in the dark, the film at the front is marked with four notches, that at the back with two. The more sensitive film (relative exposure factor 0.25) is normally used for the autoradiography. For this, a microchromatogram with ^{14}C-dansyl amino acids or dansyl ^{14}C-amino acids is carefully laid with the correct side against the sensitive X-ray film in total darkness and sandwiched between two 3×3 cm single glass plates. By tapping them briefly on the table top, the film, chromatogram, and glass plates are aligned relative to one another, and the glass plates are then bound on three sides with clear adhesive tape. The glass plates can be obtained from a glazier or can be cut from glass remnants. The glass edges must be smoothed by rubbing the edges briefly on emery paper; otherwise, the fingertips, film, or chromatogram can be easily damaged on working in the dark. The pack is wrapped in black paper and stored in the dark. It is strongly recommended to lay out all the necessary components, film, glass plates, chromatograms, and roll of adhesive tape, ready at hand, so that they can be found easily in the complete darkness when the X-ray films are taken out of their packing. It is best to practice the whole procedure in the light first.

The exposure time is determined by the intensities of the individual spots on the chromatograms. A spot which, in the scintillation counter, yields about 100 cpm, produces a good blackening of the film after 1 to $1\frac{1}{2}$ days. The films are developed in 50 ml beakers (one beaker for each film) in the dark, using developer at room temperature for 5 min (Agfa-Gevaert G230; 54.2 g compound A and 6.3 g compound B dissolved in 500 ml distilled water, or 75 ml G150 + 450 ml distilled water). The films are rinsed thoroughly in running water (holding the film only by the edges) and then fixed for 10 min with sodium thiosulfate (10% in distilled water). After this, they are washed for about 1 hr. in several changes of distilled water, and dried in air.

It is important not to damage the chromatogram, which may still be required for further estimations, or to scratch the film on removing it, as the film overlaps both glass plates by a centimetre. It is therefore better not to pull the film out, but to remove the adhesive tape first from at least one side of the glass plates. The film and the chromatogram stick to each other frequently, so one must check carefully that only the film is transferred to the developer. A chromatogram is useless if it is once dipped into the developer. If several autoradiograms are being processed simultaneously, do not forget to mark the individual chromatograms carefully on the unused side.

Autoradiograms may be evaluated quantitatively by densitometry, particularly if a standard substance of known radioactivity has been applied to the chromatogram in a suitable position prior to autoradiography. However, quantitative densitometry of such spots presents considerable difficulties because the spots are round, though often unevenly so. Quantitative estimation is possible with remission photometry in epi-illumination (see chapter 9). WEISE and EISEN-

BACH (1972) evaluated such autoradiograms with a scanning microscope photo-meter (see chapter 9) and were able to estimate as little as 10^{-14} moles of ^{14}C-dansyl amino acids quantitatively.

The purity of ^{14}C-amino acid preparations can be checked easily by treatment with unlabelled dansyl chloride, followed by autoradiography of the chromato-grams. Fig. 27 shows the autoradiogram of a micro-chromatogram of a ^{14}C-labelled protein hydrolysate which has been treated with dansyl chloride. Only 10 of the fluorescent spots denoting dansyl amino acids were recognizable under the UV lamp. The dense black spot in the right corner of the autoradiogram is due to radioactive impurities in the preparation, which did not react with dansyl chloride and were not identifiable under the UV lamp. In this case the impurities accounted for about 50% of the total radioactivity in the preparation, and, since the price of a radioactive sample is generally based on the radioactivity, this sample proved an expensive buy. By applying micro dansyl methods to the direct characterisation of transfer ribonucleic acids, NEUHOFF et al. (1969) have shown that radioactive impurities which have not been removed by the purification procedures can still be found in highly purified preparations of amino acids (see Fig. 39).

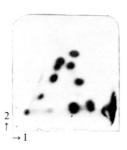

Fig. 27. Autoradiogram (original size) after reaction of ^{14}C-amino acids with unlabelled dans-Cl. The intense film-blackening in the right hand corner represents ^{14}C impurities in the preparation

Identification of Dansyl Derivatives

The sheets have similar polyamide layers on both sides, so that both the front and the back may be chromatographed simultaneously. If chromatograms are to be compared, for example with standard mixtures, care must be taken that the application points on the front and back are in identical positions. To achieve this, the sheet is held over a UV lamp after applying the sample on one side of the foil. The fluorescent dansylated sample shines through the thin foil, and the reverse side can be marked for the application of the second sample. This procedure is not advisable if the exact localization of a single spot is required, because, on chromatography in 50 ml beakers, the solvent always runs faster on the side of the foil nearest to the glass wall. This small difference in the rate of migration of the solvent causes the dansyl compounds on the different sides of the polyamide layer to migrate slightly differently. The same effect, but more pronounced, is also observed when a sample from biological material is fractionated on one side of the sheet, with a sample of a pure test mixture on the other; the differences become even more pronounced if a mixture of dansylated amino acids from biological material is compared with a single pure dansyl amino acid. Substances always run somewhat differently when pure than when in biological material, as does a substance when single or in a mixture. If it is necessary to identify a single dansyl compound in a complex mixture it is advisable to apply a quantity of the standard dansylated material, either together with, or separately from, the mixture. Preliminary experiments are required to ascertain the optimal concentration of the pure dansyl compound so that it appears as a clearly-visible minute spot when chromatographed alone. This quantity should be dissolved in the smallest possible volume, so that when it is applied together with the analysis mixture at the origin, the sample is not influenced in any way. If the mixture is correct, the spot to be identified will fluoresce about twice as strongly in the developed chromatogram as the same spot when the sample is chromatographed alone. The following conditions have been found to be best for the comparison of two compounds, especially if they normally migrate closely together. After microchromatography on one or more polyamide sheets, the dansyl compound under investigation is scraped off, put in a small tube, eluted with ethanol, and the ethanol removed in vacuo. The dried extract is dissolved in a minute volume of acetone and applied to a polyamide sheet together with the standard dansyl compound as described above. In this way no interfering compounds from the reaction mixture can impair the chromatography. In mixing experiments the addition of ^{14}C-dansyl compounds is not always necessary, although it is often valuable to use radioactive controls.

For the identification of a substance it is also often necessary to scrape off and elute a spot after chromatography of dansyl compounds on 5×5 cm plates, in order to record its fluorescence spectrum. Elution is best carried out with absolute ethanol, since this solvent also gives a high quantum yield (0.70 according to GRAY, 1964) which is suited to recording the fluorescence spectrum. If, for the further characterisation of a dansyl compound after elution from the polyamide sheet, a second treatment of the eluate with dansyl chloride is necessary, some other dansyl amino acids are often discovered on the second chromatogram (NEUHOFF, BRIEL, MAELICKE, 1971). These amino acids are contained as impurities in the polyamide sheets, are eluted with alcohol, and subsequently found on dansyl treatment. This should always be borne in mind if, for instance, dansyl peptides have been separated by chromatography on polyamide sheets, and are to be further characterised by dansylation after elution and hydrolysis. The impurities in the sheet are not important for normal chromatography of the dansylated amino acids, since they do not influence the quality of the separation.

Knowledge of the amount of ^{14}C-dansyl bound to amino acids with more than one reactive group ($-NH_2$, $-OH$) is crucial for quantitative studies, so the three dansyl histidine compounds were characterised (NEUHOFF, BRIEL, MAELICKE,

1971). Commercial bis-dansyl histidine, obtained as N,N'-di-(1-dimethylamino-naphthalene-5-sulphonyl)-histidine, could not be used for the identification of any of the three different dansyl histidine compounds, since it contained not only dans-OH and dans-NH$_2$, but also four other dansyl compounds with the typical yellow dansyl fluorescence, together with α-amino-dansyl histidine and bis-dansyl histidine. Therefore, L-histidine was obtained from Sigma (Σ-Grade), 1-methyl- and 3-methyl-histidine from Koch & Light Lab. Ltd. For dansylation, the compounds were dissolved in 0.05 M NaHCO$_3$ adjusted to pH 9 with NaOH, and mixed thoroughly with dansyl chloride dissolved in acetone (3 mg/ml). Two-dimensional chromatography was performed on 3×3 cm or 5×5 cm micro-polyamide layers. Fluorescence spectra of the three dansyl histidine compounds were obtained after separating the individual derivatives on a number of 5×5 cm layers, and then isolating and eluting them with absolute ethanol. A Turner Model 210 "spectro" spectrofluorometer (TURNER, 1964) was used for the spectra. To dansylate the isolated compounds a second time, the ethanol was evaporated and the residue was dissolved in 0.05 M NaHCO$_3$ (pH adjusted to 10.0), mixed with the solution of dansyl chloride in acetone, and incubated for 30 min at 37° C. The pH was adjusted to 10 as the formation of bis-dansyl compounds is favoured by a higher pH (NEUHOFF and WEISE, 1970). On the second chromatogram, a number of amino acids present as impurities in the polyamide sheet were also detected. Fig. 28 shows the characteristic positions of the three dansyl histidine compounds, as well as dansyl imidazole, dansyl 1-methyl-histidine, and dansyl 3-methyl-histidine, after two-dimensional micro-chromatography on a 3×3 cm polyamide layer. From the different migration rates in an aqueous phase and in an organic phase, it was to be expected that spot 1 represented the bis-dansyl histidine, and that the other spots were two different mono-dansyl histidine compounds. After isolation and a second dansylation of all three isolated histidine compounds at pH 10, spot 1 remained unchanged, and spots 2 and 3 were converted completely into a compound which migrated in an identical way to spot 1, thus indicating that they were transformed to the bis-dansyl histidine product of spot 1. After 60 h of incubation of dans-Cl with histidine at room temperature, only spot 1 is found on the chromatogram. On the basis of this finding, the fluorescence colour of spot 2 was initially puzzling since such an intense orange fluorescence was not expected from a mono-dansyl compound. However, HARTLEY and MASSEY (1956) have found dansyl-α-benzoyl-L-histidine methyl ester to be bright orange, with a UV-absorption maximum at 365 mμ.

Fig. 29 shows the excitation and emission spectrum of L-histidine in absolute ethanol. It is characterised by an excitation maximum at 248 mμ and an emission maximum at 305 mμ. The intensity of the emission maximum is increased in strong alkali and shows a second maximum at 343 mμ, with a minimum at 338 mμ (upper curve in Fig. 29). When the pH is lowered with HCl the fluorescence is the same as that in ethanol. Imidazole in ethanol has the same maxima as those of histidine.

Fig. 30 shows the excitation and emission spectrum of dansyl imidazole in absolute ethanol. If the concentration is such that concentration quenching does not occur in the imidazole peak (left), scarcely any dansyl-specific fluorescence is detectable. If the concentration is optimal for measuring the dansyl fluorescence

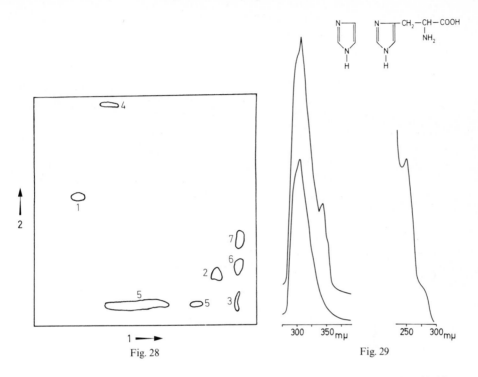

Fig. 28

Fig. 29

Fig. 28. Characteristic positions of *1* dans$_2$-histidine, *2* 1(3)-N-dans-histidine, *3* α-amino-dans-histidine, *4* dans-imidazole, *5* dans-OH. *6* α-amino-dans-1-N-methyl-histidine, *7* α-amino-dans-3-N-methyl-histidine, after two dimensional chromatography on 3 × 3 cm micropolyamide layers.
1st dimension: water/formic acid, 2nd dimension: benzene/acetic acid. (From NEUHOFF, BRIEL and MAELICKE, 1971)

Fig. 29. Emission spectrum (left) and excitation spectrum (right) of 1-histidine in abs. ethanol. Both spectra had 25 Å bandwidth on the excitation, and 100 Å bandwidth on the emission monochromator. For the emission spectrum the excitation was at 248 mμ, and for the excitation spectrum the emission monochromator was at 305 mμ

(second spectrum on the left), the imidazole shows a peak, but this is an artefact, with maxima at 291 mμ and 303 mμ and a minimum at 296 mμ. As well as the fluorescence maximum at 530 mμ, specific for the dansyl chromophore dissolved in ethanol, there is a little shoulder at 445 mμ. The excitation spectrum of dansyl imidazole, with 530 mμ at the emission monochromator and a bandwidth of 100 Å on both monochromators, has the same characteristics as other dansyl compounds: three sharp maxima at 252 mμ, 340 mμ, and 350 mμ, and three minima at 243 mμ, 297 mμ, and 345 mμ (compare Figs. 37 and 38).

Bis-dansyl histidine (spot 1), with excitation at 248 mμ, is characterized by two maxima in the emission spectrum: the histidine maximum at 305 mμ, and the dansyl maximum at 530 mμ (see Fig. 31). For this bis-dansyl compound the histidine peak is higher than the dansyl peak. The excitation spectrum taken with the same instrument settings, but at a wavelength of 530 mμ on the emission monochromator,

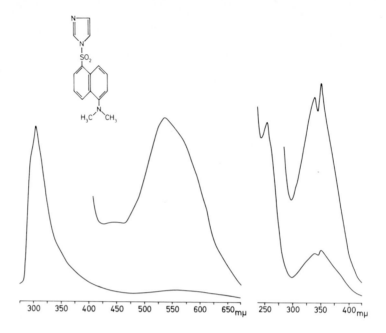

Fig. 30. Emission spectrum (left) and excitation spectrum (right) of dans-imidazole in abs. ethanol. The emission spectrum was taken at 248 mµ excitation, with 25 Å bandwidth on the excitation mono-chromator and 100 Å bandwidth on the emission monochromator. Under these conditions scarcely any dansyl fluorescence can be detected beside the imidazole peak. The second emission spectrum was taken with the same dansyl-imidazole concentration as for the excitation spectrum, and had 530 mµ on the emission monochromator, with 100 Å bandwidth on both monochromators. The second excitation spectrum was taken at the same instrument settings, but with a diluted sample

shows the three maxima typical for dansyl compounds at 252, 340, and 350 mµ. In alkali the intensity of both emission peaks increases, and in acid the dansyl peak is lost completely while the histidine peak remains unchanged.

The emission spectrum of the dansyl histidine compound III, taken with the same instrument settings as the spectrum of bis-dansyl histidine, is shown in Fig. 32. Compound III has maxima at identical wavelengths, but the relation bet-ween the histidine peak and the dansyl peak is different. The intensity of fluores-cence for the dansyl peak is increased. The characteristic maxima for this dansyl compound remain unchanged in the excitation spectrum. On polyamide, these two compounds appear yellow and yellow-orange respectively under UV light. The fluorescence spectra of the dansylated 1-methylhistidine and 3-methyl-histidine are very similar to that of compound III, showing excitation maxima at 253 and 335–340 mµ, and emission maxima at 305 and 520–530 mµ, with the double peaks normally observed for the dansyl region of these spectra.

The fluorescence spectra of the dansyl histidine compound II differ widely from those of compounds I and III (see Fig. 33). The emission spectrum, at an excitation wavelength of 245 mµ with a bandwidth of 100 Å on the emission mono-chromator and 25 Å on the excitation monochromator, shows the histidine peak

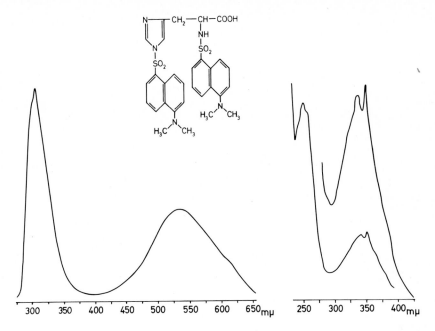

Fig. 31. Emission spectrum (left) and excitation spectrum (right) of bis-dansyl-histidine in abs. ethanol. The emission spectrum was taken at 248 mμ excitation, and the excitation spectrum at a setting of 530 mμ of the emission monochromator. The bandwidth was at 100 Å on the emission side, and 25 Å at the excitation side, for both spectra

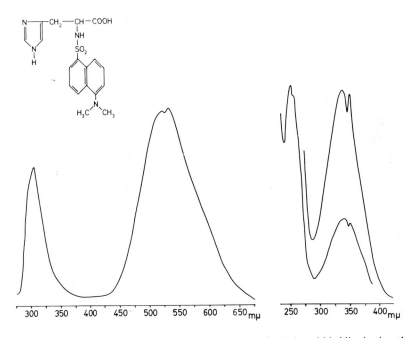

Fig. 32. Emission spectrum (left) and excitation spectrum (right) of α-N-dansyl-histidine in abs. ethanol. The instrument settings were the same as for the spectra in Fig. 31

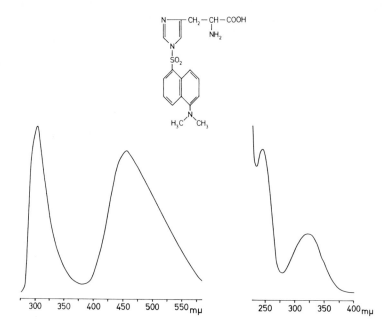

Fig. 33. Emission spectrum (left) and excitation spectrum (right) of 1-N-dansyl-histidine in abs. ethanol. The emission spectrum was taken with an excitation of 245 mμ, and the excitation spectrum at 456 mμ on the emission monochromator; for both spectra the bandwidth was 100 Å on the emission side, and 25 Å at the excitation monochromator. (Figs. 29–33 from NEUHOFF, BRIEL and MAELICKE, 1971)

at the typical wavelength of 305 mμ, but the dansyl maximum has shifted to 456 mμ, and the minimum to 382 mμ. The intensity of emission is nearly the same for both peaks. This emission spectrum corresponds very well to the red-orange fluorescence of the compound when seen under UV light on a polyamide layer. The excitation spectrum of this compound, with the same bandwidth at 456 mμ of the emission monochromator, differs widely from the normal dansyl excitation spectra; it is characterized by maxima at 245 mμ and 322.5 mμ, and by minima at 235 mμ and 277.5 mμ.

The difficulty in interpreting the fluorescence spectra is due to the fact that histidine or imidazole, as well as the dansyl chromophore, can be activated at 248 mμ. The relationship between the intensities of the emission maxima: 305 mμ for histidine or imidazole, up to 530 mμ for dansyl, suggests a special steric arrangement for both the chromophores. The dansyl maximum at 530 mμ always increases in proportion to the imidazole maximum at 305 mμ if there is a preceding, fast, radiationless energy-transfer step, whereas the intensity of the imidazole maximum corresponds to the immediate deactivation by radiation. Dansyl imidazole is deactivated completely by radiation, as the ratio of the maximum at 305 mμ to that at 530 mμ is greater than 100 : 1. The emission spectra of the dansyl histidine compounds are not so easy to interpret, since the side-chain gives more possibilities for oscillation, even if the dansyl group is only on the imidazole ring. Nevertheless the direct process of radiation will be preferred for mono-ring-substituted histidine, so that the ratio of the 305 mμ maximum to that at longer wavelength must be greater than that obtained for histidine with substitution in the side-chain. From the spectra in Figs. 32 and 33 one can conclude that compound II is the monodansyl histidine with the substitution on the imidazole ring, and that compound III is the α-amino-dansyl histidine. As mentioned earlier, conversion of compounds II and III into the bis-dansyl histidine compound I is possible by a longer or repeated dansylation. In agreement with this, only in 1(3)-N-mono-dansyl histidine (compound II) is there the possibility of reciprocal interactions between the chromophores, as demonstrated by the shifting to shorter wavelength of the dansyl maximum. A similar, but weak effect is also shown in the dansyl imidazole spectrum by the shoulder at 445 mμ.

The emission spectrum of bis-dansyl histidine cannot be compared with that of the mono-dansyl compounds because of the double substitution. In contrast to the mono-products, it shows intra-molecular quenching between the two dansyl chromophores. Hence the ratio from 305 mμ to 530 mμ is increased again.

A comparison of the excitation spectra shows that the maxima of dansyl imidazole, bis-dansyl histidine, and α-amino-dansyl histidine, are identical and are also similar to other more simple dansyl compounds (compare Figs. 37 and 38). 1(3)-N-dansyl histidine is a special amino acid insofar as the second ring-N in the 5-member ring favors the electrophilic reaction with one of the two ring-N. The excitation spectrum of this compound reflects the influence of the imidazole on the dansyl chromophore. These particular fluorometric properties make it possible to characterize histidine in a multi-component system such as biological material.

For the determination of Serotonin (5-hydroxy-tryptamine) in pico-mole quantities with dansyl chloride it is necessary to characterize the reaction products, as the reaction of serotonin with dansyl chloride can yield three fluorescent products: N-dansyl serotonin, bis-(N,5-oxy)-dansyl serotonin, and O-dansyl serotonin. To synthesize dansyl serotonin, serotonin oxalate (Sigma) was dissolved in $NaHCO_3$ (pH 9), dansyl chloride acetone solution was added, and the mixture was incubated at 37° C. The reaction mixture was then applied as a line to a 6 × 15 cm polyamide layer, and the chromatogram was developed with water/formic acid. During this process dans-OH, recognizable by its pale blue fluorescence migrated to half the distance travelled by the front. The reaction product of dansyl chloride and serotonin remained in the region of the starting line. This strip was scraped off, eluted with absolute ethanol, concentrated, applied as a line to a second 6 × 15 cm polyamide layer, and developed with benzene/acetic acid. Under UV illumination two fluorescent yellow bands could be seen, one just above the starting line and the other close to the solvent front. After drying carefully, both bands were scraped off and eluted with absolute ethanol. From the behaviour in aqueous and organic solvents, it could be predicted that the slower-migrating product of the reaction was N-dansyl serotonin, and the faster one the bis-dansyl serotonin. After the purity of these preparations had been tested on a further small two-dimensional chromatogram on a polyamide layer, the emission and excitation spectra shown in Figs. 34 and 35 were obtained, using a Turner Model 210 "Spectro" with a bandwidth setting of 25 Å for both the excitation and emission monochromators.

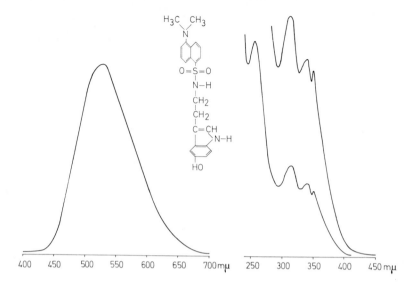

Fig. 34. Emission spectrum (left) and excitation spectrum (right) of pure N-dansyl-serotonin in abs. ethanol. Both spectra were taken with a 25 Å bandwidth on both monochromators. For the emission spectrum the excitation was at 340 mµ, and for the excitation spectrum the emission monochromator was at 530 mµ. The second excitation spectrum was taken at the same instruments settings, with a diluted sample

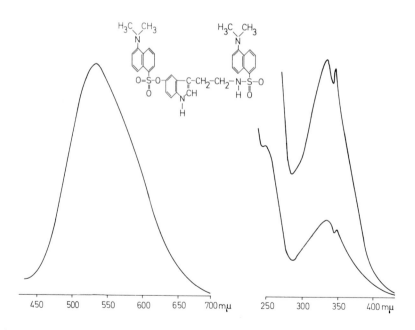

Fig. 35. Emission spectrum (left) and excitation spectrum (right) of pure bis-(N,5-oxy-dansyl)-serotonin in abs. ethanol. Instrument settings as in Fig. 34. For the emission spectrum the excitation was at 340 mµ, and for the excitation spectrum the emission monochromator at 535 mµ. The second excitation spectrum was taken at the same instrument settings, with a diluted sample

The broad emission spectrum for N-dansyl serotonin in absolute ethanol (Fig. 34), obtained with an excitation wavelength of 340 mµ, shows a maximum at 530 mµ. The excitation spectrum, with a setting of 530 mµ on the emission side, is characterized by 4 sharp peaks at 255, 313, 338 and 350 mµ, which decrease in intensity towards the longer wavelengths. Minima occur at 243, 295, 327 and 347 mµ. Absolute ethanol (G. R. Merck 972) shows no emission or excitation maxima at these wavelengths with the instrument settings used.

Bis-dansyl serotonin, under the same conditions (Fig. 35), has an emission spectrum with a maximum at 535 mµ and hardly differs from N-dansyl serotonin in this respect. In contrast, the excitation spectrum, recorded with 535 mµ on the emission side, has only three peaks, at 250, 338, and 350 mµ: the peak at 313 mµ for N-dansyl serotonin is missing. Minima occur at 243, 283, and 347 mµ. To determine the origin of the 313 mµ peak of N-dansyl serotonin, the spectra for serotonin oxalate (Sigma) in absolute ethanol, shown in Fig. 36, were recorded. The emission spectrum, with an excitation wavelength of 280 mµ and a bandwidth of 25 Å for both monochromators, shows a sharp narrow peak at 336 mµ. The excitation spectrum, with a setting of 335 mµ on the emission side, reveals peaks at 278 and 305 mµ, and minima at 250 and 295 mµ. The 313 mµ peak for N-dansyl serotonin is due to the excitation of serotonin, and the short shift from 305 mµ for pure serotonin to 313 mµ for N-dansyl serotonin is due to the presence of the dansyl group on the serotonin. The absence of this peak for bis-dansyl serotonin can be attributed to quenching of the 313 mµ maximum by the second dansyl

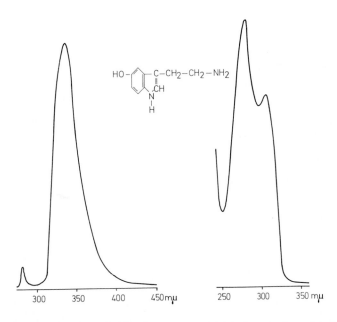

Fig. 36. Emission spectrum (left) and excitation spectrum (right) of serotonin-oxalate (Sigma) in abs. ethanol. Instrument settings as in Fig. 34. For the emission spectrum the excitation was at 280 mµ, and for the excitation spectrum the emission monochromator was at 335 mµ

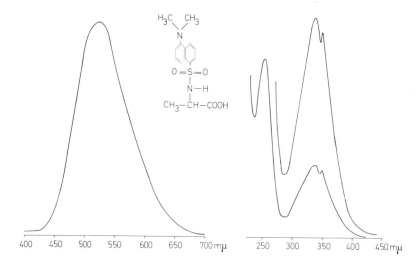

Fig. 37. Emission spectrum (left) and excitation spectrum (right) of dansyl-alanine (Sigma) in abs. ethanol. Instrument settings as in Fig. 34. The emission spectrum was taken with an excitation at 340 mμ, and the excitation spectrum with a setting of 525 mμ of the emission monochromator. The second excitation spectrum was obtained with the same instrument settings after dilution of the sample

group attached to the indole ring. As proof that bis-dansyl serotonin is the band which migrates faster on chromatography on polyamide layers with organic solvents, the slower-migrating band was eluted and treated a second time with dansyl chloride at pH 10, to give the faster-migrating compound.

A comparison between the spectra of dansyl alanine and dansyl γ-amino butyric acid (Figs. 37 and 38) reveals that the 3 peaks at 255, 338 and 350 mμ in the excitation spectrum can be attributed to the dansyl group, since in this substance, as well as in dans-NH_2 peaks are also found at these wavelengths. The very narrow, but clear and consistent, small peak at 350 mμ, which has so far not been shown in published spectra of dansylated amino acids (GRAY, 1964), is visible in the spectra shown here, because of the very high resolution of the Turner absolute spectrofluorometer using the narrow bandwidth of 25 Å.

Theoretically, it is also possible to form O-dansyl serotonin during the reaction of serotonin (3-(β-aminoethyl)-5-hydroxyindole) with dans-Cl, but under normal conditions of dansylation no O-dansyl serotonin is found. Bufotenin (3-(β-dimethyl-aminoethyl)-5-hydroxyindole) (Fluka) was treated with dans-Cl as a control. Pure O-dansyl bufotenin also has a broad emission maximum at 540 mμ. The excitation spectrum is characterized by 4 maxima at 255, 300, 338, and 350 mμ and by minima at 288, 305, and 347 mμ. In contrast to the excitation spectrum of N-dansyl serotonin, the maximum at 300 mμ is only weak for O-dansyl bufotenin. The emission and excitation spectra of bufotenin and of serotonin in ethanol are identical. O-dansyl bufotenin on a polyamide layer is characterized by its orange-red fluorescence under UV light; O-dansyl serotonin migrates nearly as far as O-dansyl bufotenin in the first dimension, but in the second dimension only half the distance of O-dansyl bufotenin. These results, and also the differences in the

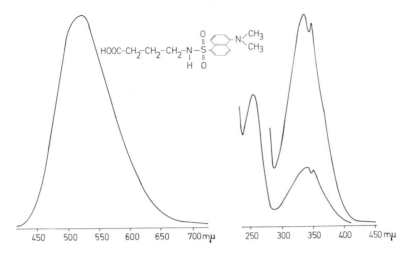

Fig. 38. Emission spectrum (left) and excitation spectrum (right) of pure dansyl-GABA in abs. ethanol. Instrument settings as in Fig. 34. The emission spectrum was obtained with an excitation at 340 mμ, and the excitation spectrum with 525 mμ on the emission monochromator. The second excitation spectrum was obtained after dilution of the sample. (Figs. 34–38 from NEUHOFF and WEISE, 1970)

excitation spectra, demonstrate again that after treatment of serotonin with dans-Cl, and two-dimensional chromatography very little migrating dansyl product is N-dansyl serotonin.

Characterization of tRNA

The reaction mixtures were fractionated after aminoacylation of the tRNA, using specially purified Sephadex G-25 columns in order to obtain the ^{14}C-aminoacyl-tRNA (NEUHOFF et al., 1969). A sample of eluate from a purified sephadex column had shown that only two impurities capable of being dansylated were present, and that, on chromatography, these migrated in positions which did not interfere with the estimation of the amino acids. Owing to presence in chromatography papers of extraordinarily high quantities of impurities which react with dansyl chloride and which cannot be removed even by very intensive washing with various solvents, the simpler preparation of ^{14}C-aminoacyl-tRNA samples by the starting spot method (HELLER, 1966; MATTHAEI et al., 1967) is not possible. The column eluate containing ^{14}C-aminoacyl-tRNA was therefore lyophilized, taken up in $NaHCO_3$, and allowed to react with unlabelled dansyl chloride as described previously. During the incubation at 37° C at the high pH of the reaction mixture, the complete hydrolysis of the aminoacyl-bond and the reaction of the liberated ^{14}C-amino acids with dansyl chloride occur. Fig. 39 shows the autoradiogram after the processing and two-dimensional chromatography of a mixture of ^{14}C-seryl-tRNA and ^{14}C-phenylalanyl-tRNA on a 3×3 cm polyamide foil. Near the blackenings due to ^{14}C-serine and ^{14}C-phenylalanine, which are found at their typical positions, a further dark spot is observed, which cannot be seen under the UV lamp. Subsequent examination of the chromatogram by scintillation counting revealed that 21% of the total radio-activity on the chromatogram was at the position of this impurity, which was introduced into the aminoacyl preparation by unspecific adsorption on the tRNA. It should therefore be borne in mind that the normal estimation procedure in amino-acylation experiments, in which radioactivity alone is measured, without direct reference to the amino acids, can lead to false results. In order to test to what degree a small amount of one aminoacyl-tRNA is detectable when another aminoacyl-tRNA is present in excess, pure ^{14}C-seryl-tRNA and ^{14}C-phenyl-alanyl-tRNA were mixed, according to their radioactivity, in the ratios ^{14}C-seryl-tRNA/^{14}C-phenylalanyl-tRNA of 8.7:1, 1:1, 1:10, and analysed. The results are shown in Table 6, and are in good agreement with the mixtures made up. The low values for mixtures with high serine contents are due to the high amounts of impurities in the preparation of ^{14}C-serine. From these experiments it is seen that it is possible to identify about 5% of a particular tRNA in a specific tRNA preparation by using this method without further refinement. The sensitivity of this method depends largely on the specific radioactivity of the amino acids used for aminoacylation.

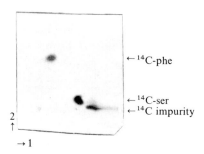

Fig. 39. Autoradiogram (original size) of a mixture of ^{14}C-seryl-tRNA and ^{14}C-phenylalanyl-tRNA. Dans-^{14}C-serine and dans-^{14}C-phenylalanine, at typical positions, are visualized under UV-light. The impurity in the right-hand corner comes from the ^{14}C-serine preparation used and is visible only on the autoradiogram. It represents 21% of the total radioactivity on the chromatogram. (From NEUHOFF *et al.*, 1969)

Table 6. Recovery after two-dimensional microchromatography of ^{14}C-seryl-tRNA and ^{14}C-phenyl-alanyl-tRNA mixtures. (From NEUHOFF *et al.*, 1969)

	Predetermined activity (relativ values)	Measured activity	
		IpM[a]	(relativ values)
Ser/Phe	8.7/ 1	423/ 52	8.2/1
Ser/Phe	1 / 1	97/121	0.8/1
Ser/Phe	1 /10	47/464	1 /9.9

[a] Corrected for background activity. $\sigma \pm 3$ IpM.

Determination of C- and N-Terminal Amino Acids

GRAY and HARTLEY (1963) were the first to use dansyl chloride for the determination of end-groups and in the sequence analysis of proteins and peptides (revue see GRAY, 1967a and b). BURTON and HARTLEY (1970) performed micro "dansyl-Edman" reactions (revue see HARTLEY, 1970) to determine the sequence at the N-terminus, and around a lysine residue in the catalytic site of methionyl-tRNA synthetase from *Escherichia coli*, using 1–2 mg of enzyme in each case.

NEUHOFF, WEISE and STERNBACH (1970) used the micro dansylation method to determine the C- and N-terminal amino acids of DNA-dependent RNA polymerase from *Escherichia coli*, using 0.2 mg in each case.

For the determination of the C-terminal amino acid, a suspension of carboxy-peptidase A (Worthington) was first dialysed for 2 days against 0.05 M $NaHCO_3$ (pH 7.3), then an equimolar quantity of $ZnCl_2$ was added. 50 µl of RNA polymerase (4 mg protein/ml) was dialysed in a microdialysis chamber (NEUHOFF and KIEHL, 1969) (see Chapter 12) against 0.05 M $NaHCO_3$ for 6 to 8 days; it was then added to 40 µl 0.125 M sodium phosphate buffer, pH 8.0, and 10 µl of the dialysed carboxypeptidase A. 10 µl were removed immediately as control, and the remaining solution was incubated at 37° C. 10 µl samples were removed after $\frac{1}{2}$, 2 and 4 hrs., and freeze-dried immediately. As another control, the same amount of carboxypeptidase A was incubated in phosphate buffer alone and treated in the same manner. The freeze-dried residues of the 10 µl portions were dissolved in $NaHCO_3$ and allowed to react with ^{14}C-dans-Cl. The mixture was passed through a Dowex 50 micro-column (see Fig. 40); in order to remove carboxypeptidase A and dans-OH, which interfere with microchromatography, the effluent was monitored for UV absorbance, freeze dried, dissolved in (1–2 µl) absolute ethanol, and chromatographed. Only serine, glycine, and alanine could be detected as dansyl derivatives on a autoradiogram. Densitometry of this autoradiogram (using an Integramat, Leitz, Wetzlar) gave a composition of 48.5% glycine, 36.2% serine and 15.3% alanine. Scintillation counting of the three spots, after incubation for different times with carboxypeptidase A, gave of 42.2% glycine, 35.4% serine, and 22.4% alanine. After incubation for two hours, the values were 48.0%, 28.7%, and 23.3% respectively; the values remained constant after incubation for longer than 4 hrs. In all cases, ^{14}C-dansyl glycine accounted for most of the radioactivity, suggesting that glycine is the C-terminal group. The C-terminal amino acid was also determined by hydrazinolysis (BRAUNITZER, 1955; FRAENKEL-CONRAT and CHUN MING TSUNG, 1967), followed by dansylation with ^{14}C-dans-Cl, and microchromatography. 0.5 mg of freeze-dried RNA polymerase was used for this analysis. Glycine was again found as the C-terminal acid, thus confirming the result with carboxypeptidase A. From the distribution of the ^{14}C-activity it seems likely that the C-terminal sequence is glycyl-seryl-alanine. The fact that no other amino acid was liberated after 2 to 4 hrs. incubation could indicate the next amino acid is arginine or proline, neither of which can be liberated by carboxypeptidase A (AMBLER, 1967; SCHROEDER, 1968).

For the determination of the N-terminal amino acid, 50 µl of purified and micro-dialysed RNA polymerase (4 mg protein/ml) was treated with 10 µl of

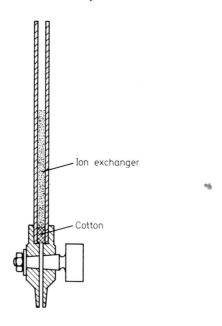

Fig. 40. Micro-column for ionexchange chromatography (inner diameter of the pyrex tube 2 mm, outer diameter 4 mm, length 4 cm). The column is closed with a piece of cotton-wool. The pyrex glass tube exactly fits in a micro tap from Dynal, and after use it can be replaced by a new glass tube

5 % sodium dodecyl sulphate for 2 hrs. at room temperature. The sample was freeze-dried and then oxidised at 0° C in a mixture of hydrogen peroxide, formic acid, and ethanol (HIRS, 1956). It was freeze-dried again, dansylated, and micro-dialysed for 4 days. The dialysate was freeze-dried and hydrolyzed in approx. 30 μl 5.7 N HCl for 12 hrs. at 100° C. The mixture was freeze-dried once again, and the residue was dissolved in 1–2 μl ethanol for chromatography on a poly-amide layer. To identify the end-group, the anticipated dansyl amino acid was chromatographed as a standard together with a portion of the sample. In addition to dans-OH, dans-O-tyrosine, and dans-ε-lysine, dansyl methionine sulfone could be detected, indicating that methionine is the N-terminal amino acid of the RNA-polymerase (this is in agreement with findings of BURGESS, 1969).

Experiments to determine the N-terminal amino acid with aminopeptidase M (Röhm u. Haas) were performed after treatment of the dialysed polymerase with ^{14}C-dans-Cl. Before incubation with aminopeptidase M (37° C, 12–24 hrs.), a control chromatogram was prepared for comparison with the chromatogram of the material after incubation, to find which amino acid had been liberated. As a further check on the action of the aminopeptidase M, a portion was treated with ethanol to precipitate the aminopeptidase after the incubation, and the clear supernatant was allowed to react with ^{14}C-dans-Cl again. Nearly all the amino acids were found on a corresponding chromatogram, and this was first taken as a sign that aminopeptidase M was able to split off the dansylated N-terminal amino acid from the polymerase. After treatment of the dansylated

RNA polymerase with aminopeptidase M, the chromatogram showed a spot with the characteristic R_f-value of dansyl glutamic acid or dans-O-tyrosine, but which did not show the typical reddish fluorescence of dans-O-tyrosine clearly, because the spots were too small for recognition of the colour with certainty. In order to distinguish between the two dansyl derivatives, a few mg of BOC-1-tyrosine (t-butyloxycarbonyl-tyrosine, Fluka A.G.) was dansylated, then treated with trifluoroacetic acid for 2 hrs. at 0° C to remove the BOC-group, and purified by chromatography on polyamide layers. Using this as a standard for dans-O-tyrosine, it could be shown that the dansyl spot obtained after treatment of the dansylated polymerase with aminopeptidase M was dans-O-tyrosine, thus indicating that dansylation of the polymerase was incomplete. Experiments with aminopeptidase M on completely dansylated albumin, pepsin, and ribonuclease did not reveal O-tyrosine or any N-terminal amino acid; these results indicate that aminopeptidase M is not able to split off a dansylated N-terminus, as has been discussed previously (NEUHOFF, 1969).

Determination of Amino Acids from Biological Material

If the dansyl method is to be used for the micro-determination of amino acids from very small biological samples, e.g. isolated nerve cells, very small cores of material from particular regions of the brain, biopsy material, etc., the usual agents for extraction, such as picric acid, ethanol, sulphosalicylic acid, acetone/hydrochloric acid, ethanol/HCl, or perchloric acid, are unsuitable, as they all affect the dansyl reaction and interfere with microchromatography, resulting in bad separations. However, extraction with 0.05 M $NaHCO_3$ solutions (pH 10) has proved very suitable. The importance of standardising the pH for the dansyl reaction which follows has already been explained. After a single micro-homogenisation (see Chapter 13) of weighed pieces of tissue in $NaHCO_3$, 85% of the free amino acids are found in the supernatant. The remaining 15% are extracted almost completely on a second homogenisation with $NaHCO_3$; no further dansyl amino acids can be detected on micro-chromatography after a third extraction (BRIEL, 1972).

The proteins contained in the supernatant after extraction and centrifugation, can interfere with the dansyl reaction, since dansyl chloride also reacts with proteins. In addition, the proteins often remain at the origin and impair the final quality of the microchromatogram. They must therefore be removed if possible before allowing the sample to react with dansyl chloride. Heat denaturation (5 min, 85° C), using a capillary is possible, but in this very drastic method, just as in precipitation with trichloroacetic acid, many of the free amino acids co-precipitate with the denatured material so that the final yield is reduced by ca. 30%. Precipitation with trichloroacetic acid is also unsuitable because any excess is difficult to remove completely, and the pH for the dansylation must be adjusted accordingly. Also trichloroacetic acid which is not been removed completely on lyophilisation and which is subsequently neutralised, causes very bad micro-chromatograms. The best method of precipitating the proteins is to add an equal volume of cold acetone and let the mixture stand for 60 min at

− 20° C to complete precipitation. Not all the proteins are precipitated if a solution contains only 50% acetone (about 90% acetone is necessary to precipitate all the proteins), but enough of them are precipitated and removed on centrifugation, so that good micro-chromatograms can be obtained after treating the remaining solution with dansyl chloride (e.g. Figs. 41, 43, 44, 47). The use of acetone as the agent for precipitating the proteins has the additional advantage of providing the right conditions for the reaction with dansyl chloride.

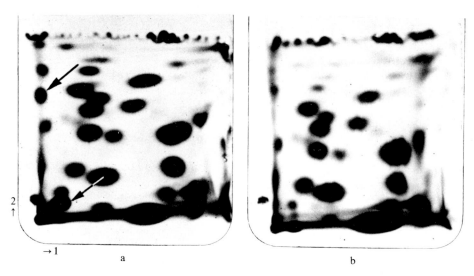

Fig. 41 *a* and *b*. Autoradiogram of ^{14}C-dansyl compounds from metacerebral cells (GSCs) *a*, and posterior buccal cells *b*, from the snail *Helix pomatia*. The position of the two dansyl serotonin compounds in the GSCs are indicated by arrows. (From BRIEL, NEUHOFF and OSBORNE, 1971)

If the optimal concentration of dansyl-Cl is used, the dans-OH formed by reaction with water is so low that its removal by micro-column chromatography is unnecessary. If too much dans-OH is present, the chromatography and the quantitative evaluation of the chromatogram can be appreciably impaired. If separation of the dans-OH, or of other impurities, is necessary, this is carried out in micro-columns (inner diameter 2 mm, length 4 cm, cf. Fig. 40) of Dowex 50 (mesh 200–400; No. 65). The Dowex is previously treated 3 times with a mixture of acetone and 25% ammonia (1:1 v/v), washed with water until neutral, and equilibrated with 0.01 M acetic acid. The sample is washed into the column with 20 ml 0.01 M acetic acid, and the dansylated amino acids are eluted with an acetone/ammonia mixture (25% acetone, 1.7% NH$_4$OH). The effluent is monitored at a wavelength of 248 nm, freeze dried, the residue dissolved in acetone/acetic acid or in ethanol, and micro-chromatographed on polyamide layers. It is strongly recommended that, for any attempt at even semi-quantitative analysis, the sample should be compared to a standardised mixture of ^{14}C-dansyl compounds similar to the sample under investigation.

The giant serotonin-containing cells (GSCs), located in the metacerebral ganglia of *Helix pomatia* and other gastropods (OSBORNE and COTTRELL, 1971), are easily identified in vivo, and in histological sections, by their large size, their particular position within each cerebral ganglion, and the absence of other large cells in the vicinity. This makes it possible to isolate individual GSCs and to perform biochemical analysis on them (revues, see OSBORNE, 1973; OSBORNE and NEUHOFF, 1973). Extracts prepared from four cells, either GSCs or posterior buccal cells (non-amine-containing buccals), of *Helix pomatia*, are sufficient for a single good microchromatogram (BRIEL, NEUHOFF, OSBORNE, 1971). Fig. 41 (*a* and *b*) shows autoradiograms of the free amino acids of the GSCs (*a*) and the posterior buccal (*b*) cells. Cells from the metacerebral ganglia show serotonin clearly, as N-dansyl- and bis-dansyl serotonin (indicated by arrows). In contrast, the buccal cells produce no blackening of the film at the positions of these two dansyl-serotonin compounds. Examination of several pairs of chromatograms showed that it was indeed possible to obtain consistent results (OSBORNE and COTTRELL, 1972). The results were confirmed by studying the *in vivo* synthesis of serotonin in the GSC, after perfusing the central nervous system of the snail with ^{14}C-5-hydroxytryptophan (OSBORNE, 1972a). Extracts from the GSCs and the buccal cells were then allowed to react with unlabelled dansyl chloride, and the dansyl derivatives were chromatographed. It is clear from the results shown in Table 7 that only the GSCs form ^{14}C-serotonin from ^{14}C-5-HTP. In no instance is ^{14}C-serotonin found in the non-amine-containing cells. From a number of experiments it was calculated that a single GSC *in vivo* formed 0.8 ng ^{14}C-serotonin in 2 hrs.

Table 7. Amount of serotonin produced in GSCs and buccal cells, after perfusing the snail's CNS with ^{14}C-5-HTP. (From OSBORNE, 1972a)

Perfusion of ^{14}C-5-HTP	GSC's ^{14}C-serotonin formed/cell	Buccal cell ^{14}C-serotonin formed/cell
0.5 hrs.	0.2 ± 0.1 ng (6)	0 ng (6)
1 hr.	0.4 ± 0.2 ng (8)	0 ng (8)
2 hrs.	0.8 ± 0.2 ng (8)	0 ng (8)
3 hrs.	0.8 ± 0.3 ng (8)	0 ng (8)
4 hrs.	0.8 ± 0.1 ng (6)	0 ng (6)

In a single experiment, 4 cells were used in each instance. The figure shown in brackets represents the number of experiments performed.

The effect of electrical stimulation on the glucose and glutamic acid metabolism of the GSCs was studied by OSBORNE (1972b). The GSCs were stimulated via the exterior lip nerve, during perfusion of the brain with snail saline containing ^{14}C-glucose or ^{14}C-glutamic acid. The extracts from fourteen GSCs were then allowed to react with unlabelled dans-Cl, chromatographed, and the radioactive metabolites formed from the labelled glucose or glutamic acid were investigated by autoradiography. Fig. 42 shows that the GSC metabolizes ^{14}C-glucose to form a number of substances; the formation of alanine (spot *B*),

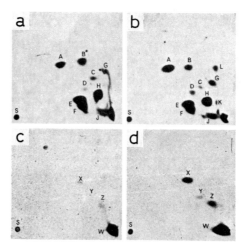

Fig. 42 *a–d.* Autoradiograms of micro-chromatograms of extracts, from GSCs dissected from snails perfused with ^{14}C-glucose, after reaction with unlabelled dans-Cl. Metabolites from ^{14}C-glucose which occur in the GSCs are shown in a. The substances are identified as alanine spot *B*, glutamic acid *E*, aspartic acid *F*, glutamine *H*, and unknowns *A*, *C*, *D*, *G*, *J*. In b, the effect of electrical stimulation on the metabolism of ^{14}C-glucose in the GSCs is shown. The concentration of spot *A* (unknown) has increased while spot *B* (alanine) and spot *C* (unknown) have decreased; in addition, two unknown metabolites appear, spot *L* and spot *K*. The metabolites from ^{14}C-glutamic acid which occur in the GSCs are shown in c. Only three substances could be identified, alanine spot *X*, unknown *Y*, and glutamine *Z*. The substance *W* is an impurity from the radioactive glutamic acid. The effect of electrical stimulation on the metabolism of ^{14}C-glutamic acid is shown in d. It is quite clear that the concentration of all three substances (alanine, glutamine and spot *Y*) is increased. (From Osborne, 1972)

glutamine (*H*), aspartic acid (*F*), glutamic acid (*E*), and the absence of γ-amino butyric acid, is consistent with the data published on the *in vitro* metabolism of glucose in snail nervous tissue (Osborne, Briel, Neuhoff, 1971). In addition, and in contrast to the *in vitro* experiments, a number of unknown substances (spots *A*, *C*, *D*, *G* and *J*) were produced from glucose *in vivo*. The effect of electrical stimulation upon the metabolism of glucose is shown in Fig. 42 *b*. In this case there is an increase in the amount of spot *A*, and a slight decrease in the amount of alanine (*B*) and spot *C*. Remarkably large quantities of two new unknown metabolites are also formed (spots *L* and *K*).

Fig. 42 *c* shows that no labelled glutamic acid was found in the GSCs after perfusion with ^{14}C-glutamic acid (compare the position of glutamic acid in a, spot *E*). It therefore appears that the neurons lack the ability to take up glutamic acid. However, an unknown contaminant (spot *W*), which is normally present as an impurity in the ^{14}C-glutamic acid used, is found in the giant neuron, together with minute amounts of alanine (spot *X*), glutamine (*Z*), and an unknown substance (*Y*). Fig. 42 *d* shows that after electrical stimulation there is no glutamic acid detectable in the cell, but that the amount of alanine, glutamine, and unknown compound (spot *Y*) increases.

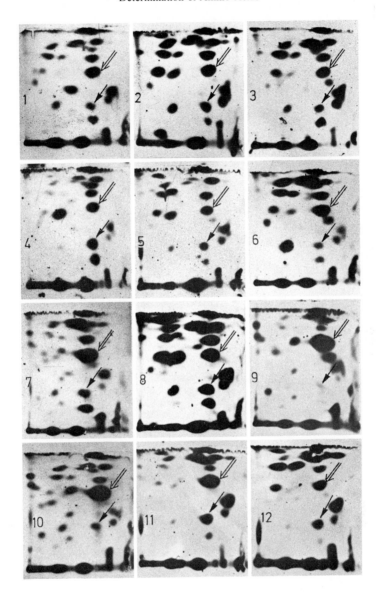

Fig. 43. Autoradiograms (original size) of microchromatograms of extracts of *Aphrodite aculeata 1*, *Lumbricus terrestris 2*, *Buccinum undatum 3*, *Eledone cirrhosa 4*, *Limax maximus 5*, *Carcinas maenas 6*, *Nephrops norvegicus 7*, *Locusta migratoria 8*, *Asterias rubens 9*, *Echinus esculentus 10*, *Scyllium canicula 11* and *Pleuronectes platessa 12*, after reaction with ^{14}C-dansyl chloride. The microchromatograms were developed in the first dimension (horizontal direction) with water/formic acid, in the second dimension (vertical direction) with benzene/acetic acid, and once more in the second dimension with ethyl acetate/methanol/acetic acid. The positions of glycine (double arrow) and glutamic acid (single arrow) are marked on each chromatogram. (From OSBORNE, 1972)

Comparatively few studies on the occurrence of free amino acids in the nervous system of invertebrates have been carried out, mainly because of a lack of techniques suitable for the quantitative analysis of amino acids in small quantities of tissues. OSBORNE (1971 a, 1972 c), using the methods described, has analyzed the free amino acids of nervous tissue for a number of different invertebrates and also for two species of fish, with special emphasis on glycine, glutamic acid (OSBORNE, 1972 c), GABA, and taurine (OSBORNE, 1971 a), since these amino acids are considered to be transmitter-substances in certain situations. Fig. 43 illustrates the varying contents of glycine (double arrow) and glutamic acid (single arrow) in a number of different species. The chromatograms also indicate variations in the proportions of other amino acids.

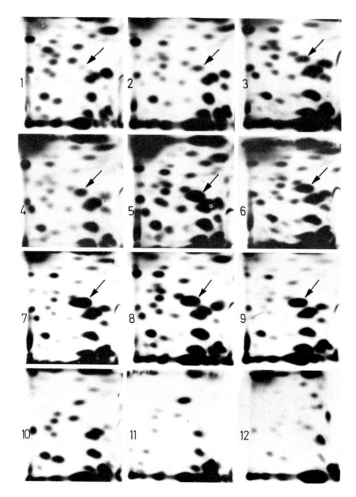

Fig. 44. Autoradiograms (original size) after reaction with ^{14}C-dans-Cl and microchromatography of free amino acids, from the optic pathway of rabbit.
1 retina, *2* nerv. opticus. *3* chiasma, *4* tractus opticus, *5* corp. gen. lat. pars dorsalis, *6* corp. gen. lat. pars ventralis, *7* colliculus anterior, *8* pulvinar, *9* area striata, *10* cornea, *11* lense, *12* corp. vitreum. The arrow indicates the position of γ-amino butyric acid (GABA). (From FREY, 1972)

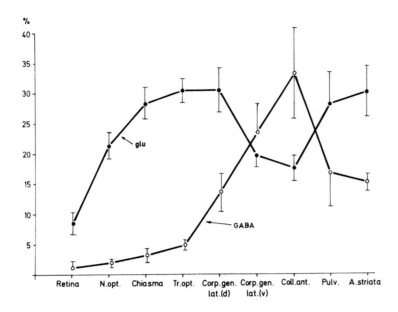

Fig. 45. Distribution of glutamic acid (—•—) and GABA (—○—) in the optic pathway of the rabbit (The Fig. represents the mean value of four experiments. σ is given as ⊢—⊣). The ordinate represents the percentage of all measured free amino acids from the tissue. (From FREY, 1972)

FREY (1972) analysed the free amino acids along the optic pathways of the rabbit. The autoradiograms of the ^{14}C-dansyl-amino acids (Fig. 44) show considerable variation that the distribution of free amino acids in the different regions of the optic pathway. Of special interest is the distribution of GABA (marked with an arrow), which is know to act as an inhibitory transmitter (ROBERTS et al., 1960; CURTIS, 1972) and which reaches its highest concentration in the anterior part of the lamina quadrigemina. Fig. 45 shows the distributions of GABA and glutamic acid in the regions analysed, as a percentage of all amino acids analysed (trypthopan, ornithine, lysine, tyrosine, 5-hydroxyindole, phenylalanine, histidine, leucine, isoleucine, γ-amino butyric acid, valine, proline, arginine, asparagine, glutamine, threonine, aspartic acid, glutamic acid, glycine, alanine, taurine). Taurine, which may also be an inhibitory transmitter (DAVISON and KACMAREK, 1971), reaches its highest concentration (25%) in the retina, and declines in the more central areas of the optic pathway (see Fig. 46). This may indicate that taurine is a specific transmitter substance in the retina. The main amino acids in the cornea are glycine (15%), alanine (15%), and taurine (25%), while the lense contains proline (16%), glycine (14%), and taurine (20%). In the corpus vitreum taurine accounts for only 10% of the content of free amino acids.

A considerable amount of data on phenylpyruvic amentia has been amassed (revue see KNOX, 1966), as a basis for the study of mental defects in phenylketonuria, which might give insight into the nature and mechanism of the development of intellectual functions. The pharmacological effects of some

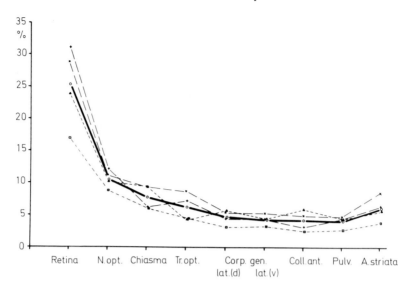

Fig. 46. Distribution of taurine in the optic pathway of four rabbits, as percentage of all amino acids counted after reaction with ^{14}C-dans-Cl and microchromatography. (Dotted lines show single experiments, the thick line represents the mean value.) (From FREY, 1972)

abnormal metabolites, especially derivatives of tyrosine and serotonin, were investigated with relation to the symptoms of this severe disease. Relatively little information is available on the amino acids in the cerebrospinal fluid of phenylketonuric babies. Using the micromethod described here, QUENTIN and NEUHOFF (1972) analysed the cerebrospinal fluid of a four-month-old male baby before therapeutic treatments, and 17 days afterwards, and compared this with the cerebrospinal fluid of a normal baby of the same age. Fig. 47 (a and b) shows autoradiograms of ^{14}C-dansylated amino acids before and after therapy, compared to those from the normal (c) cerebrospinal fluid. Maps to assist in the identification of the different dansyl compounds are also given in the figures.

Before therapy, the spot of phenylalanine (9) is the most pronounced in the autoradiogram. Fig. 47 (b) shows the decrease of this spot after successfull therapy. The appearance of serotonin (spot 43), 5-hydroxyindoleacetic acid (41), and 5-hydroxyindole (47) before therapy, and their almost complete disappearence

Fig. 47a–c. Autoradiograms after reaction with ^{14}C-dans-Cl of the amino acids from the cerebro- ▷ spinal fluid of a four month old phenylketonuric baby. *a* before therapy, *b* after 17 days of therapy, *c* cerebro-spinal fluid of a normal baby for comparison. Maps are given for each autoradiogram to assist in identification of the spots.
1 origin, *2* dans-OH, *3* dans-NH$_2$, *4* tryptophane, *5* ornithine (dans$_2$), *6* lysine (dans$_2$),*7* homocarnosine, *8a* tyrosine (dans$_2$), *8b* tyrosine, *9* phenylalanine, *10* histidine (dans$_2$), *11* leucine, *12* isoleucine, *13* proline, *14* valine, *15* GABA, *16* ethanolamine, *17* alanine, *18* glycine, *19* asparagine, glutamine, *20* methionine, *21* taurin, *22–26* arginine, threonine, α-amino histidine, α-lysine, *27* 5-HO-tryptophane, *28–42* unknown, *43* serotonin (dans$_2$), *44–46* unknown, *47* 5-HO-indol, *48* histamin, *49* unknown

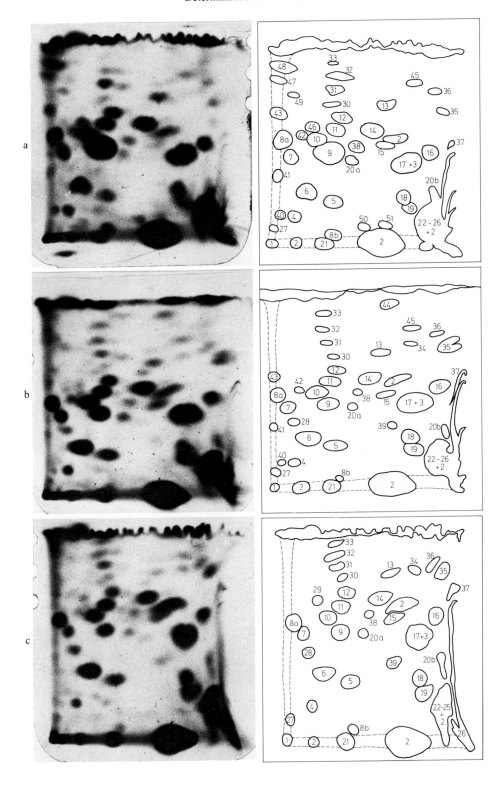

after therapy is remarkable. In the normal cerebrospinal fluid (*c*) these spots are absent. Before therapy there are also three, as yet unidentified, dansyl compounds (spots 40, 42 and 46) which disappear after therapy and which are absent in normal spinal fluid. On the other hand, spot 28 is an unknown dansyl compound which is distinctly visible in the autoradiogram from the normal cerebrospinal fluid and also in the autoradiogram of the phenylketonuric baby after therapy, but absent before the therapy. There are some other differences between the three autoradiograms, which may also reflect changes in the composition of the cerebrospinal fluid between normal and phenylketonuric humans. The same pattern of compounds reacting with dansyl chloride was also found in a second case of phenylketonuria. It will be very difficult to identify all the compounds which react with dansyl chloride, but it may be necessary to do so in order to obtain more detailed information on the disturbed metabolism of the brain during phenylketonuria.

The micromethod has been used for the analysis of several other biological materials. For example, UNGAR, DESIDERIO, and PARR (1972) used this technique for the identification of a pentadecapeptide isolated from the brains of rats taught to avoid the dark. The amine and amino acid compositions of 6 easily-identifiable giant neurons in the oesophageal ganglion of the snail *Helix pomatia* has recently been reported (OSBORNE, SZCZEPANIAK and NEUHOFF, 1973). Only 2 of the 6 neurons contained serotonin; the amino acid composition of all the neurons was similar, although they differed in detail. The predominant amino acids in the cells were ornithine, alanine, glycine, and glutamic acid. Other substances detected were tyrosine, taurine, tryptophan, lysine, phenylalanine, histidine, 5-hydroxyindole, isoleucine, leucine, valine, GABA, ethanolamine, aspartic acid, glutamine, serine, asparagine, threonine, arginine, cystine and proline. It was concluded that, because of the similar pattern of amino acids in each of the six cells, information as to the functional role of individual amino acids whithin the neuron may best be obtained by studying their metabolism.

Determination of Mono-Amines

The micro-determination of biologically active monoamines (e.g. 5-hydroxy-tryptamine, 5-hydroxytryptophan, adrenaline, dopamine, and noradrenaline), by the dansyl method after their extraction from individual nerve cells, is extraordinarily difficult since these compounds form several dansyl derivatives, depending on the number of reactive $-NH_2$ or $-OH$ groups. For quantitative determinations after micro-chromatographic fractionation, the exact location of these derivatives on a micro-chromatogram must be known. With biological materials, appreciable difficulties are caused by the overlapping of other dansyl compounds. Moreover, for quantitative evaluation it is necessary to know which spot corresponds to a mono- and which spot to a bis-dansyl derivative. It has been shown for histidine and serotonin that such identification is possible. BELL and SOMERVILLE (1966) described the reaction between certain amines and formaldehyde vapour, using conventional paper and thin-layer chromatographic techniques. OSBORNE (1971 b) described a micro-chromatographic method for the detection of biologically active monoamines in the nanogram range, using one-dimensional chromatography on polyamide layers, followed by treatment with formaldehyde vapour. Standard amounts of amine (1–1000 ng), dissolved in 50% acetone in 0.01 N HCl, were spotted with an ultra-thin capillary, 4 mm from the bottom edge of a 3 × 3 cm micropolyamide sheet under a stereomicroscope, with drying in a stream of cool air. The chromatogram was developed in a 50 ml beaker, the bottom of which was just covered with either methyl acetate/isopropanol/ammonia 25% (9:7:5 v/v) or butanol/chloroform/acetic acid (4:1:1 v/v). The beaker was covered to prevent evaporation of the solvent. When the solvent reached the upper edge of the chromatogram (after 3–6 min), it was removed, dried carefully with cold air from a hair dryer, and then placed in a sealed jar (500 ml) with about 3 g of paraformaldehyde. The paraformaldehyde must have been stored at a relative humidity of 60% for at least 5 days to produce optimum fluorescence. After heating the jar in an oven for 3 h at 80° C, the amines were located by viewing the chromatograms under UV light.

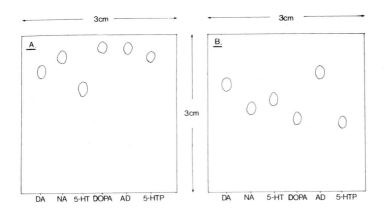

Fig. 48 A and B. Diagram of microchromatograms (polyamide layers) for the solvents: A methyl-acetate/isopropanol/ammonia 25% (9:7:5 v/v), B butanol/chloroform/acetic acid (4:1:1 v/v)
DA: dopamine, NA: noradrenalin, 5-HT: 5-hydroxytryptamine, DOPA: 3,3-dihydroxyphenylalanine, AD: adrenalin, 5-HTP: 5-hydroxytrypthophan

Fig. 48 shows the positions typical of some mono-amines after chromatography on 3 × 3 cm micropolyamide sheets with the two different solvent systems. Table 8 gives the minimum detectable amounts of biological amines and other compounds on treatment with formaldehyde. This very simple procedure for the analysis of amines in very small quantities of tissue can be applied to many neurochemical problems. In order to demonstrate the applicability of the method, 6–10 giant neurons (each cell measures 120 μm across its major axis) from the metacerebral ganglion of *Helix pomatia* (gastropod-Mollusca), which are known to contain only 5-HT were dissected and placed in a 5 μl capillary containing saline. The neurons were sedimented by centrifugation; after removing the saline and replacing it with approximately 1 μl of 70% acetone in 0.01 N HCl, the cells

Table 8. Minimum detectable amounts of biological amines and other compounds after formaldehyde treatment. (From OSBORNE, 1971 b)

Compound	Detectable amount (ng)
Adrenaline	100
Bufotenine	80
n,n-Dimethyltryptamine	1 000
3,4-Dimethoxyphenethylamine	15
3,4-Dihydroxyphenylalanine	50
Dopamine	6
5-Hydroxytryptophan	50
5-hydroxytryptamine	5
3-hydroxykynurenine	80
3-Hydroxy-4-Methoxyphenylethylamine	20
Kynurenine	100
Kynurenic acid	100
Kynuramine	1 000
Melatonin	90
Metadrenaline	1 000
Mescaline	90
α-Methyl-m-tyrosine	1 000
3-Methoxy-4-hydroxyphenylethylamine	50
Noradrenaline	7
Normetadrenaline	100
Octopamine	1 000

were micro-homogenized. Precipitation of the neuronal proteins was accomplished by placing the capillary in a freezer for 30 min. After centrifugation, the supernatants were chromatographed on 3×3 cm micropolyamide layers. Standard amines were chromatographed separately, or added to the extract, to assist in identification. From the results of 12 different experiments it was estimated that a single neuron contained 0.9 ng of 5-hydroxytryptamine. This method seems to be very suitable for the study of the amine content of neurons which are thought to contain more than one amine, and the method can also be adapted to measure the rate of incorporation of radioactivity into specific neurons, so as to measure the biosynthesis and half-lives of certain monoamines. The method has the disadvantage that amounts of proteins which would not interfere with the separation of dansyl derivative impaire the separation of the amines.

Concluding Remarks

On account of its intense fluorescence, dansyl chloride is one of the best reagents available at present for the detection of amino acids. Unfortunately, the reaction depends on a number of parameters which are difficult or impossible to control, so that quantitative analysis with dansyl chloride is possible only under certain conditions. Only when the optimal reaction conditions for each particular case have been ascertained, can even semiquantitative data be obtained.

UDENFRIEND (1973) has recently described a new reagent, fluorescamine (4-phenylspiro[furan-2(3H), 1¹-phthalan]-3,3¹-dione, RO-20-7234), for the assay in the pico-mole range of amino acids, peptides, proteins, and other primary amines. Initial tests have shown that amino acids treated with fluorescamine cannot be separated chromatographically on thin layers of polyamide or silica gel. Its suitability for the microprocedure (i.e. not only for micro-chromatography but also for forming derivatives of amino acids in very small volumes and from very small samples of material), remains to be investigated. Attempts are being made (URBAN, 1972; URBAN et al., 1973) to synthesise an intensely fluorescent compound processing better reactivity than dansyl chloride, based on the pseudosaccharine chloride described by HETTLER (1966, 1968); if these are successfull it may be possible to analyse amino groups in solution quantitatively.

Literature

AMBLER, R. P.: Enzymic hydrolysis with carboxypeptidases. In: Methods in enzymology, vol. XI, Enzyme structure, ed. by C. H. W. HIRS, p. 155–166. New York-London: Academic Press 1967.

BELL, C. E., SOMERVILLE, A. R.: A new fluorescence method for detection and possible quantitative assay of some catecholamine and tryptamine derivatives on paper. Biochem. J. **98**, 1 c–3 c (1966).

BRAUNITZER, G.: Bestimmung der Reihenfolge der Aminosäuren am Carboxylende des Tabakmosaikvirus durch Hydrazinspaltung. Chem. Ber. **88**, 2025–2036 (1955).

BRIEL, G.: Mikroanalyse von Aminosäuren als ¹⁴C-Dansyl-Verbindungen und ihre Anwendung zur Bestimmung von freien Aminosäuren des Zentralnervensystems. Inaugural-Dissertation, Göttingen 1972.

BRIEL, G., GYLFE, E., HELLMANN, B., NEUHOFF, V.: Microdetermination of free amino acids in pancreatic islets isolated from obese-hyperglycemic mice. Acta physiol. scand. **84**, 247–253 (1972).

BRIEL, G., NEUHOFF, V.: Microanalysis of amino acids and their determination in biological material using dansyl chloride. Hoppe-Seylers Z. physiol. Chem. **353**, 540–553 (1972).

BRIEL, G., NEUHOFF, V., OSBORNE, N. N.: Determination of amino acids in single identifiable nerve cells of Helix pomatia. Int. J. Neurosci. **2**, 129–136 (1971).

BRUTON, C. J., HARTLEY, B. S.: Chemical studies on methionyl-tRNA synthetase from *Escherichia coli*. J. molec. Biol. **52**, 165–178 (1970).

BURGESS, R. R.: Separation and characterization of the subunits of ribonucleic acid polymerase. J. biol. Chem. **244**, 6168–6178 (1969).

CROWSHAW, K., JESSUP, J., RAMWELL, P. W.: Thin-layer chromatography of 1-dimethylaminonaphthalene-5-sulphonyl derivatives of amino acids present in superfusates of cat cerebral cortex. Biochem. J. **103**, 79–85 (1967).

CURTIS, D. R.: Central synaptic transmitters. Proc. Aust. Assoc. Neurologists **7**, 55–60 (1970).

DAVIDSON, A. N., KACMAREK, L. K.: Taurine—a possible neurotransmitter? Nature (Lond.) New Biol. **234**, 107–108 (1971).

FRAENKEL-CONRAT, H., CHUN MING TSUNG: Hydrazinolysis, In: Methods in enzymology, vol. XI, Enzyme structure, ed. by C. H. W. HIRS, p. 151–155. New York-London: Academic Press 1967.

FREY, W.: Freie Aminosäuren der Sehbahn des Kaninchens. Inaugural-Dissertation, Göttingen 1972.

GRAY, W. R.: Ultra-micro methods for investigation of protein structure. Ph. D. Thesis, St. John's College, University of Cambridge 1964.

GRAY, W. R.: Dansyl chloride procedure, In: Methods in enzymology, vol. XI, Enzyme structure, ed. by C. H. W. HIRS, p. 139–151. New York-London: Academic Press 1967 a.

GRAY, W. R.: Sequential degradation plus dansylation. In: Methods in enzymology, vol. XI, Enzyme structure, ed. by C. H. W. HIRS, p. 469–475. New York-London: Academic Press 1967 b.

GRAY, W. R., HARTLEY, B. S.: A fluorescent end-group reagent for proteins and peptides. Biochem. J. **89**, 59 p (1963).

GROS, C., LABOUESSE, B.: Study of the dansylation reaction of amino acids, peptides and proteins. Europ. J. Biochem. **7**, 463–470 (1969).

HARTLEY, B. S.: Strategy and tactics in protein chemistry. Biochem. J. **119**, 805–822 (1970).

HARTLEY, B. S., MASSEY, V.: The active centre of chymotrypsin. I. Labelling with a fluorescent dye. Biochim. biophys. Acta (Amst.) **21**, 58–70 (1956).

HELLER, G.: Über die Hydrolyse der Aminosäure-Ester von Transfer-Ribonucleinsäuren. Dissertation, Göttingen 1966.

HETTLER, H.: Charakterisierung von primären und sekundären Aminen mit Pseudosaccharinchlorid. Fresenius Z. anal. Chem. **220**, 9–15 (1966).

HETTLER, H.: Chapman-Mumm rearrangement of pseudosaccharinesters. Tetrahedron Letters **15**, 1793–1796 (1968).

HIRS, C. H. W.: The oxydation of ribonuclease with performic acid. J. biol. Chem. **219**, 611–621 (1956).

KNOX, W. E.: Phenylketonuria. In: The metabolic basis of inherited disease, ed. by STANBURG, J. B., WYNGAARDEN, J. B., FREDERICKSON, D. S., 2. ed., part 3, chap. 11, p. 266–295. New York-Toronto-Sidney-London 1966.

MATTHAEI, J. H., HELLER, G., VOIGT, H.-P., NETH, R., SCHÖCH, G., KÜBLER, H.: Analysis of the genetic code by amino acid adapting. Genetic elements, ed. by D. SHUGAR, p. 233–250. New York-London: Academic Press 1967.

NEADLE, D. J., POLLIT, R. J.: The formation of 1-dimethylaminonaphthalene-5-sulphonamide during the preparation of 1-dimethylaminonaphthalene-5-sulphonyl-amino acids. Biochem. J. **97**, 607–608 (1965).

NEUHOFF, V.: First Harden Conf. on the Structure and Biological Role of Proteins. Wye College, 1969.

NEUHOFF, V.: Micromethods for protein and enzyme analysis. In: International Symposium VI, Chromatographie Electrophorese, p. 57–62. Bruxelles: Press Académiques Européennes S. C. 1971.

NEUHOFF, V., BRIEL, G., MAELICKE, A.: Characterization and micro-determination of histidine as its dansyl compounds. Arzneimittel-Forsch. **21**, 104–107 (1971).

NEUHOFF, V., HAAR, F. VON DER, SCHLIMME, E., WEISE, M.: Zweidimensionale Chromatographie von Dansyl-aminosäuren im pico-Mol-Bereich, angewandt zu direkten Charakterisierung von Transfer-Ribonucleinsäuren. Hoppe-Seylers Z. physiol. Chem. **350**, 121–128 (1969).

NEUHOFF, V., KIEHL, F.: Dialysiergeräte für Volumen zwischen 10 und 500 μl. Arzneimittel-Forsch. (Drug Res.) **19**, 1898–1899 (1969).

NEUHOFF, V., WEISE, M.: Determination of pico-mole quantities of γ-amino-butyric acid (GABA) and serotonin. Arzneimittel-Forsch. (Drug Res.) **20**, 368–372 (1970).

NEUHOFF, V., WEISE, M., STERNBACH, H.: Micro-analysis of pure deoxyribonucleic acid-dependent ribonucleic acid polymerase from *Escherichia coli*. VI. Determination of the amino acid composition. Hoppe Seylers Z. physiol. Chem. **351**, 1395–1401 (1970).

OSBORNE, N. N.: Occurrence of GABA and Taurine in the nervous system of the dogfish and some invertebrates. Comp. Pharmac. **2**, 433–438 (1971a).

OSBORNE, N. N.: A micro-chromatographic method for the detection of biologically active monoamines of isolated neurons. Experientia (Basel) **27**, 1502–1503 (1971b).

OSBORNE, N. N.: The in vivo synthesis of serotonin in an identified serotonin-containing neuron of Helix pomatia. Int. J. Neurosci. **3**, 215–228 (1972a).

OSBORNE, N. N.: Effect of electrical stimulation on the in vivo metabolism of glucose and glutamic acid in an identified neuron. Brain Res. **41**, 237–241 (1972b).

OSBORNE, N. N.: Occurrence of glycine and glutamic acid in the nervous system of two fish species and some invertebrates. Comp. Biochem. Physiol. **43**B, 579–585 (1972c).

OSBORNE, N. N.: The analysis of amines and amino acids in micro quantities of tissue. In: Progress in neurobiology, ed. G. A. KERKUT and J. W. PHILLIS. Oxford-New York-London-Paris: Pergamon Press 1973 (in press).

OSBORNE, N. N., BRIEL, G., NEUHOFF, V.: Distribution of GABA and other amino acids in different tissues of the gastropod mollusc *Helix pomatia*, including in vitro experiments with ¹⁴C glucose and ¹⁴C glutamic acid. Int. J. Neurosci. **1**, 265–272 (1971).

OSBORNE, N. N., COTTRELL, G. A.: Distribution of biogenic amines in the slug, *Limax maximus*. Z. Zellforsch. **112**, 15–30 (1971).

OSBORNE, N. N., COTTRELL, G. A.: Amine and amino acid microanalysis of two identified snail neurons with known characteristics. Experientia (Basel) **28**, 656–658 (1972).

OSBORNE, N. N., NEUHOFF, V.: Neurochemical studies on characterized neurons. Naturwissenschaften **60**, 78–87 (1973).

OSBORNE, N. N., SZCZEPANIAK, A. C., NEUHOFF, V.: Amines and amino acids in identified neurons of *Helix pomatia*. Int. J. Neurosc. **5**, 125–131 (1973).

PETZOLDT, U., BRIEL, G., GOTTSCHEWSKI, G. H. M., NEUHOFF, V.: Free amino acids in the early cleavage stages of the rabbit egg. Developmental Biology **31**, 38–46 (1973).

QUENTIN, C.-D., NEUHOFF, V.: Unpublished observations (1972).

ROBERTS, E., BAXTER, C. F., HARREVELD, A. VAN, WIERSMA, C. A. G., ADEY, W. R., KILLAM, K. F. (eds.): Inhibition in the nervous system and gamma-aminobutyric acid. Oxford-New York-London-Paris: Pergamon Press 1960.

SCHROEDER, W. A.: The primary structure of proteins. New York: Harper and Row Publ. 1968.

SEILER, N.: Use of dansyl reaction in biochemical analysis. In: Methods of biochemical analysis, ed. by D. GLICK, vol. 18, p. 259–337. New York-London-Sydney-Toronto: Intersci. Publ. 1970.

SEILER, N., WIECHMANN, M.: Quantitative Bestimmung von Aminen und von Aminosäuren als 1-Dimethylamino-naphthalin-5-sulfonsäureamid auf Dünnschichtchromatogrammen. Z. analyt. Chem. **220**, 109–127 (1966).

SEILER, N., WIECHMANN, M.: TCL analysis of amines as their dans-derivatives. In: Progress in thin-layer chromatography and related methods, ed. by A. NIEDERWIESER and G. PATAKI, vol. I, p. 95–144. Michigan, USA: Ann Arbor science publishers 1970.

SPIVAK, V. A., SCHERBUKHIN, V. V., ORLOV, V. M., VARSHAVSKY, LA. M.: Quantitative ultramicroanalysis of amino acids in the form of their DNS-Derivatives, II. On the use of the dansylation reaction for quantitative estimation of amino acids. Analyt. Biochem. **39**, 271–281 (1971).

TANG, J., HARTLEY, B. S.: Amino acid sequences around the disulphide bridges and methionine residues of porcine pepsin. Biochem. J. **118**, 611–623 (1970).

TURNER, G. K.: An absolute spectrofluorometer. Science **146**, 183–189 (1964).

UDENFRIEND, S., STEIN, S., BÖHLEN, P., DAIRMANN, W., LEIMGRUBER, W., WEIGELE, M.: A reagent for assay of amino acids, peptides, proteins and other primary amines in the pico-mole range. Science **178**, 871–872 (1972).

UNGAR, G., DESIDERIO, D. M., PARR, W.: Isolation, identification and synthesis of a specific-behaviour-inducing brain peptide. Nature (Lond.) New Biol. **238**, 198–202 (1972).

URBAN, I.-S.: Versuche zur Synthese eines fluoreszierenden Reagenz zum Nachweis von Aminosäuren. Diplomarbeit, Göttingen 1972.

URBAN, I.-S., JASTORFF, B., HETTLER, H., NEUHOFF, V.: Synthesis of a new reagent for microanalysis of amino acids and related compounds. (In preparation.)

WANG, K.-T., WEINSTEIN, B.: Thin-layer chromatography on polymide layers. In: Progress in thin-layer chromatography and related methods, ed. by A. NIEDERWIESER and G. PATAKI, vol. III, p. 177–231. Michigan, USA: Ann Arbor science pub. inc. 1972.

WEBER, G.: Polarization of the fluorescence of macromolecules. 2. Fluorescent conjugates of ovalbumin and bovine serum albumin. Biochem. J. **51**, 155–167 (1952).

WEISE, M., EISENBACH, G. M.: Quantitative determination of amino acids in the 10^{-14} molar range by scanning microscope photometry. Experientia (Basel) **28**, 245–247 (1972).

WOODS, K. R., WANG, K. T.: Separation of dansyl-amino acids by polyamide layer chromatography. Biochim. biophys. Acta (Amst.) **133**, 369–370 (1967).

ZANETTA, J. P., VINCENDON, G., MANDEL, P., GOMBOS, G.: The utilisation of 2-dimethylamino-naphthalene-5-sulphonyl-chloride for quantitative determination of free amino acids and partial analysis of primary structure of proteins. J. Chromat. **51**, 441–458 (1970).

Chapter 3

Micro-Determination of Phospholipids

General Remarks

As phospholipids are determined fluorometrically, rigorous standards of purity and cleanliness for the solvents, reagents, and cuvettes employed must always be observed. Any fluorometer can be used, so long as it is possible to record the spectra either directly or by a coupled pen-recorder. No fluorometric measurement, even if it is only a routine measurement with a well-established procedure, should be carried out without recording the spectrum. It is not enough simply to measure and record a single pen deflection at a known fluorescence maximum; under such conditions an alteration in a spectrum, due perhaps to an impurity which could falsify the measurement, can be overlooked very easily.

A salutary example from our own laboratory experience serves to illustrate this. In the quantitative estimation of 3-acetylpyridine, an anti-metabolite of nicotinamide, large numbers of extractions of organs were carried out, with methyl ethyl ketone, NaOH, and finally with 5 N HCl, to isolate an intensely fluorescent product (NEUHOFF and HERKEN, 1962; HERKEN and NEUHOFF, 1963; WILLING, NEUHOFF and HERKEN, 1964). At times, series of 50 to 100 preparations were made. The 5 N HCl was prepared freshly each week in a 1 l measuring flask with a ground-glass stopper. In one of the excitation spectra, recorded routinely for each single measurement, an appreciable broadening of the spectrum in one of the experimental series was observed, which brought all the measured values into question. On investigation it was found that the ground-glass stopper of the HCl-flask had accidently come into contact with the cement of a tiled bench surface, and an unknown substance had been introduced into the HCl. This substance had an emission maximum at a similar wavelength to that of the actual reaction product, and had falsified the readings appreciably. This impurity would not have been noticed if the spectra had not been recorded.

Recording the spectrum is necessary not only to detect accidental impurities, but also to detect fluorescent compounds in inadequately purified extracts, as is shown by the following example. On recording the excitation spectrum of extracts of rat kidney, after treatment with 6 N NaOH for NAD and NADP determination according to LOWRY et al. (1957, 1961), an indication of an additional maximum at the excitation wavelength of 300 mμ (uncorrected) was observed; no such maximum was detectable after treatment of pure NAD or NADP with 6 N NaOH (Fig. 1). This additional maximum was not present in the excitation spectra of other organ extracts of rat produced similarly (brain, liver, spleen, lung, heart, muscle, pancreas, aorta, blood); it was also not demonstrable in kidney extracts of man, rabbit, guinea pig, or mouse. The substance responsible for this fluorescence maximum had therefore to be traced. Since the substance interfered with the quantitative estimation of NAD or NADP in rat kidney, it was isolated chromatographically. The excitation spectrum of the purified substance is shown in Fig. 1; it possesses three distinct maxima, of which two practically coincide with those of NAD and NADP after treatment with 6 N NaOH. Contamination of NAD or NADP solutions with this substance results in spuriously high measurements for the concentration of nucleotides because

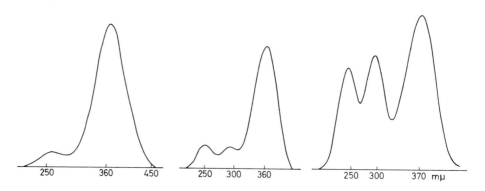

Fig. 1. Excitation spectra of pure NAD or NADP (left) and NAD in a $HClO_4$ extract from rat kidney (middle) after reacting with 6 N NaOH according to Lowry et al. (1957). The excitation spectrum on the right belongs to an unknown compound isolated from the $HClO_4$ extracts of rat kidney. (From Herken and Neuhoff, 1964)

of superposition. It was shown that about 15% of the observed total fluorescence of an extract of rat kidney, after treatment with 6 N NaOH, may be traced to this substance, which is at present unidentified (Herken and Neuhoff, 1964). Such incidents emphasize the correctness of the first basic rule, that in every kind of fluorometric analysis the spectra must be recorded, and only such spectra should be used for evaluation.

Only the purest quality (spectrally pure) solvents should be used, and in addition the emission and excitation spectra of each solvent should be recorded; the spectra should be recorded at the wavelength at which the actual measurements are to be carried out.

If water is used at any stage of the assay, only tap water which has been distilled four times from quartz should be used; this is easy to produce and has very little fluorescence. Water from ion-exchangers is unsuitable, as, even after several distillations from quartz it often contains traces of fluorescent impurities which co-distill on distillation. Pure distilled water was already a qualitative concept at the turn of the century, as illustrated by the ice-delivery wagon shown in Fig. 2; this is now one of the rarities housed in the National Museum of History and Technology, Smithsonian Institution, U.S.A.

Fig. 2. Ice delivery wagon, at the National Museum of History and Technology, Smithsonian Institution, U.S.A.

Quartz cuvettes are best stored in 12 N HNO₃ (prepared from fuming HNO₃ by dilution (1 : 1 v/v) with four-times quartz-distilled mains water), and renewed occasionally. Before use the cuvettes are rinsed, first with mains water and then with quartz-distilled water, and dried with purest-grade acetone. After use, they are again rinsed carefully with mains water and placed in nitric acid. The cuvette stoppers are rinsed and cleaned in exactly the same way as the cuvettes. Further valuable advice on the measurement of fluorescence, along with the theoretical basis of fluorescence may be found in various text books, e.g. UDENFRIEND (1962, 1969), GUILBAULT (1967), BECKER (1969), WHITE and ARGAUER (1970), LOWRY and PASSONNEAU (1972).

Calibration Curves

The procedures in use for the determination of phospholipids in microgram quantities necessitate the digestion of organic material, followed by colorimetric determination of inorganic phosphate (BARTLETT, 1959; CHALVARDJIAN and BUDNICKI, 1970; EIBL and LANDS, 1970). These methods are precise, but they are time-consuming and are not sensitive enough for the analysis of minute amounts of tissues or membraneous material. KLEINIG and LEMPERT (1970), and SVETASHEV and VASKOSKY (1972), have described micromethods for the determination of phospholipids after chromatography on 6 × 6 cm silica gel thin-layer plates; the limit of detection reaches the nanomole range, but this is still not sensitive enough for many requirements.

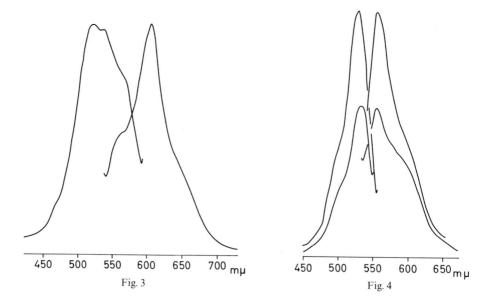

Fig. 3   Fig. 4

Fig. 3. Excitation spectrum (left) and emission spectrum (right) of rhodamine. The cuvette contained 2 ml cyclohexane and 20 μl rhodamine solution (10 mg/25 ml chloroform). For the emission spectrum the excitation wavelength was set at 527 nm, for the excitation spectrum the emission wavelength was set at 608 nm. Bandwidth for both monochromators was 100 Å

Fig. 4. Excitation spectrum (left) and emission spectrum (right) of phosphatidylethanolamine (11.6 μg) before (lower curve) and after (upper curve) the addition of acetic acid. The cuvette contained 2 ml cyclohexane, 20 μl rhodamin solution and 2 μl acetic acid (upper curve). The spectra were measured using a 25 Å bandwidth on both monochromators. For the emission spectrum the excitation wavelength was set at 532 nm, for the excitation spectrum the emission wavelength was set at 558 nm

The fluorescence spectrum of rhodamine 6 G O dissolved in chloroform and diluted with cyclohexane is shown in Fig. 3. Rhodamine has excitation maxima at 523 and 537 nm, and an emission maximum at 608 nm; acetic acid shifts the maxima to 527 and 555 nm. The same effect is seen on the addition of phosphatides, as shown in Fig. 4. The spectral maxima of each phospholipid-rhodamine complex differ slightly (see Table 1), due perhaps to differences in the fatty acids responsible for the hydrophobic interaction between the dye and the phospholipid. The addition of acetic acid (Fig. 4) increases the sensitivity of the determination, and the sharpness of the emission spectra, possibly by its influence on the ionic and steric arrangements of the dye and the phosphatide. Oleic acid, cholesterol, and triglyceride neither form a fluorescent rhodamine complex, nor change the shape or position of the maxima of the fluorescence spectrum for the phospholipid-rhodamine complex. Only triglyceride in high concentrations exerts a slight quenching effect.

Each phosphatide requires individual reaction conditions for maximum sensitivity. The reaction mixture in the cuvette contains cyclohexane, rhodamine

6 GO in chloroform, and acetic acid in varying concentrations depending upon the lipid and its concentration (see Table 1). The cuvettes are stored in HNO_3 and rinsed successively with tap water, absolute ethanol, chloroform, and cyclohexane. Afterwards 2 ml of cyclohexane and the required volumes of rhodamine stock solution (10 mg/25 ml chloroform) and acetic acid are added (see Table 1). The addition of acetic acid is indispensable for the determinations of sphingomyelin and cardiolipin, and also increases the sensitivity of the determination. The rhodamine 6 GO reagent should be prepared daily and stored in the dark. The reaction mixture is photosensitive, so the experiments must be performed with dim lighting. Even small amounts of methanol and water impair the analysis: the emission maxima change and the fluorescence of the blank solution increases.

It was possible to obtain calibration data by making consecutive measurements in one cuvette, by the stepwise addition of known volumes of a single phosphatide, or the phospholipid mixture, dissolved in chloroform, until the rhodamine was exhausted. Single determinations gave identical results. The low basic fluorescence of rhodamine is increased by the addition of the phospholipid in chloroform. The "blank" contained no lipid, but always the same amount of chloroform that had been added to the other cuvettes. Each value was obtained from a complete emission spectrum, the exciting beam being set at maximum wavelength. The fluorescence was measured in arbitrary units at the emission maximum and was corrected for volume and apparatus amplification according to the formula:

$$F = \frac{(\text{test value} - \text{blank}) \cdot \text{volume}}{\text{amplification}}.$$

The fluorescence data were plotted on a log-log chart, and the equations of the standard curves were calculated according to the method of linear regression: $\lg y = \lg b + a \cdot \lg x$. The calculation of the standard curves gave values for "a" (the factor for the steepness of the curve), greater than 1.0. This is due to the fluorescence-quenching of free rhodamine, which is lowered on binding of the dye to the phospholipid added. The calibration curve for the total lipid extract from guinea-pig brain is the only one with a factor lower than 1.0; this is probably due to quenching substances in the brain extract, e.g. gangliosides (HARRIS, SAIFER, and WEINTRAUB, 1961).

Figs. 5–11 give the standard curves of phosphatidylcholine, phosphatidylethanolamine, phosphatidylserine, phosphatidylinositol, sphingomyelin, cardiolipin, and of a total phospholipid extract from guinea-pig brain. Cardiolipin, phosphatidyl-ethanolamine, -serine, -inositol, and sphingomyelin, were purchased from Koch-Light, and phosphatidylcholine from Fluka. The phosphatides were found to be pure on two-dimensional thin-layer chromatography. Weighed quantities of the substances were dissolved in chloroform and the solutions stored at $-20°$ C until use. Karlsberg microlitre pipettes were used for pipetting volumes. The dry preparations were stored in a desiccator over blue silica gel at $-20°$ C. If the preparations become damp they must be dried, because otherwise not only is the weight incorrect, but also the samples are not completely soluble in chloroform.

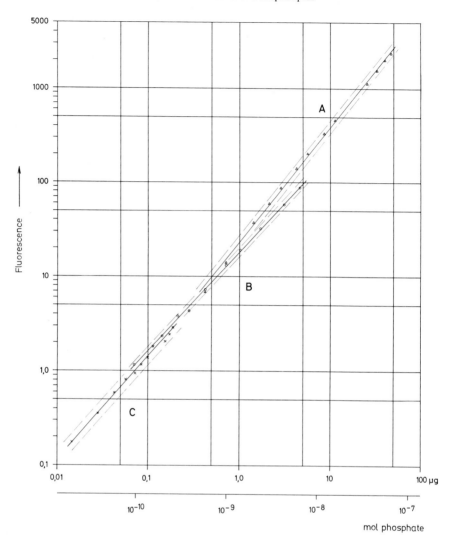

Fig. 5. Standard curves for the fluorometric determination of phosphatidylcholine. The excitation wavelength was set at 535 nm. The fluorescence was measured at the emission maximum (563 nm).

Curve A: 2 ml cyclohexane, 20 µl rhodamine solution, 1 µl acetic acid.

Calculations: $\lg y = (1.352 \pm 0.015) + (1.220 \pm 0.015) \cdot \lg x$
$\sigma = \pm 0.034, \ r = 0.9992$

Curve B: 2 ml cyclohexane, 5 µl rhodamine solution, 1 µl acetic acid.

Calculations: $\lg y = (1.259 \pm 0.010) + (1.047 \pm 0.016) \cdot \lg x$
$\sigma = \pm 0.029, \ r = 0.9990$

Curve C: 2 ml cyclohexane, 5 µl rhodamine solution, 1 µl acetic acid.

Calculations: $\lg y = (1.250 \pm 0.493) + (1.096 \pm 0.044) \cdot \lg x$
$\sigma = 0.053, \ r = 0.991$

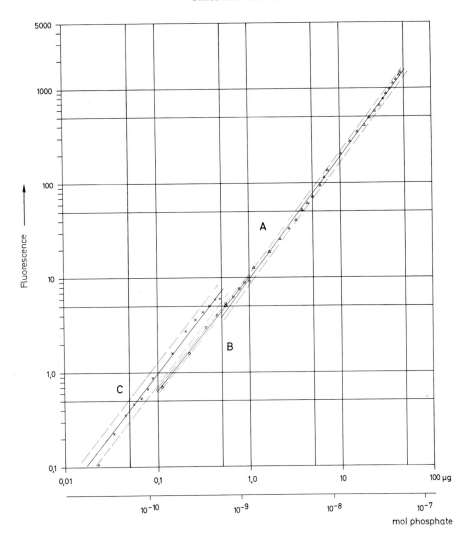

Fig. 6. Standard curves for the fluorometric determination of phosphatidylethanolamine. The excitation wavelength was set at 532 nm. The fluorescence was measured at the emission maximum (558 nm).

Curve A: 2 ml cyclohexane, 20 µl rhodamine solution, 0.2 µl acetic acid.
$\qquad$ Calculations: $\lg y = (0.967 \pm 0.014) + (1.302 \pm 0.012) \cdot \lg x$
$\qquad\qquad\qquad\qquad \sigma = \pm 0.033, \; r = 0.9990$

Curve B: 2 ml cyclohexane, 5 µl rhodamine solution, 0.2 µl acetic acid.
$\qquad$ Calculations: $\lg y = (0.996 \pm 0.015) + (1.193 \pm 0.033) \cdot \lg x$
$\qquad\qquad\qquad\qquad \sigma = \pm 0.029, \; r = 0.997$

Curve C: 2 ml cyclohexane, 2 µl rhodamine solution, 0.2 µl acetic acid.
$\qquad$ Calculations: $\lg y = (1.237 \pm 0.039) + (1.257 \pm 0.037) \cdot \lg x$
$\qquad\qquad\qquad\qquad \sigma = \pm 0.068, \; r = 0.994$

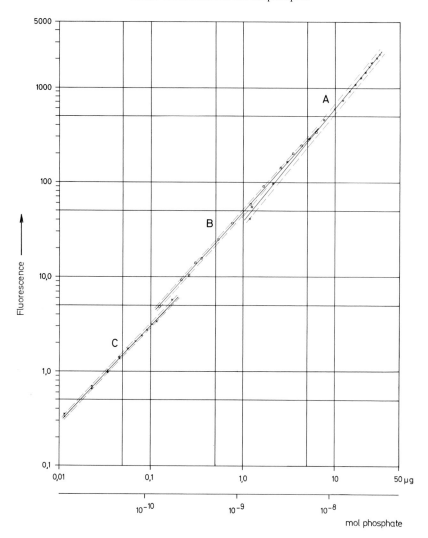

Fig. 7. Standard curves for the fluorometric determination of phosphatidylserine. The excitation wavelength was set at 532 nm. The fluorescence was measured at the emission maximum (560 nm).

Curve A: 2 ml cyclohexane, 20 µl rhodamine solution.
$$\text{Calculations:}\quad \lg y = (1.582 \pm 0.019) + (1.195 \pm 0.017) \cdot \lg x$$
$$\sigma = \pm 0.028,\ r = 0.998$$

Curve B: 2 ml cyclohexane, 5 µl rhodamine solution.
$$\text{Calculations:}\quad \lg y = (1.689 \pm 0.005) + (1.090 \pm 0.009) \cdot \lg x$$
$$\sigma = \pm 0.017,\ r = 0.9996$$

Curve C: 2 ml cyclohexane, 2 µl rhodamine solution.
$$\text{Calculations:}\quad \lg y = (1.496 \pm 0.017) + (1.015 \pm 0.012) \cdot \lg x$$
$$\sigma = \pm 0.016,\ r = 0.9991$$

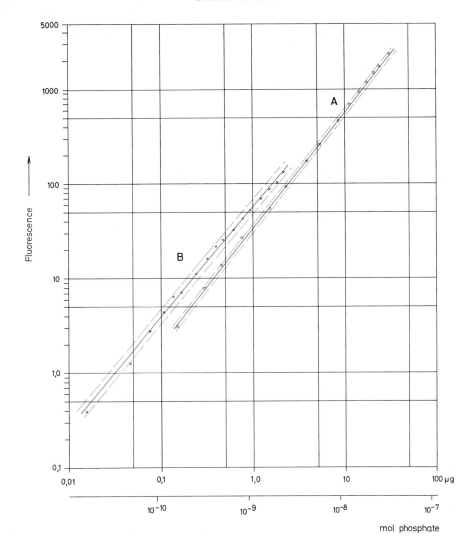

Fig. 8. Standard curves for the fluorometric determination of phosphatidylinositol. The excitation wavelength was set at 528 nm. The fluorescence was measured at the emission maximum (557 nm).

Curve A: 2 ml cyclohexane, 20 µl rhodamine solution, 2 µl acetic acid.
     Calculations: $\lg y = (1.522 \pm 0.007) + (1.240 \pm 0.008) \cdot \lg x$
                              $\sigma = \pm 0.022, \ r = 0.9998$

Curve B: 2 ml cyclohexane, 5 µl rhodamine solution, 1 µl acetic acid.
     Calculations: $\lg y = (1.746 \pm 0.014) + (1.165 \pm 0.019) \cdot \lg x$
                              $\sigma = \pm 0.045, \ r = 0.998$

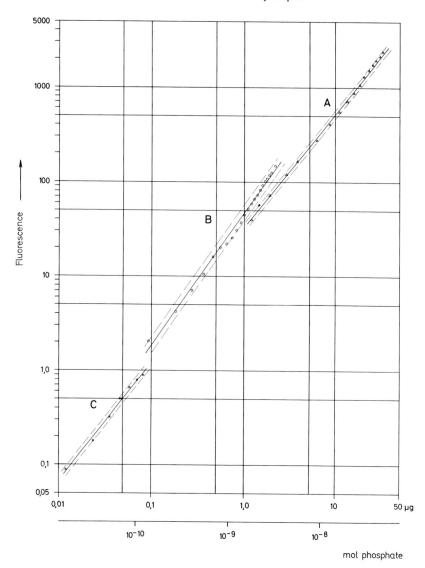

Fig. 9. Standard curves for the fluorometric determination of sphingomyelin. The excitation wave-length was set at 533 nm. The fluorescence was measured at the emission maximum (565 nm).

Curve A: 2 ml cyclohexane, 20 µl rhodamine solution, 2 µl acetic acid.

$$\lg y = (1.495 \pm 0.015) + (1.230 \pm 0.014) \cdot \lg x$$
$$\sigma = \pm 0.027, \; r = 0.9990$$

Curve B: 2 ml cyclohexane, 10 µl rhodamine solution, 1 µl acetic acid.

$$\lg y = (1.643 \pm 0.009) + (1.407 \pm 0.026) \cdot \lg x$$
$$\sigma = \pm 0.043, \; r = 0.996$$

Curve C: 2 ml cyclohexane, 2 µl rhodamine solution, 0.2 µl acetic acid.

$$\lg y = (1.307 \pm 0.061) + (1.240 \pm 0.043) \cdot \lg x$$
$$\sigma = \pm 0.031, \; r = 0.997$$

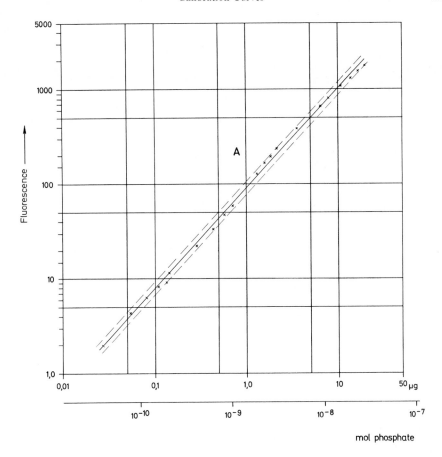

Fig. 10. Standard curve for the fluorometric determination of cardiolipin. The excitation wavelength was set at 530 nm. The fluorescence was measured at the emission maximum (537 nm).

Curve A: 2 ml cyclohexane, 20 µl rhodamine solution, 2 µl acetic acid.

$$\text{Calculations:} \quad \lg y = (1.945 \pm 0.009) + (1.061 \pm 0.010) \cdot \lg x$$
$$\sigma = \pm 0.042, \ r = 0.9990$$

Conditions for obtaining the standard curves are given in Table 1. The measurements were carried out with a Turner 210 absolute spectrofluorometer (TURNER, 1964). It is possible to use the luminescence or fluorescence mode of the instrument, since the changes in optical density during the experiment are so low that the electronic correction system of the instrument does not affect the energy of the exciting light beam significantly. The sensitivity is increased, for the same excitation band width (25 Å), by using the fluorescence mode. The excitation band width must be kept constant, espicially during the preparation of the calibration curves. A greater, or lesser, band width allows more, or less, energy into the cuvette, resulting in higher, or lower, readings. The emission band width can be varied depending on the fluorescence emitted, as each band width setting has a known

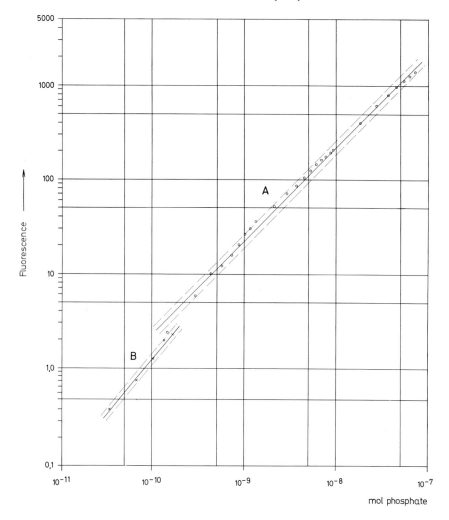

Fig. 11. Standard curves for the fluorometric determination of total phospholipid extract from guinea-pig brain. The excitation wavelength was set at 534 nm. The fluorescence was measured at the emission maximum (558 nm).

Curve A: 2 ml cyclohexane, 20 µl rhodamine solution. 2 µl acetic acid.

Calculations: $\lg y = (1.362 \pm 0.012) + (0.987 \pm 0.012) \cdot \lg x$

$\sigma = \pm 0.046, \ r = 0.998$

Curve B: 2 ml cyclohexane, 10 µl rhodamine solution, 1 µl acetic acid.

Calculations: $\lg y = (1.289 \pm 0.070) + (1.165 \pm 0.064) \cdot \lg x$

$\sigma = \pm 0.036, \ r = 0.995$

(Figs. 3–11 from SCHIEFER and NEUHOFF, 1971)

amplification factor which can be taken into account when calculating the calibration values by the above formula. Even if the amplification factors are quoted by the makers for their instruments, it is always advisable to carry out an independent

Table 1. Reaction mixtures and conditions for the determination of phosphatides

Lipid	Wave-length of excitation max. [nm]	Wave-length of emission max. [nm]	Conc. range [µg]	Cyclo-hexane [ml]	Rhodamine (10 mg/25 ml CHCl$_3$) [µl]	Acetic acid [µl]
Phosphatidyl-choline	535	563	0.01– 5	2	5	1
			1 – 50	2	20	1
Phosphatidyl-ethanolamine	532	558	0.01– 0.5	2	2	0.2
			0.1 – 1	2	5	0.2
			0.5 – 50	2	20	0.2
Phosphatidyl-serine	532	560	0.01– 1	2	2	—
			0.1 – 5	2	5	—
			1 – 30	2	20	—
Phosphatidyl-inositol	528	557	0.01– 2	2	5	1
			0.1 – 30	2	20	2
Sphingomyelin	533	565	0.01– 0.09	2	2	0.2
			0.09– 2	2	10	1
			1 – 30	2	20	2
Cardiolipin	530	557	0.01– 20	2	20	2
Total lipid extract from guinea pig brain	534	558	0.03– 0.2 nM P	2	10	1
			0.15–130 nM P	2	20	2

measurement, using solutions of fluorescent substances at suitable concentrations as it is worthwhile to have a really exact standard curve.

It is always advisable to determine one's own standard curves, rather than to use those published by others. Different fluorometers do not agree exactly in their wavelength settings; if very large discrepancies from the given maxima are observed, then either the substances, the solvents, or the fluorometer are to be suspected. Furthermore, the determination of a standard curve is the best means of mastering the method. If the total phospholipids are to be determined in a series of experiments, the standard curve for each organ investigated must be determined separately, since all organs show a different combination of phospholipids. It is also recommended that the phosphate content of such extracts be determined by the method of STOFFEL and SCHEID (1967), both as a check on the standard curve, and to express the result in moles of phosphate.

As little as 0.02 µg of phospholipids can be determined quantitatively by the same procedure as that used to derive the standard curves. These measurements are carried out in a volume of 2 ml in fluorescence quartz cuvettes (all four sides of polished quartz) of 1 cm path length. It is possible to obtain greater sensitivity, without changing the experimental procedure, by using microcuvettes. However, these should only be employed in exceptional cases, as the increase in sensitivity is accompanied by a similar increase in the susceptibility to errors.

Analysis of Biological Material

Extraction of Phospholipids

It is often difficult to punch out a cylinder of fresh tissue with a glass capillary, owing to the tendency of the punch to mash the tissue. This can be overcome by freezing the tissue to $-20°$ C; a slice of frozen tissue, about 1 mm thick, is then cut with a razor blade and placed on a clean, fat-free, microscope slide which has been precooled on a block of dry ice. Very small pieces of tissue (0.01–1 mg) can then be punched out with the open end of a 5 µl or 10 µl Drummond microcap, heat-sealed at one end and filled with water to within approximately 1 mm of the capillary top. During punching, the capillary is moved firmly in a vertical direction with a screw-like rotation. After centrifugation, the wet weight of the tissue is determined as described in Chapter 14. The supernatant is then removed carefully, and the capillary is filled with chloroform/methanol (2 : 1 v/v). The tissue is homogenized in the capillary, using a dentist's nerve-canal drill driven at about 24000 rpm (see Chapter 13) and allowed to stand at room temperature for 1 hr. After centrifugation, the total lipid extract is removed with a very fine capillary pipette; the extraction procedure is repeated twice. For the fluorometric determination of the total phospholipids, the extracts are combined in a cuvette, dried under a stream of nitrogen, and analysed after addition of cyclohexane, rhodamine, and acetic acid as summarized in Table 1.

If larger quantities of tissue (1–2 mg) are available, a cylindrical piece of tissue can be obtained by using a punch made from a needle of the appropriate diameter (1–1.5 mm). The tissue is then introduced into a larger weighed capillary tube by means of a stilette and the capillary is reweighed. This capillary is made from Pyrex glass (inner diameter 2 mm, outer diameter 4 mm, length approx. 40 mm), and must be heat-sealed carefully at one end and well annealed before use. If this is not done, the tubes break when placed in the liquid nitrogen. Provided that the tissue sample is placed at the bottom of the tube, loss of weight owing to evaporation is negligible during the short time necessary for weighing (less than a minute).

After weighing, the tube is half-filled with quartz-distilled water, and frozen in liquid nitrogen. After allowing the solution to thaw, 2 or 3 splinters of razor blade (prepared as described in Chapter 6), pre-cleaned carefully in chloroform/methanol, are introduced and the contents of the tube are agitated on a whirlmix until the tissue is homogenised. The homogenate is frozen in liquid nitrogen and lyophilized; it is advisable to put the frozen tube in a small beaker containing liquid nitrogen. If this is not done the sample at the bottom of the tube can thaw out sufficiently to expel the homogenate from the tube when the vacuum is applied. As the aqueous phase is not removed in this procedure, phospholipids which may be absorbed onto proteins are not lost.

After lyophilization, 10–20 µl of freshly-prepared chloroform/methanol (2 : 1 v/v) are added to the residue in the tube, and agitated on a whirlmix. The splinters of razor blade still in the tube aid the extraction procedure. The tube is then centrifuged (see Chapter 5) at 15000 rpm for 5 min at 0–4° C. The clear supernatant is transferred by a capillary pipette to a round-bottomed micro-flask (see Fig. 12), which is then supported on a cushion of plasticine (see Fig. 13) in a lyophilizor.

The neck of the flask is at an angle of approximately 45°, to avoid loss of material during evaporation of the chloroform/methanol. The organic solvent is removed in 6 to 10 min. The chloroform/methanol extraction procedure is repeated twice and the extracts are combined in the micro-flask. After the final evaporation, the residue is dissolved in chloroform/methanol, 10 μl of the solvent mixture being used for each milligram of fresh tissue taken. Aliquots of 1–2 μl are removed with calibrated capillary pipettes (see Chapters 6 and 16), and the total phospholipids are either determined as described or are applied to a micro-chromatoplate for separation into the individual phospholipid components.

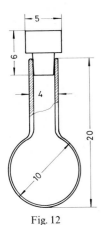

Fig. 12

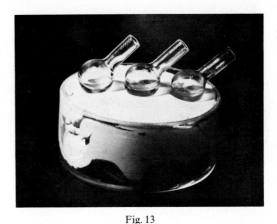

Fig. 13

Fig. 12. A round-bottomed micro-flask suitable for the evaporation of small volumes of organic solvents

Fig. 13. Micro flasks supported on a plasticine cushion for lyophilization

Micro-Chromatography

Two-Dimensional Micro-Chromatography

For two-dimensional micro-chromatography, cover-glasses, 32 × 24 mm, are spread with a thin slurry of silica gel (Camag D-0) as follows: a homogeneous suspension of silica gel is prepared by shaking 30 g silica gel vigorously with 130 ml chloroform in a screw-cap glass bottle. Alternatively the suspension may be prepared by homogenisation for a short time, using an Ultra-turrax. If the silica gel/chloroform suspension is too thick, then the layer, which should be about 50 μm, is shed too easily from the plate. The cover glasses are rinsed with chloroform, dipped into the suspension with cross-action tweezers (see Fig. 21 in Chapter 2), removed immediately, and dried (2 min). TLC plates prepared in this way should be used within ca. 30 min of preparation, since on longer storage the silica layer absorbs water and the separating qualities are impaired. The plates are prepared without a binder, and, since the layer is very thin, they are very suspectible to mechanical damage, and also to loss of the silica gel, by its being blown away during the drying necessary between the two chromatographic runs.

The layer of silica gel can be stabilized by the addition of gypsum (30 g silica gel Camag D-0, 2 g $CaSO_4 \cdot 2 H_2O$, and 154 ml chloroform/Methanol/water 130 : 20 : 4 v/v). The suspension is homogenized with the Ultra-turrax, the plates are coated as described, held ca. 3–4 sec in steam, and air-dried for 10–15 min. ALTHAUS and NEUHOFF (1973) have recently found that impregnation by pre-chromatography with sodium silicate (Na-silicate/water, 1 : 19 v/v) stabilises the silica gel layer on microplates. A solution of Na-silicate/water/conc. HCl (15 : 70 : 1 v/v) can also be used as a spray for the fixation of silica plates. The impregnation with Na-silicate (water-glass) aids in the application procedure and does not impair the quality of separation. This stabilization process has the advantage in allowing autoradiograms from chromatoplates containing fractionated radioactive extracts to be prepared.

These TLC plates are stable enough to allow the use of spray reagents for the detection of phospholipids. The mixture described by HANES and ISHERWOOD (1969) is well suited for this purpose: 0.5 g ammonium molybdate in 5 ml H_2O + 1.5 ml 25% HCl + 2.5 ml 70% perchloric acid, made up to 50 ml with acetone after cooling. The plate is sprayed with a spray gun from a distance of about 30 cm, under a fume hood, in such a way that the silica layer is evenly dampened. The plate is dried first in warm air, then under an infra-red lamp (about 20 cm distant) for 2–3 min, and irradiated 5 min with long-wave UV light. In this way, about 50 ng of phospholipid may be visualized as a blue spot on a white background. On keeping for longer periods the background becomes bluish. This method is specific for phospholipids; other spray reagents (see STAHL, 1967) colour all lipids unspecifically. Similarly, 50–80 ng of phospholipid can be detected using the spray reagents described by JONES et al. (1966), and by POPOV and STEFANOV (1968), in which a rhodamine-lipid complex is produced and can be detected as a fluorescent spot under UV light.

The chromatography plates can be developed with the following solvents: for plates without binder, chloroform/methanol/water 65 : 25 : 4 (v/v) in the first dimension, and n-butanol/acetic acid/water 60 : 20 : 20 (v/v) in the second dimension; for plates with gypsum binder: chloroform/methanol/cyclohexane/water 65 : 30 : 10 : 4 (v/v) in the first dimension, and n-butanol/chloroform/acetic acid/water 60 : 10 : 20 : 20 (v/v) in the second dimension. There is no visible difference in the chromatographic behaviour of phospholipids on gypsum-containing or gypsum free plates, provided that the solvent systems described are used.

After chromatographic fractionations, the plates are dried carefully and put into a chamber saturated with iodine vapour. The solvent front is stained clearly with iodine vapour even if a lipid extract has not been applied. This can be a problem if a substance migrates either near, or at, the solvent front. In such cases it is advisable, before applying the sample, to pre-run the plate in a mixture of all the solvents which will be used for the chromatographic separation of the extracts. This cleaning chromatography must travel to the extreme edge of the chromatoplate. With this technique the solvent front is no longer marked with the iodine vapour.

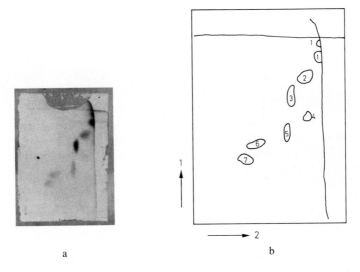

a   b

Fig. 14. *a* Two-dimensional thin-layer chromatogram on a microplate (original size, 32 × 24 mm) of the total lipid extract from tissue punched from the anterior column of cat's spinal cord (iodine staining). Solvent systems: 1st dimension: chloroform/methanol/water 65 : 25 : 4 (v/v), 2nd dimension: n-butanol/acetic acid/water 60 : 20 : 20 (v/v).

b Copy showing localization of the spots: *1* neutral lipids, *2* cerebroside, *3* phosphatidylethanolamine, *4* sulfatide, *5* phosphatidylinositol and -serine, *6* phosphatidylcholine, *7* sphingomyelin

SKIDMORE and ENTENMAN (1962) described a solvent system which is also well suited for the two dimensional micro-chromatography of phospholipids: first dimension, chloroform/methanol/7 N ammonium hydroxide 60 : 35 : 5 (v/v); second dimension, chloroform/methanol/7 N ammonium hydroxide 35 : 60 : 5 (v/v). The solvent system described by HORROCKS (1968), HORROCKS and SUN (1972), is also suited for the two-dimensional micro-chromatography of phospholipids. Preparation of micro-chromatoplates and application of the sample is performed as described above. Development in a lined 50 ml beaker is performed in the first dimension with chloroform, methanol, conc. NH_4OH (60 : 25 : 4 v/v). After careful drying, the plate is held for 3 min in HCl vapour and then dried for 3–5 min with a hair dryer. The second dimension is developed with chloroform, methanol, acetone, acetic acid, water (75 : 15 : 30 : 15 : 7.5 v/v). Figs. 14, 15 and 16 show photographs of two-dimensional chromatograms of phospholipids with the solvent systems described.

The critical factor in this whole micro-chromatographic method is the quality of the silica gel. The finest suitable silica gel obtainable is D-0 quality from Camag (Muttenz, Switzerland). However, even for this material, different batches are not identical in quality, and alterations in the manufacture are not recorded on the bottles. It is thus advisable, if one obtains a gel of suitable quality, to order the total quantity needed, from the same specific batch immediately. If the silica gel has too inconsistent a particle size, and if the particle size is, in the main, too large, the gel may be treated as follows:

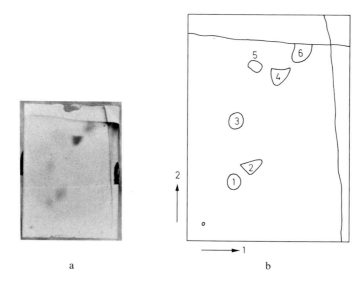

Fig. 15. *a* Two-dimensional thin-layer chromatogram on a microplate (original size) of the total lipid extract from tissue punched from the anterior column of cat's spinal cord (iodine staining). Solvent systems: 1st dimension: chloroform/methanol/7 N NH$_4$OH 60 : 35 : 5 (v/v), 2nd dimension: chloroform/methanol/7 N NH$_4$OH 65 : 60 : 5 (v/v).
b Copy showing localization of the spots: *1* sphingomyelin, *2* phosphatidylcholine, *3* phosphatidylserine, *4* phosphatidylethanolamine, *5* phosphatidylinositide, *6* cerebrosides

The material is extracted for 24 hrs. in a large soxhlet apparatus with chloroform/methanol (2 : 1 v/v), and dried. The effectiveness of this purification step is seen in the brownish colour extracted into the solvent. Next, the dry silica gel is placed in a measuring vessel, and a homogeneous suspension is prepared by shaking it vigorously with distilled water. The suspension is allowed to stand until half the silica gel has settled. The remaining suspension is decanted, the water removed in a rotary evaporator, and the powder dried in a drying cupboard. The last traces of water are removed by shaking the gel with absolute methanol. The methanol is then evaporated and the residue dried in high vacuum. If larger amounts of silica gel are prepared by this method, it is necessary to repeat the methanol wash. Traces of water remaining in the gel cause the upper surface of a chromatography plate to be inhomogeneous and uneven, covered with fine ripples. Such plates give very bad separations.

A maximum of 2 µl of phospholipid solution is applied to one corner of the plate, 3 mm from the glass edges, by means of a very fine capillary pipette. A stereomicroscope is used to help in the accurate application of the spot. For two-dimensional chromatography, a total weight of not more than 10 µg phospholipids should be applied. The quality of separation depends critically on the way in which the sample is applied. The area of the application spot should be as small as possible, and should in no case have a diameter larger than 1.5 mm. For this reason, the capillary is applied slowly and carefully to the fragile layer of silica gel. On contact with the silica gel, the chloroform-methanol-lipid solution is automatically absorbed into the silica layer. This process must be carefully

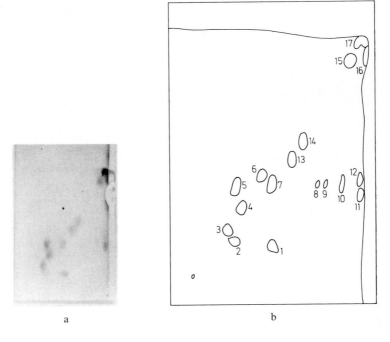

Fig. 16. *a* Two dimensional thin-layer chromatogram on a microplate (original size) of the total lipid extract from tissue punched from the anterior column of cat's spinal cord (iodine staining). Solvent systems: 1st dimension: chloroform/methanol/conc. NH_4OH 60 : 25 : 4 (v/v), 2nd dimension: chloroform/methanol/acetone/acetic acid/water 75 : 15 : 30 : 15 : 7.5 (v/v).

b Copy showing localization of the spots: *1* phosphatidylserine, *2* phosphatidylinositide, *3* sphingo-myelin, *4* phosphatidylcholine, *5* acyl phosphatidylethanolamine, *6* cerebroside sulfates, *7* acid-stable phosphatidylethanolamine, *8, 9* unknown, *10* cardiolipin, *11, 12* unknown, *13, 14* cerebrosides, *15* un-known, *16, 17* neutral lipids

controlled, and the flow of chloroform regulated by closing the mouthpiece of the silicone tube intermittently with the tongue. If the chloroform spot reaches a diameter of ca. 1 mm, the capillary must be removed from the silica gel and further application withheld until the chloroform has evaporated. On replacing the capillary, care must be taken that contact with the plate is at exactly the same point as for the first application. If the capillary is pressed into the silica gel, its orifice can easily become blocked by the gel. This usually means that quantitative application is no longer possible, since the silica gel can only be blown out again with relatively high pressure, and either the extract is lost, or too large an applica-tion spot results. When the contents of the capillary have been applied, a little chloroform/methanol is taken up in the capillary to remove any traces of phospho-lipid material. This solution is then applied to the plate. It is not advisable to apply volumes larger than 2 µl, otherwise the process is lengthened unnecessarily, and the danger of damaging the plate is increased. Plates which are damaged during application so that there is a small hole in the silica gel, give bad chromato-grams and should be discarded. An experienced worker needs about 5–10 min for the application of 2 µl.

For development, the plates are placed in 50 ml beakers containing a 1 mm layer of the solvent to be used. The point of application of the phospholipids must not dip into the solvent. One third of the glass wall of the beaker should be covered with filter paper (thickness ca. 0.2 mm) soaked in the solvent, so that the atmosphere in the vessel is saturated. The filter paper must always be renewed when it dries out; a paper should not be re-used once it has dried out, since it does not take up the solvent correctly the second time. Such papers, once used, should be thrown away immediately, since there is a danger that they might be used for chromatography with another solvent, with a resulting deterioration in the quality of separation. The filter paper should not be higher than the rim of the beaker, and should in no circumstances be bent over the rim, since the resulting wick effect disturbs the chromatography. A correct assembley is drawn in Fig. 17. Fresh developer should be placed in the beaker for each chromatogram. The solvent mixtures must be renewed regularly and should not be kept longer than a few hours, as the changes in the concentrations of the components, due to repeated opening of the storage vessel, cause a rapid deterioration in the quality of the separation.

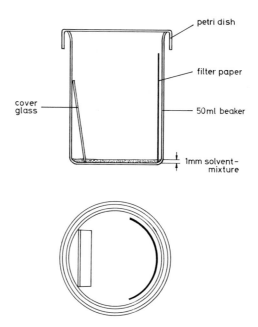

Fig. 17. Scheme of the assembly for chromatography on micro-chromatoplates

The chromatographic plate is supported, vertically if possible, against a part of the beaker wall which is not covered with filter paper, so that no solvent from the sides can travel through the filter paper into the silica layer. The beaker is covered with a Petri dish during chromatography. The plates are developed in both dimensions until the solvent front is about 5 mm from the edge of the glass, a process taking about 10 min. After each development the plate should be dried carefully; after chromatography, it is placed in a Petri dish which contains some iodine crystals. It is advisable to use the plastic lid of a snap-top sample-vial as a support under the plate, so that it can be lifted easily out of the dish after the phospholipids have shown up as brown iodine spots. The plate should only stay in the iodine vapour until the phospholipids are clearly visible (ca. 5 min). If the plate is developed with the ammoniacal system of SKIDMORE and ENTENMAN (1962) the phospholipid spots are not clearly defined after iodine development and are coloured brown on a brown background. The spots are sharper if the plate is exposed briefly (2–3 sec) in hydrochloric acid vapour before iodine development; also, the background stays bright longer after this treatment.

After development with iodine, the single spots are marked with a very fine needle and the iodine allowed to evaporate. An area somewhat larger than the spot should be marked with a fine row of dots; if a continuous line is used to outline the spots, fragments of silica gel are displaced from the layer and can be lost. The iodine must be removed completely from the plate, since the least trace of iodine quenches the fluorescence reaction with rhodamine which is eventually used for quantitative estimation. It is recommended that the colouring with iodine be carried out for as a short time as possible, since otherwise removal of the iodine takes a very long time. After correct colouring, evaporation in air takes ca. 2–3 hrs., and in a desiccator connected with a vacuum pump ca. 20–30min.

After complete removal of the iodine the silica gel is taken up quantitatively from the marked spots, with a micro-shovel (Fig. 18), and placed in Pyrex tubes (inner diameter 2 mm, length ca. 40 mm). It is necessary to use a second micro-shovel as a pusher to pick up the last grains of silica gel. Instead of using micro-shovels, the spots can be removed from the plate in one of two ways: Using a razor blade, or a suitable scalpel, the silica gel surrounding the marked spots can be scraped off carefully. The spots can then be transferred easily using a glass micro-funnel which fits tightly into the top of the tube. This technique of collecting the material is easier and more rapid than the method using micro-shovels. Alternatively, when the plate is first prepared, the slurry of silica gel can be spread

Fig. 18. Micro-shovel for removal of silica gel spots from micro-chromatoplates

onto plastic foil of the type used to support polyamide layers used in amino acid separation (see Chapter 2). After separation and marking, the spots can be cut out with scissors. The silica gel layer should first be moistened by putting the chromatoplate in a beaker containing a small quantity of moist cotton wool. The water is absorbed into the gel and binds the gel sufficiently strongly to enable the spots to be cut out with scissors.

The test-tube containing the silica gel from the spots is then half-filled with chloroform/methanol (2 : 1 v/v), agitated on a whirl-mix, centrifuged briefly at 5000 rpm in a capillary rotor, and the supernatant transferred to the measuring cuvette. In principle, this can be carried out with Pasteur pipettes; however, these are not recommended, because the suction cannot be regulated so accurately with the rubber teat as in the case of the capillary which has a silicone tube to the mouth. The elution sequence is repeated another four times to ensure complete elution of the phospholipid from the very fine silica. Control experiments have shown that 90–105% of the amount of lipids applied can be recovered from the plate (Schiefer and Neuhoff, 1971). Small traces of silica gel in the cuvettes do not influence the final measurement so long as they are not in suspension.

The combined extracts in the cuvette are evaporated carefully to complete dryness by means of a well-regulated stream of nitrogen delivered through a capillary. Care must be taken that none of the extract is sprayed out of the cuvette! Complete drying is necessary because traces of methanol interfere with the measurement. 2 ml of the mixture of cyclohexane, acetic acid, and rhodamine, are added according to Table 1. Measurement is carried out immediately after shaking the cuvette carefully. The quartz cuvettes best suited for these measurements have a fitted Teflon stopper which allows the cuvette to be shaken without loss of the contents. Normal cuvettes are unsuitable, as are those which must be sealed with aluminium foil or parafilm before they can be shaken. No seal with aluminium foil is completely safe, and Parafilm is soluble in chloroform. The "analytical thumb"[1], otherwise frequently employed for closure, is for obvious reasons not permissible here.

One-Dimensional Micro-Chromatography

Good separation of phospholipids can also be achieved by one-dimensional chromatography on thin layers of silica gel on 24×48 mm cover glasses. The preparation of the plates and application of the samples are identical to those described for the two-dimensional technique. One-dimensional development has the advantage that either several samples can be separated simultaneously, or comparison samples can be run under the same conditions. However, it should be noted that, as in almost all methods of chromatographic separation, a substance applied on its own runs somewhat differently to when it is applied in a mixture. Care must be taken when drawing conclusions about substances which run close together. The following solvent system has proved best for the separation of phospholipids on one-dimensional chromatography: chloroform/methanol/acetone/acetic acid/water (60 : 26 : 6 : 10 : 6). Fig. 19 shows a one-dimensional fractionation of phospholipids extracted from the anterior horn of cat's spinal cord and for

---

[1] Obtainable on the experimenter's right hand.

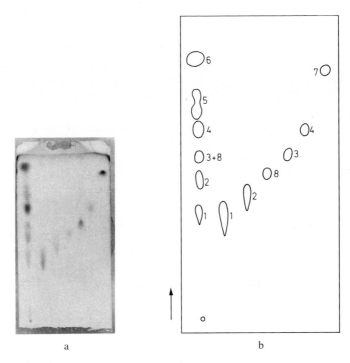

Fig. 19. *a* One-dimensional thin-layer chromatogram on a microplate (original size, 48 × 24 mm) of the total lipid extract from tissue punched from the anterior column of cat's spinal cord (iodine staining). Solvent system: chloroform/methanol/acetone/acetic acid/water 60 : 26 : 6 : 10 : 6 (v/v).
b Copy showing localization of the spots: *1* sphingomyelin, *2* phosphatidylcholine, *3* phosphatidylserine, *4* phosphatidylethanolamine, *5* cerebroside, *6* neutral lipids, *7* cardiolopin, *8* phosphatidylinoside

comparison, single commerciable phospholipids (sphingomyelin, 1-α-lecithin, phosphatidylinositol, phosphatidylserine, phosphatidylethanolamine, and cardiolipin, from Koch-Light, Colnbrook, G.B.).

A good separation of phospholipids can be achieved on silica gel microplates impregnated with water-glass by one-dimensional chromatography. Even phosphatidylinositide and phosphatidylserine, which have proved difficult to separate an micro plates, can be resolved with a stepwise chromatographic technique. The first development is allowed to procede until the solvent front has migrated a distance of 20 mm, employing the solvent mixture chloroform/methanol/water/acetic acid (65 : 35 : 3.6 : 2 v/v). After carefully drying the plate, a second development in the same direction with chloroform/methanol/water/acetic acid (60 : 25 : 5.3 : 10 : 5 v/v) is performed until the front has migrated a distance of 30 mm. Thereafter the plate is once again carefully dried and the third development again in the same direction but with n-hexane/diethyl ether/acetic acid (70 : 30 : 1 v/v) is carried out. The last solvent mixture seperates neutral lipids. The spots may then be detected either with iodine vapour, HANES reagent or other suitable spray reagents like sulfuric acid or fluorescent dyes.

Two-Dimensional, Two-Step, Technique for Micro-Chromatography of Lipids

HUBMANN and NEUHOFF (1973) have developed a method for the separation of neutral fats, free fatty-acids, cholesterol, cholesterol esters, and phospholipids (lecithin, sphingomyelin, phosphatidylethanolamine, phosphatidylcholine, phosphatidylinositol and -serine), on a single micro-chromatoplate. A normal microscope slide is cut down to 45×26 mm. After cleaning the slide carefully, a slurry of silica gel (Camag D-0, without gypsum) is spread as described previously, and air-dried. It is advisable to handle the plate at one corner only, with forceps (compare Fig. 20).

The plate is pre-run in n-butanol/acetic acid/water/chloroform/methanol/hexane $= 6:2:2:6:10:8$ (v/v), so as to avoid staining of the solvent front with iodine vapour. It is advisable to use a closed flat-bottomed chamber containing the smallest possible volume of solvent mixture; this ensures that the silica gel remains attached to the edge of the plate. The sides of the chamber are covered with filter paper. The plate is held by the left upper corner when being introduced into the chamber.

The chromatoplate is dried by means of a hair dryer, turned through 180°, and the sample applied carefully to the left corner at a distance of 5 mm from each edge of the plate. The first chromatographic separation is performed, using chloroform/methanol/water ($65:25:4$ v/v), until the solvent front is 1.5 cm below the upper edge of the plate. It is advisable to mark this distance before starting the chromatography. Then the plate is dried carefully with hot air from a hair dryer held underneath it.

The second separation takes place in the same direction as the first, but is allowed to continue until the solvent reaches the upper end of the plate. The solvent system used is n-hexane/diethyl ether/acetic acid $=85:20:2$ (v/v). During this sepa-

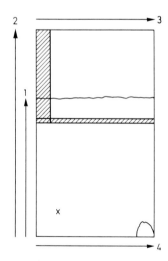

Fig. 20. Scheme for the two-dimensional two-step chromatography of lipids on a micro-chromatoplate

ration the apolar lipids, which migrate as one spot at the solvent front during the first separation, travel to the upper part of the plate. This chromatography is performed in a closed chamber from which the filter paper lining has been removed.

After drying the plate, the third separation is performed in the second dimension with the solvent mixture used for the second separation (n-hexane/diethyl ether/acetic acid). After drying the plate, a transverse line, 1 mm wide, is drawn in the silica gel with the micro-shovel at a distance of 2 cm from the upper edge. A band of silica gel 3 mm wide is also scraped off the left side of the plate from the top of the plate down to the transverse line (see Fig. 20).

The fourth chromatographic separation is performed in the same direction as the third, in a closed chamber without the filter paper lining, using the solvent: n-butanol/acetic acid/water = 60 : 20 : 20 (v/v). After drying, staining with iodine is performed as described previously. Fig. 21 shows a separation of lipids extracted with chloroform/methanol (2 : 1 v/v) from human serum.

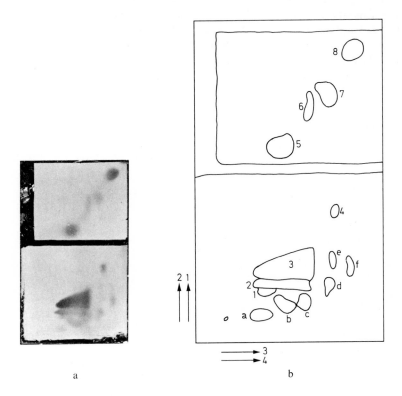

Fig. 21. *a* Two-dimensional two-step thin layer chromatography on a microplate (original size, 45 × 26 mm) of the total lipid extract from human serum (iodine staining). Solvent systems are used as described in the text.

b Copy showing localization of the spots: *1* lysolecithin, *2* sphingomyelin, *3* lecithin, *4* phosphatidylethanolamine, *5* cholesterol, *6* free fatty acids, *7* triglycerides, *8* cholesterol esters, *e–f* unknown compounds

Biological Applications

The micromethod described is sensitive enough to determine the total phospholipids of a single nerve cell. Single nerve cells (perikarya, plus parts of their dendrites and axons) are isolated from beef brain-stem Nucleus hypoglossus by freehand dissection. In order to avoid bacterial contamination, which would interfere with the lipid analysis, all solutions are prepared freshly, and care is taken that the instruments, glass-ware, and wires are fat-free and sterile. The stainless steel wire used for the isolation of single cells is refluxed in a chloroform/methanol mixture (2 : 1 v/v) first. The single cells are isolated by the method described in detail in Chapter 6. To extract the total lipids, the wire pieces, together with the adhering isolated cells, are dried in air and then transferred to cuvettes containing 1 ml chloroform/methanol (2 : 1 v/v). After 2 hrs. at room temperature, the solvent is removed with a stream of purified nitrogen, and the fluorescence spectra are measured (Fig. 22). The total phospholipid content is obtained by comparison of the result with the standard curve for the total phospholipid mixture of guinea-pig brain (Fig. 11). The average dry weight of these cells, as determined by X-ray absorption, is 2–10 ng. This method of dry weight determination does not include the long dendrites and axons of the cells. From the dry weight of the cells one would expect 0.5–2.5 ng phospholipids per cell, assuming that 25% of the cellular dry weight is phosphatides

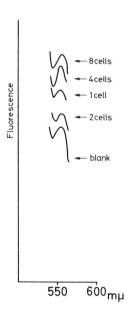

Fig. 22. Emission spectra of the total phospholipid extract from isolated single nerve cells. The cuvettes contained 2 ml cyclohexane, 2 µl rhodamine solution, and 1 µl acetic acid. The excitation wavelength was set at 534 nm, the bandwidth of the emission monochromator at 100 Å, the bandwidth of the excitation monochromator at 25 Å, and the apparatus amplification at 10 ×. (From SCHIEFER and NEUHOFF, 1971)

[as determined from the analysis of grey matter (BRANTE, 1969)]. The higher values found (see Table 2) are presumed to be due to the technique used for the isolation of nerve cells, which always have many long dendrites and axons (compare Fig. 3 in Chapter 6). As axons and dendrites both contain large amounts of phospholipids, and as their dry weight is not taken into account, it is not surprising that the values found by this analytical procedure are higher than those calculated by other methods. The differences in the phospholipid content per single cell are due at least partially to differences both in the cell size of the single perikarya, and in the lengths of their dendrites and axons. The high value found for the phospholipid content of the single cell is due to the large size of this cell, as assessed under the microscope. Furthermore, one must consider the possibility that the composition of phospholipids of single nerve cells may differ from the composition of the brain extracts, which were used for the calibration curve. The results, however, demonstrate the high sensitivity of these fluorometric micro-assay procedures, which allow the study of phosphatide metabolism and changes on the cellular level.

Table 2. Total phospholipid content of isolated single nerve cells from beef brain stem Nucleus hypoglossus. The mean molecular weight of phospholipids was taken to be 750

Number of cells	Fluorescence	Phospholipid content		Phospholipid content
		[nMol]	[ng]	per one cell [ng]
1	0.5154	0.043	32.2	32.2
2	0.1955	0.019	14.3	7.2
4	0.8708	0.069	51.7	12.9
8	1.0308	0.082	61.5	7.7

For the determination of single phospholipids by micro-chromatography, 60–100 µg fresh weight of tissue, depending on the type of tissue, is sufficient starting material (SCHIEFER and NEUHOFF, 1971). OSBORNE, ALTHAUS, and NEUHOFF (1972) have shown that, for the phospholipids in the nervous system of the gastropod mollusc, Helix pomatia, the incorporation of ^{32}P in the individual phospholipids of the brain and single nerve cells can be investigated by micro-chromatography. In these experiments the anterior aorta of the snail was canulated just before it enters the brain, and perfused with snail saline (MENG, 1960) containing $NaH_2{}^{32}PO_4$ (from Radiochemical Amersham Buchler, specific activity 221 mCi/mMol; the concentration of perfused $NaH_2{}^{32}PO_4$ was 80 µCi/ml of perfused saline). After perfusion with the labelled substance (1 ml in 90 min) the brain was rinsed in snail saline for 5 min. In some experiments the labelled phospholipids of the brain were extracted into chloroform/methanol, separated by micro-chromatography, and the substances visualised with iodine vapour. In the other experiments, giant neurons from the metacerebral ganglia of three snails, whose brains had been perfused with $NaH_2{}^{32}PO_4$, were dissected, placed in a 10 µl Drummond microcap containing chloroform/methanol, and their phospholipids isolated as described. The phospholipids were applied to silica gel microplates, together with unlabelled phospholipids for comparison, chromatographed, and the individual substances were localised with iodine vapour. The spots were removed, suspended in 10 ml scintillation liquid, and the radioactivity associated with each phospholipid of the brain and identified neurons was counted in a Packard Tricarb liquid scintillation spectrometer. Incorporation of ^{32}P into the individual phospholipids of the snail brain and identified neurons are shown in Table 3. It can be seen that the radioactivity associated with phosphoinositides and phosphatidylserine in the neurons is almost twice that in the brain, while the radioactivity in the other phospholipids is approximately the same in both types of nervous tissue. The importance of phosphoinositides in nervous tissue is well established. They appear to be involved, not only in post-synaptic events accompanying transmission of nerve impulses, but also in the axonal membrane, possibly because of their affinity with divalent cations (HAWTHORNE and KAI, 1970). In addition, acetylcholine, noradrenaline, and electrical stimulation are known to increase the incorporation of ^{32}P into the phosphoinositides (HOKIN, 1969; FRIEDEL and SCHANBERG, 1971).

Table 3. Incorporation of ^{32}P into the phospholipids of the brain (an aliquot separated and identified on a single microchromatogram) and identified neurons (a total of 6 neurons). Results are expressed in cpm, with background activity of 35 cpm already substracted

	Brain (cpm)	Neurons (cpm)	Ratio of radioactivity in brain : neurons
Sphingomyelin	47.2	10.9	4.33 : 1
Phosphatidylcholine	99.8	24.6	4.1 : 1
Phosphoinositides and phosphatidylserine	739	364.1	4.02 : 2
Phosphatidylethanolamine	172	36.4	4.7 : 1

REIFENRATH (1973) has used micro-chromatography on silica gel for the determination of the lipid content of the lung alveolar surfactant. After fractionation, the plates were soaked with amido black 10B in 7.5% acetic acid for 5 min, and then with acetic acid to remove the unbound excess of dye. A piece of filter paper is put onto the upper edge of the plate to give a greater surface for evaporation of the acetic acid which soaks continuously through the plate. Finally the dried plate is soaked with liquid paraffin and evaluated by micro-densitometry. The lower limit for quantitative measurements using this technique is between 2 and 5×10^{-8} g of lecithin, lysolecithin, sphingomyelin, or cephalin.

Literature

ALTHAUS, H.-H., NEUHOFF, V.: One-dimensional micro chromatography of phospholipids and neutral lipids on sodium silicate impregnated silica gel layers. Hoppe-Seylers Z. physiol. Chem. in press.

BARTLETT, G. R.: Phosphorus assay in column chromatography. J. biol. Chem. **234**, 466–468 (1959).

BECKER, R. S.: Theory and interpretation of fluorescence and phosphorescence. New York-London-Sydney-Toronto: Wiley Interscience 1969.

BRANTE, G.: Studies on lipids in the nervous system with special reference to quantitative chemical determination and topical distribution. Acta physiol. scand. **18**, Suppl. 63 (1949).

CHALVARDJIAN, A., RUDNICKI, E.: Determination of lipid phosphorus in the nanomolar range. Analyt. Biochem. **36**, 225–226 (1970).

EIBL, H., LANDS, W. E. M.: A new, sensitive determination of phosphate. Analyt. Biochem. **30**, 51–57 (1970).

FRIEDEL, R. O., SCHANBERG, S. M.: Incorporation in vivo of intracisternally injected ^{32}P into phospholipids of rat brain. J. Neurochem. **18**, 2191–2200 (1971).

GUILBAULT, G. G.: Fluorescence, theory, instrumentation, and practice. New York: Marcel Dekker, Inc. 1967.

HANES, C. S., ISHERWOOD, F. A.: Separation of the phosphoric esters on the filter paper chromatogram. Nature (Lond.) **164**, 1107–1112 (1949).

HARRIS, A. F., SAIFER, A., WEINTRAUB, S. K.: Fluorescence quenching of acridines by strandin. Arch. Biochem. Biophys. **95**, 106–113 (1961).

HAWTHORNE, N., KAI, M.: Metabolism of phosphoinositides. In: Handbook of neurochemistry, ed. by A. LAJTHA, vol. III, p. 491–505. New York: Plenum Press 1970.

HERKEN, H., NEUHOFF, V.: Mikroanalytischer Nachweis von Acetylpyridin-adenin-dinucleotid und Acetylpyridin-adenin-dinucleotid-phosphat im Gehirn. Hoppe-Seylers Z. physiol. Chem. **331**, 85–94 (1963).

HERKEN, H., NEUHOFF, V.: Spektrofluorometrische Bestimmung des Einbaus von 6-Aminonicotinsäureamid in die oxydierten Pyridinnucleotide der Niere. Naunyn-Schmiedebergs Arch. exp. Path. Pharmak. **247**, 187–201 (1964).

HOKIN, L. E.: Effect of norepinephrine on ^{32}P incorporation into individual phosphatides in slices from different areas of the guinea pig brain. J. Neurochem. **16**, 127–134 (1969).

HORROCKS, L. A.: The alk-l-enyl group content of mammalian myelin phosphoglycerides by quantitative two-dimensional thin-layer chromatography. J. Lipid Res. **9**, 469–472 (1968).

HORROCKS, L. A., SUN, G. Y.: Ethanolamine plasmalogens. In: Research methods in neurochemistry, ed. by N. MARKS and R. RODNIGHT, vol. 1, p. 223–231. New York-London: Plenum Press 1972.

HUBMANN, F.-H., NEUHOFF, V.: In preparation (1973).

JONES, D., BOWYER, D. E.: GRESHAM, G. A., HOWARD, A. N.: An improved spray reagent for detecting lipids on thin-layer chromatograms. J. Chromatogr. **23**, 172–174 (1966).

KLEINIG, H., LEMPERT, U.: Phospholipid analysis on a micro scale. J. Chromatogr. **53**, 595–597 (1970).

LOWRY, O. H., PASSONNEAU, J. V.: A flexible system of enzymatic analysis. New York-London: Academic Press 1972.

LOWRY, O. H., PASSONNEAU, J. V., SCHULZ, D. W., ROCK, M. K.: The measurement of pyridine nucleotides by enzymatic cycling. J. biol. Chem. **236**, 2746–2755 (1961).

LOWRY, O. H., ROBERTS, N. R., KAPPHAN, J. I.: The fluorometric measurement of pyridine nucleotides. J. biol. Chem. **224**, 1047–1064 (1957).

MENG, K.: Untersuchungen zur Störung der Herztätigkeit bei *Helix pomatia*. Zool. Fahr. **68**, 539–566 (1960).

NEUHOFF, V., HERKEN, H.: Mikromethode zum Nachweis von 3-APAD (3-Acetylpyridin-Adenin-nukleotid) in tierischen Geweben. Naturwissenschaften **49**, 519 (1962).

OSBORNE, N. N., ALTHAUS, H. H., NEUHOFF, V.: Phospholipids in the nervous system of the gastropod mollusc *Helix pomatia*, and the in vivo incorporation of ^{32}P into the phospholipids of identified neurons. Comp. Biochem. Physiol. **43 B**, 671–679 (1972).

POPOV, A. D., STEFANOV, K. L.: Über einen neuen kontrastfähigen Fluoreszenzindikator für die Dünnschichtchromatographie der Lipide. J. Chromatogr. **37**, 533–535 (1968).

REIFENRATH, R.: In preparation (1973).

SCHIEFER, H.-G., NEUHOFF, V.: Fluorometric microdetermination of phospholipids on the cellular level. Hoppe-Seylers Z. physiol. Chem. **352**, 913–926 (1971).

SKIDMORE, W. D., ENTENMAN, C.: Two-dimensional thin-layer chromatography of rat liver phospholipids. J. Lipid Res. **3**, 471–475 (1962).

STAHL, E.: Dünnschichtchromatographie. Ein Laboratoriumshandbuch, 2. Aufl. Berlin-Heidelberg-New York: Springer 1967.

STOFFEL, W., SCHEID, A.: Zur Polyfettsäure- und Phospholipidsynthese in der Gewebekultur von HeLa Zellen. Hoppe-Seylers Z. physiol. Chem. **348**, 205–226 (1967).

SVETASHEV, V. I., VASKOVSKY, V. E.: A simplified technique for thin-layer microchromatography of lipids. J. Chromatogr. **67**, 376–378 (1972).

TURNER, G. K.: An absolute spectrofluorometer. Science **146**, 183–189 (1964).

UDENFRIEND, S.: Fluorescence assay in biology and medicine. New York-London: Academic Press 1962 and 1969.

WHITE, C. E., ARGAUER, R. J.: Fluorescence analysis, a practical approach. New York: Marcel Dekker, Inc. 1970.

WILLING, F., NEUHOFF, V., HERKEN, H.: Der Austausch von 3-Acetylpyridin gegen Nicotinsäureamid in den Pyridinnucleotiden verschiedener Hirnregionen. Naunyn-Schmiedebergs Arch. exp. Path. Pharmak. **247**, 254–266 (1964).

Chapter 4

Micro-Diffusion Techniques

Two-Dimensional Micro-Immunodiffusion

Two-dimensional micro-immunoprecipitation is equivalent to the Ouchter-lony test used on the macro-scale. The theoretical basis of immunoprecipitation and descriptions of the current macro procedures may be obtained from the appropriate text books (e.g. STAFSETH, STOCKTON and NEWMAN, 1956; KABAT and MAYER, 1961; CAMPBELL *et al.*, 1963; ACKROYD, 1964; STEFFEN, 1968; WILLIAMS and CHASE, 1967, 1968, 1971). The micro-method has the advantage that it can be performed with extremely small quantities of material and, because of the very short diffusion distances, the result is obtained rapidly. For example, on the macro-scale the test takes 24 to 48 hrs. whereas the micro-method takes several minutes to a few hours.

Micro-immunoprecipitation is performed on microscope slides which have been cut into three pieces and covered with a 1–2 mm thick layer of 1 % agarose. The plates (26 × 76 mm) are prepared by scoring a slide with a diamond glass-writer and breaking it on the edge of the table. The edges of the plate are then smoothed off with fine emery paper. The plates are cleaned by moistening them with water, rubbing well with scouring powder applied by the finger tip, rinsing with water, and drying. They are then laid horizontally on the heating-cooling plate[1] shown in Fig. 1; this can be adjusted by means of a levelling screw so as to be horizontal. A thermostat prevents the temperature rising above 50° C. After the warm glass plates have all been spread with the hot agarose solution, they are cooled rapidly and evenly, by turning on the water supply to the apparatus. If cooling is slow, the agarose layer may dry unevenly on the edges of the glass plate. The agarose plates are then transferred to Petri dishes lined with wet filter papers, and stored in the refrigerator until needed.

To prepare the agarose solution, 100 mg of agarose (purest grade) are weighed into a small Erlenmeyer flask (25 or 50 ml), 10 ml of 0.037 M potassium phosphate buffer (pH 7.1) are added, and the mixture is warmed carefully almost to boiling, until the agarose has dissolved completely. The solution is shaken constantly during this procedure. A 1 or 2 ml pipette is warmed by filling it with the hot solution several times, and then used to spread ca. 0.5 to 0.8 ml of the solution evenly on the warm glass plates. The tip of the pipette is held just over the plate, and the agarose solution is allowed to flow evenly over the surface of the plate. (If the hands have been washed immediately before pipetting,

---

[1] Obtainable from E. Schütt jr., Göttingen, Germany.

Fig. 1. Heating-cooling plate for the preparation of agarose plates

the index finger may be too dry to maintain any even rate of flow of the solution from the pipette. In this case, the index finger should be rubbed briefly on the side of the nose; this transfers sufficient fat to the fingertip to guarantee efficient sealing of the pipette.)

DAMES (1973) confirmed the observations of KOSTNER and HOLASEK (1972), who found that the sensitivity of the method is increased four to five fold if 4% Dextran T70[2] (MW 70000) is added to the agarose. However, the addition of 4% polyethylene glycol of molecular weight 20000 to the 1% agarose solution (KOSTNER and HOLASEK, 1972; HARRINGTON, FENTON, and PERT, 1971) gives such soft agarose layers that they are unsuitable for micro-immunoprecipitation.

On cooling, the 1–2 mm thick layer of agarose becomes stiff and somewhat turbid. The plates may be used immediately although it is advisable to cut the holes in the agarose after they have been stored over night in a moist chamber. It is possible to store the plates in the moist chamber in a refrigerator for several days before use without adding a sterilising agent [e.g. 0.12 M sodium azide (NaN$_3$), 0.012 ethyl-mercurithiosalicylic acid (Merthiolate)].

The rosette of holes (see Fig. 3) is punched into the agarose layer, using the tool shown in Fig. 2. This tool is prepared by bending a 10 µl Drummond microcap over a small flame, and sealing it onto a short piece of polyethylene tubing (external diameter 2 mm) with Harvard glass wax[3]. To make the seal, a small quantity of wax is put on the blade of a pre-warmed preparation knife, the wax allowed to melt completely, and the capillary then fixed to the polyethylene tubing with a small drop of the heated wax. The handle is made from a piece of pyrex glass tubing, external diameter 4–5 mm, ca. 7 cm long one end of which has been heated with even rotation until its orifice matches that of the polyethylene tubing exactly (allow the glass to cool before testing it!). The soft silicone rubber tubing is fitted to the other end of the glass handle and connected

---

[2] Pharmacia, Upsala, Sweden.
[3] Richier und Hoffmann, Harvard Dental Ges., West Berlin, Germany.

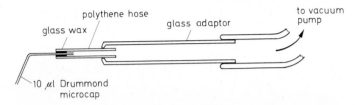

Fig. 2. Tool for punching 0.5 mm holes in agarose layers

to a vacuum pump. The vacuum should not be too high, or the agarose layer will be damaged easily or pulled away from the glass plate when the holes are punched. The vacuum should therefore be tested on the edge of the agarose plate first. The vacuum is correct when it is just sufficient to suck the small punched-out pieces of agarose into the handle.

To punch the holes in the agarose, the 10 µl capillary is held vertically above the layer. This is important, otherwise slanting holes are cut in the plates. On placing antigen and antibody solutions in such holes, the diffusion distances differ at different heights in the agarose layer, so that the precipitate appears as a double band rather than a single precipitation line expected for a known mixture. The capillary punch is held 1–2 mm above the agar layer and then pushed with a gentle, even movement down into the agar as far as the glass plate; it is then withdrawn. There should be no pause between the insertion and withdrawal of the capillary, otherwise the hole is frayed and uneven at the bottom. Punching and withdrawal should therefore be a single, continuous movement. In this way, round, even holes with a diameter of 0.5 mm are formed. The middle hole should be punched first. The distance between the holes should normally be 1–2 mm, but can be larger for special diffusion problems.

After a little practice, two-dimensional micro-diffusion plates can be prepared quickly and reproducibly by this method. An alternative method for punching the holes is to use a mechanical micro-punch which enables the holes to be punched simultaneously. Another method is to place a stencil, on which the rosette of holes has been drawn with Indian ink, under the agarose plate. The holes can then be punched after the plate has been placed under the stereo microscope using front- and back lightning. Four to six rosettes of this type can be punched into the agarose layer on the same portion of a microscope slide.

If a vacuum pump is not available, sufficiently low pressure can be obtained by strong, even suction on a mouthpiece attached to the tube used for punching the holes. Trying to re-punch the same hole always results in an increase in the diameter of the hole, and should be avoided. A hole of 0.5 mm diameter requires 0.5 to 1 µl of solution, depending on the thickness of the agarose layer. Filling the holes with solutions is carried out under the stereomicroscope, using fine capillary pipettes (see Chapter 16).

To fill a hole it is advisable to put the tip of the capillary pipette, containing the sample solution, at the bottom of the hole, and then fill it slowly by blowing gently through the tube attached to the pipette. If attempts are made to fill the

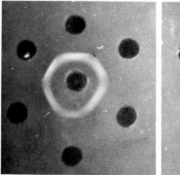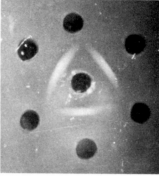

Fig. 3. Two-dimensional micro-diffusion of anti human serum (in central holes) and human albumin and beef albumin. In the left plate, all outer holes are filled with human albumin; in the right plate, human and beef albumin are filled alternately in the outer holes. Micro-photo at 10-fold magnification, without staining or fixating, in lateral incident light

hole by inserting the tip of the pipette into the upper part of the hole, or if the orifice of the pipette is to large, air bubbles can easily be trapped and prevent adequate filling.

Fig. 3 shows the result of a two-dimensional micro-immunodiffusion of anti-serum to human albumin, with human and beef albumin respectively. The anti-serum was placed in both the central holes. In the right-hand rosette of Fig. 3, human and beef albumin were placed alternately in the outer holes. Fig. 3 is a micro-photo at $10 \times$ magnification which was taken, without staining or fixing, with a Zeiss operation microscope in laterally incident light. It is advisable to avoid light reflections, by putting the agarose plate in a chamber filled with 0.9% NaCl solution. Fig. 4 shows a two-dimensional micro-diffusion plate (containing 4% dextran in 1% agarose) after filling the central hole with 0.2 µl anti human serum from horse[4] and the outer holes with 0.8 µl human serum diluted to 6 mg protein/ml. At least 9 precipitation lines are clearly visible.

The appropriate dilutions for the related antigen and antibody solutions should be determined in test series. Depending on the concentrations of antigen and antibody, the precipitates can be recognized within 5–20 min as distinct precipitation lines, in diffuse incident light under the stereomicroscope. At very low concentrations, it is necessary to wait somewhat longer. However, when the quantity of albumin which was eluted from a 20% separating gel, as described in Chapter 1 (after fractionation of 1 µl of human serum diluted 1:2000, which corresponds to ca. 0.02 µg albumin), was tested against anti human serum, a distinct precipitation line was recognizable after 20–30 min. In a similar experiment with 1 µl of serum diluted 1:4000, the precipitation line was detectable after 2 hrs. (NEUHOFF and SCHILL, 1968). However, when complex mixtures of antigen and antibody solutions are used, as for the immunoprecipitation shown in Fig. 4, it is advisable to store the plate for several hours or over-night in the moistened chamber in the refrigerator.

---

[4] Behring Werke, Marburg, Germany.

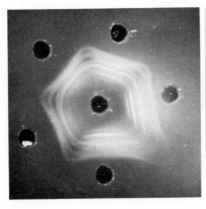 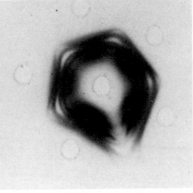

Fig. 4. Two-dimensional micro-diffusion of anti human serum from horse (in the central hole) and human serum (outer holes). Micro-photo at 10-fold magnification; left side unstained, right side stained with amido black 10B

Assuming that the human serum contains 7.5 g protein/100 ml, and that 60% of this is albumin, 1 µl of the 1:4000 diluted serum contains ca. 10^{-8} g albumin, and this can be detected by micro-immunoprecipitation. This limit of detection can be further lowered by a factor of 5, when 4% dextran T70 is added to the agarose. The limit of detectability can be brought down to ca. 10^{-9} g albumin if the antiserum is first placed in the outer holes of an agarose/dextran layer, and the dilute albumin solution introduced into the central hole after ca. 15 min diffusion time. After 15–30 min diffusion in the moist chamber, a well-defined precipitation band is already visible under the stereomicroscope, and is fully developed after 2–3 hrs (DAMES, 1973).

It is possible to elute an albumin band completely from an excised slice of polyacrylamide gel after it has been stained for 5 min with bromophenol blue or amido black in 7.5% acetic acid and cleared for 45 min with 7.5% acetic acid. If such albumin is used in the immune test, the same precipitation is observed as with native albumin; even using an albumin band eluted from a gel which had been stained with amido black and stored for three months in 7.5% acetic acid, a normal immunoprecipitation band was formed within 10 min of diffusion (NEUHOFF and SCHILL, 1968). The finding that even protein treated with acetic acid is capable of giving normal immunoprecipitation is in good agreement with the results of TER MEULEN and MÜLLER (1968), who were able to carry out normal immunological reactions on histological sections of brain tissue after fixation in the Carnoy system (chloroform, ethanol, acetic acid). The groups responsible for precipitation do not therefore appear to be affected by acetic acid.

BECKER, BONACKER, and SCHWICK (1970) have used a slight modification of the two-dimensional micro-immunodiffusion method described, for a rapid and sensitive assay of Australia-antigen (Au/SH-antigen). With an antiserum of high titre, the immunoprecipitation results can be obtained within 20 min to 2 hrs., depending on the antigen concentration.

The sensitivity of the method is not improved by staining with amido black (see Fig. 4); if, however, the preparations are to be kept, staining with amido black is advisable. For this, the agarose plates are soaked overnight in the cold, in sufficient 0.9% NaCl solution to cover them. They are then stained for 10–15 min in 1% amido black in 7.5% acetic acid, and cleared with 7.5% acetic acid. To accelerate clearing, the magnetic stirrer described in Chapter 15 is recommended. If the agarose plates are kept in 7.5% acetic acid without amido black, the immunopre-cipitates dissolve in acetic acid and disappear completely within several minutes. Stained agarose plates may either be stored for a prolonged period in the dark, in 7.5% acetic acid slightly colored with amido black, or be preserved by slow drying in air on a cover glass. To do this it is advisable to cut the agarose plate with a razor blade so that only the rosette with the stained precipitate is transferred to the glass plate. The agarose layers can be transferred with a broad spatula or a microscope slide; handling, even with soft tweezers, always damages the soft agarose layer.

Characterization of a Polymerase-Template Complex by Two-Dimensional Micro-Diffusion

To prepare the complex between DNA-dependent RNA polymerase from *E. coli* and the template poly d(A-T) or DNA, 1–2 mm thick layers of agarose are prepared as described, using 0.75% agarose in TMA buffer (10 mM Tris acetate, pH 7.4, containing 22 mM NH_4Cl, 10 mM magnesium acetate and 1 mM dithiothreitol). Before use, the plates are stored overnight in a moist chamber in the refrigerator. Next, a rosette of 0.5 mm holes is punched in the layer, and the outer holes are filled with template solution. When the template solution has diffused into the gel (ca. 15–20 min), the polymerase solution (4–5 mg protein/ml) is placed in the central hole. During diffusion the plate is stored in the moist petri dish in the refrigerator. After 1 or 2 hrs., the complex between template and polymerase can be seen in oblique incident light without staining or fixing. For complete formation of the complex, the agarose plate is kept in the moist chamber in the refrigerator for 10–15 hrs. Fig. 5 shows the micro-photo of an unstained and unfixed enzym-poly d(A-T) complex in 0.75% agarose. The aggregate is always found nearer the holes which are charged with the template; this is due to the slower diffusion of the high molecular weight matrices.

The diffusion of RNA polymerase against poly d(A-T) or DNA also gives clear aggregates when carried out in Mg^{++}-free agarose plates. However, if NH_4^+ is absent as well as Mg^{++}, and if plates have been made up only with tris-acetate buffer (pH 7.4), no complex formation is observed at all, even after diffusion for longer than 24 hrs. Also, there is no aggregate formation in agarose plates containing 0.5 M NH_4Cl; at this high concentration of ammonium chloride, enzymatic activity of the polymerase is not demonstrable under normal test conditions (ZILLIG, FUCHS, and MILETTE, 1966), which may be explained by the fact that no enzyme-template complex is formed under these conditions.

The specificity of aggregation has been tested with various proteins. On diffusion of RNA-polymerase or poly d(A-T) against human albumin, rabbit

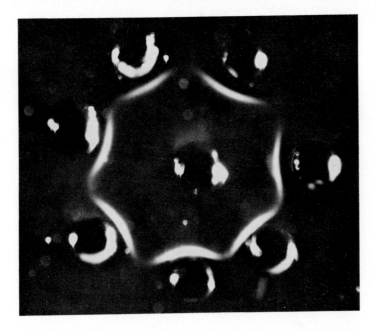

Fig. 5. Micro-photo (magnification 24 fold) of an unstained and unfixed complex of DNA-dependent RNA polymerase from *E. coli*; poly d(A-T) used as template. Plate prepared by two-dimensional micro-diffusion, 0.75 % agarose

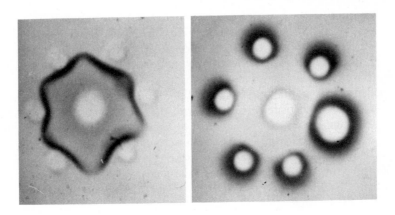

Fig. 6. Amido black staining (left) and pyronine staining (right) of agarose plates containing the complex of RNA-polymerase and poly d(A-T)

serum, tRNA$_{yeast}^{Phe}$, and various RNA synthetases, no aggregates were observed (NEUHOFF, SCHILL, and STERNBACH, 1969). According to its composition, the enzyme-matrix complex can be stained with pyronine or amido black (Fig. 6). Furthermore, the complex can be destroyed by treatment with deoxyribonuclease; this is illustrated in Figs. 7 and 8. In Fig. 7, the agarose plate was divided into

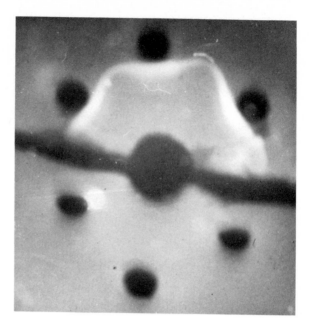

Fig. 7. Complex of RNA-polymerase and DNA (calf thymus) (upper half) is completely dissolved by applying deoxyribonuclease in the holes of the lower half of the agarose plate. Microphoto, without fixation or staining, at 18 fold magnification

two after formation of the complex was complete, and deoxyribonuclease solution was placed in the holes on one side, resulting in complete disappearence of the complex. Fig. 8 shows that the breakdown of the complex can be followed by microdensitometry. For quantitative analysis, optimum curves from unstained and unfixed complexes were first obtained, using a Joeyce Loebl double beam microdensitometer containing suitable filter combinations. Subsequently, all holes on the agarose plate were filled with deoxyribonuclease[5] and the same part of the complex was measured at fixed time intervals, without altering the microdensitometer settings. In experiments of longer duration, drying of the agarose, and the formation of artifacts, were prevented by placing a cover slip over the preparation. Fig. 8 shows the microdensitometer curves from such an experiment, in which the time course for the lysis of an enzyme-DNA complex is clearly demonstrated.

A further proof of the formation of an enzyme-template complex in the agarose layer is the ability of the complex to carry out nucleotide synthesis. This can be demonstrated as follows: after formation of the complex, solutions of ^{14}C-labelled substrates are put into all the holes in the agarose plate, and allowed to diffuse with TMA buffer; the plate is subsequently incubated in a moist chamber at 37° C. Synthesis is stopped by transferring the plate to 7.5% acetic acid; under these conditions, the complex remains unchanged. The plates are then washed for some hours in several changes of acetic acid, to remove

---

[5] Pancreatic deoxyribonuclease, D 8 KH, Worthington Biochemical Corp., Freehold, N. Y., U.S.A.

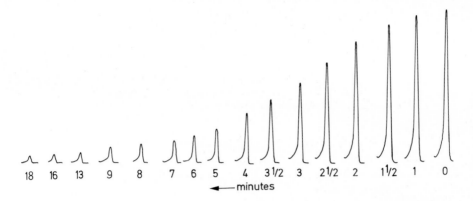

Fig. 8. Original microdensitometer curves of an unfixed and unstained complex of RNA-polymerase with calf thymus DNA, in 0.75 % agarose in TMA buffer. After recording the first curve, 1 µl of deoxyribonuclease solution (0.25 mg of protein/ml) was added to each hole in the agarose plate and further curves were recorded at the stated intervals, without altering the filter and slit-setting of the microdensitometer. The decay of the curve indicates the lysis of the complex. (From NEUHOFF, SCHILL, and STERNBACH, 1970)

excess substrates. Next, the fixed and easily visible complex is cut from the plate under the stereomicroscope, using the micro-knife described in Chapter 6. The small fragments of agarose are then dried in absolute ethanol or acetone. When the fragments are completely dehydrated (ca. 30 min, with 2 or 3 changes of the solvents) they are transferred to toluene scintillation solution. The agarose layers are then so transparent that virtually quenching-free counting of the radioactivity is possible. As a blank, a fragment of agarose of approximately the same size, which has been excised from the immediate neighbourhood of the agarose holes and treated under the same conditions, is used. If incompletely dehydrated fragments of agarose are transferred to the toluene, they shrink and are easily recognized by their milky appearance; also, due to quenching, the total counts are diminished. Table 1 shows the result of an experiment in which the synthesis of an enzyme-poly d(A-T) complex was studied, using ^{14}C-ATP and ^{14}C-UTP (each 0.001 M, specific activity 2.5 Ci/mole); the incubation time was 20 min at 37° C.

Table 1 shows further that, in the presence of the specific inhibitor of *E. coli* polymerase, rifampicin (HARTMANN *et al.*, 1967, 1968; WEHRLI *et al.*, 1968; SIPPEL and HARTMANN, 1968), no de novo synthesis from poly d(A-T) takes place. However, as shown by Fig. 9, rifampicin does not affect the formation of the enzyme-template complex; the complex is produced in the same form and at the same rate, regardless of whether rifampicin is present or not. Fig. 10 shows (on the left) the autoradiograms of a ^{3}H-rifampicin-enzyme-poly d(A-T) complex in an agarose plate. The plate, previously washed with acetic acid and then air-dried, was placed on the X-ray film from a radiation dosage badge (see Chapter 2). The radioactivity of the ^{3}H-rifampicin is detected at the same positions at which the complex was seen in incident light and visualized by staining with amido black. It was necessary to expose the film for an extended

Table 1. Enzyme activity test using a polymerase-template complex formed in a 0.75% agarose layer. In (a) a mixture of enzyme (5 mg protein/ml) and TMA-buffer (1 : 1) was diffused against poly d(A-T) (20.8 E_{260} units/ml). In (b) a mixture of the polymerase (5 mg protein/ml) and rifampicin solution (300 µg/ml (1 : 1) was diffused against poly d(A-T). After formation of the complex was complete, all holes of the agarose plates were loaded with ^{14}C-ATP and ^{14}C-UTP (0.001 M, spec. activity 2.5 Ci/Mol) and after diffusion the plates were incubated for 20 min at 37° C. Synthesis was stopped by transferring the plates into 7.5% acetic acid; radioactivity was measured after washing and preparing the pieces of agarose containing the visible complex, as described in the text. (From NEUHOFF, SCHILL, and STERNBACH, 1969)

Assay	Activity ^{14}C cpm
a) Enzyme-poly d(A-T) complex	
$+ ^{14}$C substrates	1 436
agarose blank	48
b) Rifampicin-enzyme-poly d(A-T) complex	
$+ ^{14}$C substrates	296
agarose blank	57

length of time to obtain blackening, because of the weak ^{3}H-radiation energy. It may be seen from Fig. 10 (right hand side) that, when ^{35}S-heparin is added to the pre-formed complex, all the radioactivity which is not bound to the protein is removed during the wash with acetic acid. The radioactivity is only associated with the region containing protein; no radioactivity can be detected near the holes.

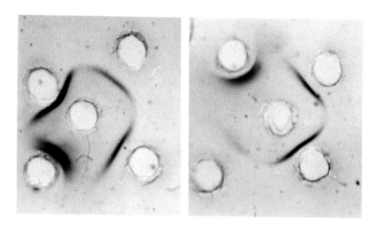

Fig. 9. Amido black staining of complexes formed between RNA polymerase and poly d(A-T). The outer holes of both plates were loaded with the same amount (0.7 µl) of poly d(A-T) solution 20.8 E_{260}-Units/ml). On the left plate the central hole was filled with 0.7 µl of a mixture of RNA polymerase and rifampicin (Enzyme: 5.2 mg protein/ml, rifampicin 30 µg/ml, 1 : 1 mixture). On the right plate the central hole was filled with 0.7 µl enzyme solution (5.2 mg protein/ml, 1 : 1 diluted with TMA buffer). After formation of the complexes, the plates were stained with amido black 10 B. Note the dependence of complex formation on the distance between the agarose holes. (From NEUHOFF, SCHILL, and STERNBACH, 1969)

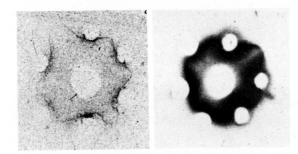

Fig. 10. Autoradiograms (produced on the film of a radiation-dosage badge) of complexes of RNA polymerase and poly d(A-T), loaded with ^{3}H-rifampicin (left) and ^{35}S-heparin (right) (magnification 8 ×). The exposure time was 6 days. The experiment with ^{35}S-heparin was terminated by fixation in acetic acid a short time after the addition of heparin, before lysis of the complex was completed. The breakdown of the complex, demonstrated by the diffuse blackening and the binding of the heparin to the protein, should be noted. No radioactivity is demonstrable in the surrounding layer of agarose

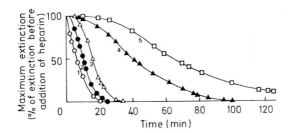

Fig. 11. Analysis of microdensitometer curves from experiments of the type shown in Fig. 8. Complexes of RNA polymerase and calf thymus DNA were subjected to lysis with 1 µl of heparin solution of different concentrations. Heparin concentrations: 1 (o) 600 µg/ml; 2 (•) 400 µg/ml; 3 (△) 200 µg/ml; 4 (▲) 100 µg/ml; 5 (□) 20 µg/ml. Note the dependence of the rate of lysis on the concentration of heparin. The same results are obtained when enzyme-poly d(A-T) complexes are used

It is known that heparin is also a potent inhibitor the synthesis of RNA by DNA-dependent RNA polymerase (DOERFLER et al., 1962; WALTER et al., 1967). As shown by Fig. 11, the mechanism of action of heparin is completely different from that of rifampicin. In contrast to rifampicin, which has no effect on the formation of the enzyme-template complex (whether the enzyme is treated with rifampicin first or the complex is formed first and then treated with rifampicin), the addition of heparin to a pre-formed complex results in complete dissociation of the complex. The period of dissociation of the complex depends on the concentration of heparin (see Fig. 11). A complex formed from calf thymus-DNA and RNA polymerase was prepared in the agarose plate, and subsequently 1 µl of heparin solutions of different concentrations were placed in each hole in the agarose plate. Determinations were carried out, as described previously for the effect of deoxyribonuclease, by determining the degree of lysis of the complex by means of a microdensitometer at different intervals of time. The curves obtained were evaluated by planimetry; the value obtained before the addition of heparin was taken as 100%. It is obvious that the dissociation of the complex with time depends on the concentration of heparin. The inhibitory effect of heparin on RNA synthesis via the polymerase is thus due to the polyanion displacing the template from its binding site on the enzyme. The autoradiogram on the right in Fig. 10 shows that the enzyme-DNA complex is already dissociating shortly after the addition of ^{35}S-heparin. Further evidence that the inhibitory effect of heparin is due to the lysis of the enzyme-template complex is shown by the fact that an enzyme-template complex is not formed at all when enzyme treated with heparin is diffused against the template, or when the diffusion is carried out in an agarose layer which contains heparin.

As shown in Table 2, ^{35}S-heparin is taken up by the enzyme after it has been treated with ^{3}H-rifampicin. This indicates that rifampicin and heparin are attached to different sites on the polymerase. LILL et al. (1969), using the gel filtration method, came to the same conclusion. Fig. 12 shows that heparin also causes the enzyme-template complex to dissociate when the latter is charged with rifampicin, or when a synthesising complex is formed by the addition of

Table 2. A complex consisting of polymerase and poly d(A-T), prepared in a micro-diffusion plate, was loaded first with ^{3}H-rifampicin, then with TMA-buffer, and finally with excess of ^{35}S-heparin. Afterwards it was fixed, and rinsed with 7.5% acetic acid. A similar complex was loaded with excess of ATP and UTP (each 1 mM), TMA buffer, and the heparin solution, then rinsed, and fixed in acetic acid. Values for the background radioactivity were obtained from areas of agarose of the same size and in close proximity to the complex

Treatment	^{3}H radioactivity (cpm)	^{35}S radioactivity (cpm)
^{3}H-rifampicin, then ^{35}S heparin	4485	1442
ATP + UTP, then ^{35}S-heparin	–	916
Agarose blank	137	110

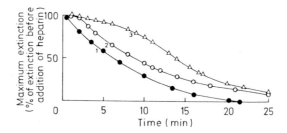

Fig. 12. Analysis of microdensitometer curves from experiments of the type shown in Fig. 8. RNA polymerase-poly d(A-T) complexes were treated: 1 (●) with 1 µl of heparin solution (600 µg/ml); 2 (o) first with an excess of 1 mM ATP + 1 mM UTP solution, then with heparin as for 1. 3 (△), the complex was first saturated with excess of rifampicin, rinsed with TMA buffer, and then treated with the heparin solution as for 1. After treatment with rifampicin, 50% lysis of the complex is only visible after twice the time necessary for an untreated complex

Table 3. Two-dimensional micro-diffusion of RNA polymerase against poly d(A-T) or calf thymus DNA in 0.75% agarose plates in TMA buffer. The complexes were either loaded only with excess of ^{3}H-rifampicin or, after rinsing with TMA buffer, either ^{14}C-ATP (1 mM) or ^{14}C-UTP (1 mM) was also added. After diffusion they were fixed, and rinsed in acetic acid. The complexes and control segments of agarose of equal area and from the vicinity of the complexes, were removed under a stereomicroscope, dehydrated in ethanol and subsequently dissolved in the toluene scintillation fluid

Complex	[^{3}H]Rifampicin		[^{3}H]Rifampicin, then [^{14}C]ATP		[^{3}H]Rifampicin, then [^{14}C]UTP	
	^{3}H radio-activity (cpm)	^{14}C radio-activity (cpm)	^{3}H radio-activity (cpm)	^{14}C radio-activity (cpm)	^{3}H radio-activity (cpm)	^{14}C radio-activity (cpm)
Enzyme–poly d(A-T)	1963	34	2030	981	1876	1663
(background)	(35)	(10)	(59)	(30)	(341)	(158)
Enzyme–DNA	2225	26	2040	837	1009	1093
(background)	(67)	(14)	(235)	(57)	(92)	(63)

the substrates. In the presence of rifampicin or substrates, the time for dissociation is prolonged. This supports the hypothesis that the conformation of the enzyme is modified (LILL et al., 1970) to a more stable form (STEAD and JONES, 1967).

As shown in Table 3, RNA polymerase has binding sites for ATP and UTP, in addition to those for rifampicin and heparin or the template, which can be demonstrated by the two-dimensional micro-diffusion technique. In addition to the use of ^{3}H radioactivity of rifampicin, it is also possible to demonstrate that ^{14}C-ATP and ^{14}C-UTP binding occurs. An identical result can be obtained with γ-^{32}P-ATP, showing that these substrates are bound as triphosphates. For these experiments the enzyme-polymerase complex is first produced, then ^{3}H-rifampicin is added to the holes in the agarose. After the ^{3}H-rifampicin has diffused into the gel, the holes are filled with TMA buffer to facilitate the diffusion of rifampicin. The labelled substrate, ^{14}C-ATP or ^{14}C-UTP, is then added to the holes, followed by TMA buffer. After diffusion the plates are fixed, the unbound radioactivity washed out with acetic acid, and the agarose pieces are dehydrated and evaluated in scintillator toluene. Table 3 shows that with both the enzyme-poly d(A-T) and enzyme-DNA complexes, ^{14}C-ATP or ^{14}C-UTP is also bound besides the ^{3}H-rifampicin.

In further experiments, enzyme-template complexes, with and without unlabelled rifampicin, were treated with ^{3}H-UTP and ^{14}C-ATP, and their radioactivity determined as described. Here, too, comparable labelling with ^{14}C and ^{3}H could be demonstrated. The simultaneous attachment of ^{3}H-rifampicin and ^{14}C-ATP to the polymerase can be confirmed by a different technique (see Table 4, lines 14 and 15). From this it may be deduced that pure RNA polymerase from E. coli has further binding sites for the substrates ATP and UTP in addition to a specific binding site for the template (or heparin); these are not identical with the binding site for rifampicin. The results are in good agreement with those of LILL et al. (1970).

Table 4. Enzyme activity tests of RNA polymerase, using the method of ZILLIG *et al.* (1966) and the binding of ^{3}H-rifampicin to the enzyme-template complex on its addition to the synthesizing medium at different times. ^{3}H-labelled or unlabelled rifampicin was added to the different media at concentrations approx. 100 times that of the normal inhibitory concentration. The reaction mixture contained 30 mM tris-acetate buffer, pH 7.9, 130 mM NH_4Cl, 30 mM magnesium acetate, 1 mM each of ATP, GTP, UTP and CTP, and 50 µg of calf thymus DNA or 0.25 E_{260} units of poly d(A-T) or poly dA, as appropriate. (Tables 2–4 from NEUHOFF, SCHILL, and STERNBACH, 1970)

	Radioactivity	
	^{14}C (cpm)	^{3}H (cpm)
1. Enzyme + DNA + [^{14}C]ATP, CTP, GTP, UTP: synthesis 0–8 min	5950	—
2. Enzyme + DNA + [^{14}C]ATP, CTP, GTP, UTP: synthesis 0–12 min	7960	—
3. Enzyme + rifampicin + DNA + [^{14}C]ATP, CTP, GTP, UTP: synthesis 0–12 min	70	—
4. Enzyme + DNA + [^{14}C]ATP, CTP, GTP, UTP, then at 8 min rifampicin: synthesis 0–12 min	7800	—
5. Enzyme + [^{3}H]rifampicin + DNA + ATP, CTP, GTP, UTP: synthesis 0–12 min	—	1406
6. Enzyme + DNA + ATP, CTP, GTP, UTP, then at 8 min [^{3}H]rifampicin: synthesis 0–12 min	—	1460
7. Enzyme + [^{3}H]rifampicin + poly d(A-T) + ATP, UTP: synthesis 0–12 min	—	2170
8. Enzyme + poly d(A-T) + ATP, UTP, then at 8 min [^{3}H]rifampicin: synthesis 0–12 min	—	2140
9. Enzyme + [^{3}H]rifampicin + poly dA + UTP: synthesis 0–12 min	—	1750
10. Enzyme + poly dA + UTP, then at 8 min [^{3}H]rifampicin: synthesis 0–12 min	—	1759
11. [^{3}H]Rifampicin alone	—	54
12. [^{3}H]Rifampicin + DNA	—	47
13. [^{3}H]Rifampicin + poly d(A-T)	—	86
14. Enzyme + [^{3}H]rifampicin + DNA	—	2085
15. Enzyme + [^{3}H]rifampicin + DNA + [^{14}C]ATP + UTP	445	2068

It is known that total inhibition by rifampicin of RNA synthesis by polymerase only occurs if it is added to the medium before synthesis commences. Synthesis already in progress is inhibited little or not at all, depending on the time of addition. This is true if the course of synthesis is only followed for 5–10 min after addition of rifampicin; however, if the course of synthesis is followed for a longer period, inhibition is observed (STERNBACH, NEUHOFF, and SCHILL, 1969; LILL *et al.*, 1970; NEUHOFF, SCHILL, and STERNBACH, 1970). Table 4 shows that the same amount of ^{3}H-rifampicin is bound to the enzyme irrespective of the time at which it is added, even if this is at a time when synthesis is not inhibited by rifampicin (line 1–4). LILL *et al.* (1969, 1970) have independently published similar findings. This effect is independent of the template used (lines 5–10).

The fact that rifampicin is taken up equally strongly before and during synthesis, and also that it is bound in presence of the substrates, leads us to a hypothesis for the mechanism of the inhibition (Fig. 13), namely that rifampicin is attached very firmly to a specific site on the enzyme, in close proximity to the site for synthesis. If rifampicin is added before the start of synthesis, it blocks the site of synthesis, but does not prevent the attachment of the substrate to the polymerase, i.e. rifampicin does not inhibit the formation of the initiation complex. This conclusion was also reached by LILL *et al.* (1970). However, if a growing RNA chain is already on the site of synthesis, the rifampicin may still become attached to its own binding site, but does not block the continuing growth of the chain. Only when the completed chain is detached can rifampicin inhibit synthesis.

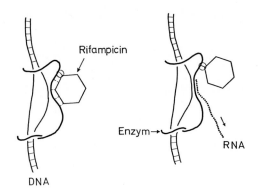

Fig. 13. Model for mechanism of inhibition of RNA synthesis by rifampicin. (Figs. 10–13 from NEUHOFF, SCHILL, and STERNBACH, 1970)

Radial Micro-Diffusion

The radial micro-diffusion method is a simple procedure for the qualitative and quantitative investigation of the attachment of ^{3}H-rifampicin, ^{35}S-heparin, or labelled substrates to the polymerase. For this, a layer of 0.75% agarose in TMA buffer is applied to slides as described for the two-dimensional micro-diffusion method. After storage in a moist chamber at 4° C for 12–15 hrs., 0.5 mm holes are punched in the layer. Polymerase is then introduced into the holes with a suitable micropipette under the stereomicroscope. For quantitative analysis, a 1 µl or 2 µl Drummond microcap is filled with the enzyme solution and this is then transferred completely to the holes with a capillary pipette. After diffusion of the enzyme is complete, the holes are filled two or three times with TMA buffer, after which the test solution is applied and the holes are filled with TMA buffer again. The plate is fixed in 7.5% acetic acid and all radioactivity not bound to the enzyme is rinsed out with several changes of acetic acid. The protein can then be seen easily (Fig. 14, left side), even without staining, as a milky white region around the hole in the agarose. The agarose, containing protein, is removed with a micro-knife, dehydrated in ethanol or acetone, and dissolved in toluene scintillation solution in the counting vials. By this method it is possible to show that 2–3 moles of heparin are bound per mole of *E. coli* RNA polymerase (NEUHOFF, SCHILL, and STERNBACH, 1970).

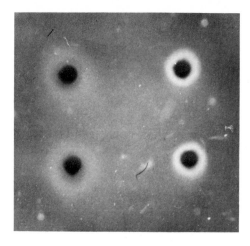

Fig. 14. Radial micro-diffusion of RNA polymerase (left). On the right side the holes are filled with poly d(A-T) followed, after rinsing with TMA buffer, by RNA polymerase. The complex is clearly visible without staining. (Magnification 10 ×)

The radial micro-diffusion technique can also be used to prepare the enzyme-template complex and to show the binding of substrates to the complex. For this, the template is first placed in the hole, and allowed to diffuse into the agarose; TMA buffer is then applied to the holes several times to ensure that the template has completely diffused out into the agarose layer. Incomplete diffusion of the matrix away from the hole leads to the formation of a complex between the enzyme and template in the hole itself, resulting in a precipitate which would prevent quantitative assessement. To avoid this, it is advisable, after the last application and diffusion of TMA buffer, to remove any residues of non-diffused template by rinsing and aspirating the hole with TMA buffer. The polymerase solution is then applied. The enzyme-template complex forms a whitish ring around the hole, which is clearly visible with the naked eye (see Fig. 14, right). To ensure complete diffusion of the enzyme, rinsing with TMA buffer is repeated several times. The binding of substrates, rifampicin, or any other radioactively labelled substance of interest, can be measured by diffusion into the agarose, rinsing with buffer, and finally washing in acetic acid, as described above. This experimental procedure is particularly advantageous if large series of experiments, e.g. testing for binding with various modifications of rifampicin, are to be carried out.

SCHILL and SCHUMACHER (1972), and SCHUMACHER and SCHILL (1972) used radial diffusion in agarose gels for the micro-determination of enzyme, e.g. muramidase, α-amylase, DNase I, RNase A, acid phosphatase, alkaline phosphatase, plasminogen activator, elastase, and non-specific proteases. The method of radial diffusion for the quantitative determination of enzymes is performed on layers of agar or agarose gel on microscope slides. Five to ten holes, with a diameter of 1.5 mm, holding 2 μl of solution, are usually punched on one slide.

For qualitative screening (e.g. of chromatographic fractions) more samples can be tested on one slide. The hydrolyzing action of enzymes during radial diffusion is shown by the formation of clear zones in the opaque substrate-containing agar or by the formation of coloured insoluble product of the enzyme reaction. The zone of diffusion of the enzyme can also be made visible by specific staining procedures. The diameters of the zones are measured in two perpendicular directions and averaged, using a millimeter scale on the screen with a magnifying projector. The diameters of the zones of diffusion for standard solutions are plotted against their concentration on semi-logarithmic paper, to obtain reference curves for the samples under investigation.

Micro Antigen-Antibody Crossed Electrophoresis

RESSLER (1960) and LAURELL (1965) have shown that the resolution of immunoelectrophoresis can be greatly improved by replacing the immuno-diffusion which normally follows antigen electrophoresis, by a second electro-phoresis of the antigen into an gel of agarose which contains antibody. In this case, the direction of migration is perpendicular to the first one. This gives sharp precipitation peaks ("rockets") instead of overlapping precipitation arcs as in conventional immunoelectrophoresis. In addition to reducing the time of the immunoreaction considerably, these peaks can be quantitatively evaluated, as the peak area is proportional to the amount of antigen (LAURELL, 1966; KRØLL, 1969; BINDER and AUERSWALD, 1972). DAMES, MAURER, and NEUHOFF (1972) have adapted this method to the micro-scale, using a polyacrylamide micro-gel as the medium for the first electrophoresis and a thin layer of agarose for the second electrophoresis.

Micro-disc electrophoresis of proteins is performed as described in Chapter 1. For agarose electrophoresis, barbital buffer, pH 8.6, ionic strength 0.02, is pre-pared by dissolving 4.12 g of sodium barbital and 0.8 g of diethyl barbituric acid in 1 l water. For the second electrophoresis, glass cells of the type described by MAURER and DATI (1972) are used. These are made from two microscope slides, separated by two glass strips glued onto one slide as a spacer. For this, the edges of normal 1 mm thick microscope slides are stuck onto microscope slides which are 0.7 mm thick with a two-component adhesive, and allowed to dry over night (see Fig. 15). The assembly of slides stuck to one another, are

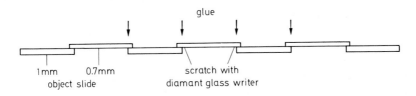

Fig. 15. Arrangement of microscope slides for preparation of chambers with an effective mould space of 75 × 18 × 0.7 mm

turned over, scored along the long edges with a diamond glass-writer, and broken on the edge of the table. The broken edges are smoothed with fine emery paper. (For routine work it is recommended that chambers of this type should be ordered from a firm which produces cuvettes for spectroscopy.) For the preparation of chambers with an effective mould space of $75 \times 18 \times 0.7$ mm, a second microscope slide is laid on the glass flanges (Fig. 16), pressed down firmly, and

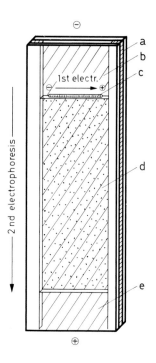

Fig. 16. Arrangement for vertical micro antigen-antibody crossed electrophoresis in an agarose gel slab containing antiserum. *a* Glass cell of two microscope slides separated by two glass strips as spacers, *b* 1% agarose top gel, *c* micro-disc gel, *d* 1% agarose gel containing antiserum, *e* agarose sealing gel. The figure gives the original size

the long edges dipped in molten paraffin wax (m · pt 56° C) to ensure a tight seal. Next, ca 75 µl of a 1% solution of agarose in barbital buffer, warmed to 45° C, are placed in the vessel to form a seal at the bottom; this may be done either with a pre-warmed pipette, or by dipping the chamber into a beaker filled to the appropriate height with the warm agarose solution. The latter method has the advantage that the vessel can be allowed to stand in the same place for further filling. In order to prevent air bubbles entering the gel it is advisable to incline the chamber slightly.

Warm (45° C) 2% antiserum containing 1% solution of agarose in barbital buffer is then added by means of a pre-warmed pipette, leaving about 15 mm from the top of the cell free. It is recommended to warm the glass vessel slightly beforehand, either by placing it in an incubater or by using a warm fan. If the antibody-agarose layer is not horizontal at the top of the cell, surface irregularities can be corrected by carefully sucking out a little agarose. After the gel is formed, an electrophoresed 5 µl polyacrylamide gel is placed on the surface of the antibody-agarose gel, using a fine, flexible teflon spatula or the fine soft forceps of the type shown in Fig. 10 in Chapter 1, and covered to the top of the cell with a warm 1% solution of agarose. Instead of performing the second electrophoresis horizontally (JOHANSSON and STENFLO, 1971), it is preferable to use vertical migration at room temperature, using the apparatus described by MAURER and DATI (1972) and shown in Fig. 17; barbital buffer is placed in the electrode reservoirs.

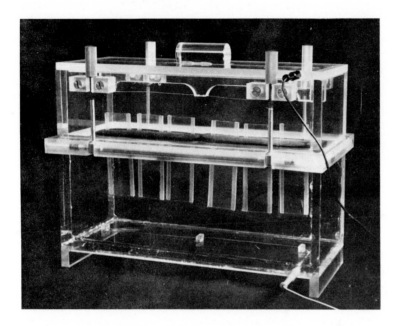

Fig. 17. Chamber for vertical electrophoresis in slab gels, constructed according to MAURER and DATI (1972)

Vertical agarose immunoelectrophoresis has several advantages over the ordinary horizontal methods. Excellent contact between the polyacrylamide gel and the agarose gel is guaranteed, making the cumbersome cutting of troughs or slits into the agarose gel unnecessary. Wicks or protruding pieces of gel used to provide contact between the agarose gel and the electrode buffer are not required. Moreover, the closed glass cell combined with gravity probably counteracts the endosmotic accumulation of liquid towards the cathode. In addition, uniformly thin (< 1 mm) layers of agarose can only be produced easily in closed cells. To test whether endosmosis, which is a major objection to the use of agarose, affects the migration of the proteins, a layer of 1% agarose gel, containing antibodies, was placed on top of the micro-disc gel. No immunoreaction was observed. This suggests that the endosmotic back-flow of proteins is negligible under these experimental conditions. Following electrophoresis, the agarose gels are removed from the glass cells after first lifting one slide from the other by means of a spatula. In so doing, care must be taken not to damage the soft agarose gels; the procedure is best carried out in a large Petri dish filled with 0.9% NaCl solution. The agarose gels are washed in 0.15 M NaCl for 24–36 hrs. to remove excess antibody, and stained with 0.2% (w/v) Coomassie Brillant Blue G-250 in ethanol/acetic acid/water (4:1:5 v/v) for 30 min. They are then destained with dye-free solvent and kept in 7% acetic acid or air dried.

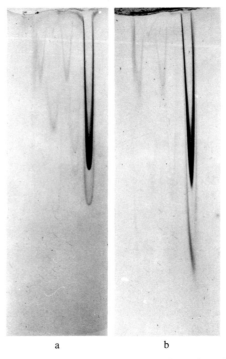

a   b

Fig. 18 a and b. Proteins of 0.9 μg human serum, analysed by electrophoresis in 8% micro-disc gels, and vertical antigen-antibody crossed electrophoresis in 1% agarose slab gels containing 2% rabbit antihuman serum. Immunoelectrophoresis for a 3 h, b 12 h. For other conditions see legend to Fig. 19

Fig. 18 shows that the peak shape varies according to the time of immuno-electrophoresis. Peaks which have stopped migration are sharp (cf. albumin peaks) while those still migrating are diffuse (left side). This suggests that sharp peaks may indicate complete, and diffuse peaks incomplete, immunoreactions. The weak peak in front of the albumin peak must be due to a protein other than albumin, since it does not occur with anti-albumin as the immunoreagent.

Immunoprecipitation peaks of albumin continue to migrate towards the anode even after 12 hrs. of electrophoresis, when human serum proteins, separated in 20% micro-disc gels, were subjected to cross-electrophoresis in a 1% agarose-

3% polyacrylamide composite gel containing 1% anti-albumin serum (Fig. 19, curve *a*). This could be due to too low a concentration of antigen in the agarose-polyacrylamide gel. To prepare agarose-polyacrylamide composite gels, equal volumes of a warm (50° C) 2% solution of agarose in barbital buffer are mixed with a warm (45° C) 6% solution of acrylamide in barbital buffer containing 0.3% bisacrylamide and 0.1% TEMED. Antiserum and ammonium persulfate are added to give final concentrations of 1% and 0.035% respectively, and the solution, after mixing carefully, is put into the glass cells as described above.

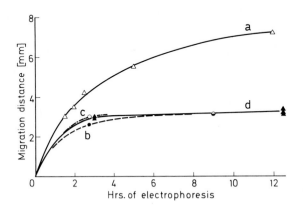

Fig. 19. Effect of concentration of antiserum on the distance of migration of the immunoprecipitation peak of albumin, following vertical antigen-antibody crossed micro-electrophoresis in agarose gel. *a* 20% micro-disc gel (1st dimension), 1% agarose −3% polyacrylamide composite gel containing 1% anti-albumin serum (2nd dimension), *b* 13% micro-disc gel (1st dimension), 1% agarose gel containing 2% anti-albumin serum (2nd dimension), *c* as *b*, but 1% agarose gel containing 2% anti-human serum, d 8% micro-disc gel (1st dimension), 1% agarose gel containing 2% anti-human serum (2nd dimension). General conditions for electrophoresis in 1st dimension: 0.6 µl human serum diluted 1 : 40 (= 0.9 µg protein), 60 V, current from 80 µA at the beginning to 20 µA at the end, 45 min room temperature. Electrophoresis in 2nd dimension: 40 V, 1 mA, room temperature. (Figs. 16, 18, and 19 from Dames, Maurer, and Neuhoff, 1972)

If the proteins are subjected to cross-electrophoresis into 1% agarose gels containing 2% antiserum, the immunoreaction of albumin is complete after 3 hrs. as shown by curves *b* and *c* in Fig. 19. Decreasing the gel concentration of the micro-disc gels from 13% to 8% for the first separation did not affect the distance of migration of the albumin peak (curve *d* in Fig. 19). Control agarose gels, i.e. without the incorporation antiserum, did not show any peaks.

On testing various compositions of the slab gels, it was found that 1% agarose gels yielded the best immunopeaks, although the gels are much more fragile and difficult to handle than polyacrylamide gels. Agarose-polyacrylamide composite gels have the disadvantage of retaining a high level of antibody, which cannot be removed even after 36 hrs. of washing, and which results in a high background staining. 2.6% polyacrylamide gels, with incorporated antiserum but without agarose, also have this disadvantage.

GIEBEL and SAECHTLING (1973) combined micro-disc electrophoresis with antigen-antibody crossed electrophoresis in horizontal agarose gel slabs. Using human standard serum, they obtained 21 individual peaks, of which 18 could be correlated with distinct protein fractions. An "optimal" polyvalent antiserum against human serum was prepared by mixing commercially available antisera, so that it was possible to evaluate all 21 protein peaks in one experiment. Individual proteins were determined quantitatively with about 3 µg of antigen protein and 40 µl of antiserum, this being the minimum sensitivity level.

One-Dimensional Multi-Stage Micro-Immunoelectrophoresis

RÜCHEL et al. (1973) adapted the polyacrylamide gradient micro-electro-phoresis as described in Chapter 1 to a multi-stage micro-immunoelectrophoresis. The principle of the method is based on the fractionation of antigens and anti-bodies successively in the same gel. This is only possible with gradient gels because the proteins do not, in practice, migrate beyond the region of the gel corresponding to their molecular weight. Fig. 20 shows that, even when proteins are applied successively to a gel for fractionation, they are concentrated into a single band after electrophoresis has been carried out for some time. In this experiment, ferritin was first applied to 8 identical gradient gels and zone-electrophoresis was carried out for 30 min at 60 V with gel-buffer (pH 8.8). The electrophoresis was then interrupted, each gel loaded a second time with the ferritin solution, and further discontinuous electrophoresis was carried out with tris/glycine electrode buffer (pH 8.4). At various time intervals, electrophoresis was stopped, and a gel was stained with amido black 10 B. It can be seen that, after 2 hrs., the second band of ferritin has migrated to the same distance as the first band, and only one combined band of ferritin is observed. Therefore, this method enables the use of the gradient system for multi-stage micro-immuno-electrophoresis.

The gradient gels are prepared as described in Chapter 1, with the important exception that the gel mixture does not contain Triton X-100. This is important because Triton X-100 either prevents the formation of the complex, or causes the antigen-antibody complex to be washed out during the re-electrophoresis stage necessary to remove excess protein.

The first electrophoresis is carried out with barbital buffer, pH 8.6, as de-scribed above, instead of with Tris/glycine buffer. The barbital electrode buffer gives an equally good fractionation of, e.g. serum proteins, to that obtained with tris/glycine buffer. The voltage applied and the time of electrophoresis is as in Chapter 1. When the first proteins are fractionated, electrophoresis is stopped, and the buffer solution at the top of the gel is removed by aspiration.

The second sample is then applied (after the necessary dilution) and the second electrophoresis is carried out for the same time using tris/glycine as electrode buffer. Next, either the gel capillary is turned round, or the current is reversed at the voltage regulator. This "wash-electrophoresis" is carried out in Tris/glycine buffer for as long as is necessary for all proteins except the antigen-antibody complex to migrate out of the gel; this takes ca. 3 hrs. at 60 V. Finally, no protein bands are seen in the gel after amido black staining other

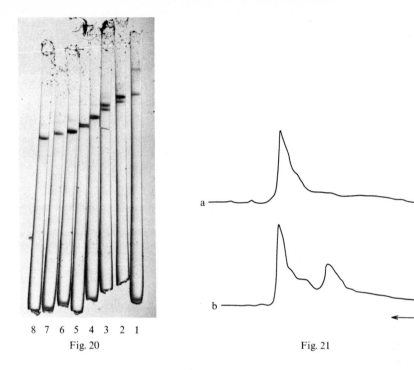

8 7 6 5 4 3 2 1
Fig. 20   Fig. 21

Fig. 20. Double fractionation of pure ferritin (MW 540000, Boehringer, Mannheim, Germany). First, all 8 gradient gels were loaded with the same amount of ferritin and zone electrophoresis was performed with gel-buffer solution. After 30 min of electrophoresis at 60 V, each gel was loaded a second time with the ferritin solution and disc electrophoresis was carried out with tris/glycine buffer. After 10 min (Gel 1), 15 min (2), 20 min (3), 30 min (4), 40 min (5), 60 min (6), 2 hrs. (7), and 4 hrs. (8), of electrophoresis at 60 V, the gels were stained with amido black 10B. In gel 6 it can be seen that the two ferritin bands have migrated together so closely that they appear as one single band; this is complete after 2 hrs. of electrophoresis

Fig. 21. a Microdensitometer curve after amido black staining of a immuno-complex from rabbit serum against human γG^* and human "standard" serum*. The first electrophoresis a was performed using rabbit antiserum for 30 min at 60 V; with barbital buffer (pH 8.6) as electrode buffer. Then the same gel was loaded with the human "standard" serum and electrophoresis performed with tris/glycine buffer (pH 8.4) for 45 min at 60 V. Finally the capillary was inverted and wash-electrophoresis performed with tris/glycine buffer for 3 hrs. at 60 V. The volume of the samples applied (approx. 0.5 µl) and the protein concentration (approx. 1 mg protein/ml) were the same in both cases.

b Using the same conditions as described for curve a, the first electrophoresis was performed with rabbit serum against γM^* and the second electrophoresis with human "standard" serum*. In these experiments, two distinct immuno-complexes are visible after staining with amido black

than the immunocomplex, as shown in Fig. 21. The non-complexed protein which is eluted can be fractionated in a second gel for further analysis. To do this, a second gel capillary is prepared as described in Chapter 1, and the elution of the protein is carried out as described above. This protein is then refractionated directly in the second gel.

—————————
* Obtained from Behringwerke, Marburg, Germany.

The use of two buffer systems (barbital buffer for the first electrophoresis and tris/glycine for the second) ensures that a moving boundary, recognizable by the bromophenol blue front, is visible in both electrophoreses, and a good fractionation is obtained. It thus becomes possible to carry out repeated electrophoresis on the gradient gels, with the possibility of producing and identifying complexes with a complicated composition. Quantitative evaluation may be attempted by means of amido black staining and microdensitometry.

Literature

ACKROYD, J. F.: Immunological methods. Philadelphia: F. A. Davis Comp. 1964.

BECKER, W., BONACKER, L., SCHWICK, H. G.: Ein einfacher immunologischer Schnelltest zum Nachweis des Au/SH-Antigens. Klin. Wschr. **48**, 887–888 (1970).

BINDER, B., AUERSWALD, W.: Methode zum Nachweis von polymerformen in Plasmaproteinlösungen. Quantitative zweidimensionale Gelwanderungs-Immun-Praecipitation (GIP). Kongreß der Dtsch. Ges. für Bluttransfusion, Gießen 1972.

CAMPBELL, D. H., HARVEY, J. S., CREMER, N. E., SUSSDORF, D. H.: Methods in immunology. New York-Amsterdam: W. A. Benjamin, Inc. 1963.

DAMES, W.: (1973) unpublished observation.

DAMES, W., MAURER, H. R., NEUHOFF, V.: Micro antigen-antibody crossed electrophoresis in vertical agarose gels following micro-disc electrophoresis. Hoppe-Seylers Z. physiol. Chem. **353**, 554–558 (1972).

DOERFLER, W., ZILLIG, W., FUCHS, E., ALBERS, M.: Untersuchungen zur Biosynthese der Proteine. Die Funktion von Nucleinsäuren beim Einbau von Aminosäuren in Proteine in einem zellfreien System aus *Escherichia coli*. Hoppe-Seylers Z. physiol. Chem. **330**, 96–123 (1962).

GIEBEL, W., SAECHTLING, H.: Quantitative serum micro-disc-immuno-electrophorese. Hoppe-Seylers Z. physiol. Chem. **354**, 673–691 (1973).

HARRINGTON, J. C., FENTON II, J. W., PERT, J. H.: Polymer-induced precipitation of antigen-antibody complexes: Precipiplex reactions. Immunochemistry **8**, 413–421 (1971).

HARTMANN, G., BEHR, W., BEISSNER, K. A., HONIKEL, K., SIPPEL, A.: Antibiotica als Hemmstoffe der Nucleinsäure- und Proteinsynthese. Angew. Chem. **80**, 710–718 (1968).

HARTMANN, G., HONIKEL, K., KNÜSEL, F., NÜESCH, J.: The specific inhibition of the DNA-directed RNA synthesis by rifampicin. Biochim. biophys. Acta (Amst.) **145**, 843–844 (1967).

JOHANSSON, B. G., STENFLO, J.: Polyacrylamide slab electrophoresis followed by electrophoresis into antibody-containing agarose gel. Analyt. Biochem. **40**, 232–236 (1971).

KABAT, E. A., MAYER, M. M.: Experimental immunochemistry, 2 ed. Springfield, U.S.A.: Ch. C. Thomas Publ. 1961.

KOSTNER, G., HOLASEK, A.: Influence of dextran and polyethylene glycol on sensitivity of two-dimensional immunoelectrophoresis and electroimmunodiffusion. Analyt. Biochem. **46**, 680–683 (1972).

KRØLL, J.: Immunochemical identification of specific precipitin lines in quantitative immunoelectrophoresis patterns. Scand. J. clin. Lab. Invest. **24**, 55–60 (1969).

LAURELL, C.-B.: Quantitative estimations of proteins by electrophoresis in agarose gel containing antibodies. Analyt. Biochem. **15**, 45–52 (1966).

LAURELL, C.-B.: Antigen-antibody crossed electrophoresis. Analyt. Biochem. **10**, 358–361 (1965).

LILL, H., LILL, U., SIPPEL, A., HARTMANN, G.: The inhibition of the RNA polymerase reaction by rifampicin. 1st Lepetit Colloquium, RNA polymerase and transcription, ed. by L. SILVESTRI, p. 55–64. Amsterdam-London: North Holland Publishing Co. 1970.

LILL, U., SANTO, R., SIPPEL, A., HARTMANN, G.: Inhibitors of the RNA polymerase reaction. In: Inhibitors as tools in cell research, ed. by TH. BÜCHNER and H. SIESS, p. 48–59. Berlin-Heidelberg-New York: Springer 1969.

MAURER, H. R., DATI, F. A.: Polyacrylamide gel electrophoresis in Micro slabs. Analyt. Biochem. **46**, 19–32 (1972).

NEUHOFF, V., SCHILL, W.-B.: Kombinierte Mikro-Disk-Elektrophorese und Mikro-Immunpräzipitation von Proteinen. Hoppe-Seylers Z. physiol. Chem. **349**, 795–800 (1968).

NEUHOFF, V., SCHILL, W.-B., STERNBACH, H.: Mikro-disk-elektrophoretische Analyse reiner DNA-abhängiger RNA-Polymerase aus *Escherichia coli*. III. Differenzierung zwischen Syntheseort und Bindungsort für die Matrize. Hoppe-Seylers Z. physiol. Chem. **350**, 335–340 (1969).

NEUHOFF, V., SCHILL, W.-B., STERNBACH, H.: Micro-analysis of pure deoxyribonucleic-acid-dependent ribonucleic acid polymerase from *Escherichia coli*. Action of heparin and rifampicin on structure and function. Biochem. J. **117**, 623–631 (1970).

RESSLER, N.: Two-dimensional electrophoresis of protein antigens with an antibody-containing buffer. Clinica chim. Acta **5**, 795–800 (1960).

RÜCHEL, R., WOLFRUM, D.-I., MESECKE, S., NEUHOFF, V.: In preparation.

SCHILL, W.-B., SCHUMACHER, G. F. B.: Radial diffusion in gel for micro determination of enzymes. I. Muramidase, Alpha-amylase, DNase I, RNase A, acid phosphatase, and alkaline phosphatase. Analyt. Biochem. **46**, 502–533 (1972).

SCHUMACHER, G. F. B., SCHILL, W.-B.: Radial diffusion in gel for micro determination of enzymes. II. Plasminogen activator, elastase, and nonspecific proteases. Analyt. Biochem. **48**, 9–26 (1972).

SIPPEL, A., HARTMANN, G.: Mode of action of rifampicin on the RNA polymerase reaction. Biochim. biophys. Acta (Amst.) **157**, 218–219 (1968).

STAFSETH, H. J., STOCKTON, J. J., NEWMAN, J. P.: A laboratory manual for immunology. Minneapolis: Burgess Publ. Comp. 1956.

STEAD, N. W., JONES, O. W.: The binding of RNA polymerase to DNA: stabilization by nucleotide triphosphates. Biochim. biophys. Acta (Amst.) **145**, 679–685 (1967).

STEFFEN, C.: Allgemeine und experimentelle Immunologie und Immunpathologie sowie ihre klinische Anwendung. Stuttgart: Thieme 1968.

STERNBACH, H., NEUHOFF, V., SCHILL, W.-B.: Kinetic analysis of pure DNA-dependent RNA polymerase from *E. coli*. FEBS-Abstracts of Communications 1969, p. 224.

TER MEULEN, V., MÜLLER, D.: Immunhistologische, feingewebliche, und neurochemische Untersuchungen bei Encephalitiden. I: Die Immunfluoreszenz an fixiertem und gefriergetrocknetem Hirngewebe. Acta neuropath. (Berl.) **10**, 74–81 (1968).

WALTER, G., ZILLIG, W., PALM, P., FUCHS, E.: Initiation of DNA-dependent RNA synthesis and the effect of heparin on RNA polymerase. Europ. J. Biochem. **3**, 194–201 (1967).

WEHRLI, W., NÜESCH, J., KNÜSEL, F., STAEHELIN, M.: Action of rifampicin on RNA polymerase. Biochim. biophys. Acta (Amst.) **157**, 215–217 (1968).

WILLIAMS, C. A., CHASE, M. W.: Methods in immunology and immunochemistry. New York and London: Academic Press, vol. I 1967, vol. II 1968, vol. III 1971.

ZILLIG, W., FUCHS, E., MILETTE, R. L.: DNA-dependent RNA polymerase (E. C. 2.7.7.6). In: CANTONI, G. L., and D. R. DAVIES, Procedures in nucleic acid research, p. 323–339. New York: Harper and Row 1966.

Chapter 5

Capillary Centrifugation

Preparative Capillary Centrifugation

When using micro methods, it is frequently necessary to centrifuge very small volumes. The arrangement shown in Fig. 1 is suitable for the centrifugation of small volumes of liquid; it also has the advantage of being independent of special adaptors (NEUHOFF, 1968b). A Pyrex glass tube of suitable dimensions is sealed carefully by rounding it off at one end and is then annealed. It is filled with the solution by means of a capillary pipette, and is sealed with Parafilm. Some fine quartz sand is placed in a suitable centrifuge bucket and cotton wool is packed well down on top of it to form a layer 2–4 mm thick. Next, a layer of quartz sand is poured on top of the cotton wool; the capillary is inserted into this layer so that it stands centrally on the cotton wool. Then quartz sand is added from a pipette with a wide nozzle, until the glass tube projects about 3 mm out of the layer of sand. The parafilm prevents the solution from contamination during the addition of the sand. The layer of cotton wool ensures that the glass tube does not touch the bottom of the centrifuge bucket even during high speed centrifugation.

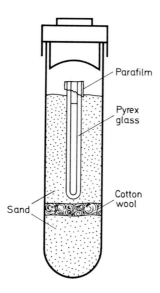

Fig. 1. Simple arrangement for the centrifugation of small volumes. (From NEUHOFF, 1968b)

Fig. 2. Special capillary rotor for capillary centrifugation up to 20000 rpm

The layer of cotton wool can be adjusted according to the length of the glass tube. Using this simple set-up, even very small volumes can be centrifuged at up to 15000–20000 rpm in an angle-head rotor for any length of time. The only precautions to be observed are that the centrifuge bucket is filled up almost to the sealed tip of the tube, and that the opposite bucket is tared with quartz sand. The various rotors are limited to the maximum rpm permissible for the centrifugation of solutions of high density.

A refinement is to use special capillary rotors. Fig. 2 shows the capillary rotor (diameter 23.5 cm) of a Heraeus-Christ[1] Zeta 20 refrigerated centrifuge. The rotor (Nr. 11299) is of aluminium, discoid in form, and can be used up to 20000 rpm. The capillaries lie horizontally in the rotor, which has holders for capillaries of 4 different diameters. The smallest holder is for the 5 µl or 10 µl Drummond Microcaps, and the larger for capillaries of external diameters of 1.5, 3, and 4 mm.

All capillaries used for centrifugation must be made out of Pyrex glass. It must be stressed once again that the capillaries must be sealed carefully, that there must

---

[1] Osterode, Harz, Germany.

be no detectable thickening at the sealed end, and that they must be annealed very carefully. Breakage of the glass during centrifugation (which can be performed at up to 20000 rpm without further precautions) is caused only by incorrect sealing or annealing of the capillaries. Only the 5 µl and 10 µl capillaries, which are sealed in a spirit flame, do not need to be annealed, since they are sealed at relatively low temperatures. Fig. 3 shows two capillaries which are correctly sealed, with external diameters of 4 mm and 3 mm; they have been filled with a solution of a dye so as to show them more clearly.

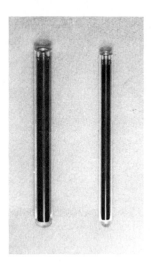

Fig. 3. Pyrex capillaries heat-sealed correctly and filled with a solution of dye for clearer display. Notice that the diameter at the sealed end is the same as that along the rest of the capillary

A capillary rotor of the type shown in Fig. 2 is also obtainable for the Heraeus-Christ Hematocrit centrifuge. This centrifuge, which runs uncooled, achieves up to 13000 rpm and is sufficient for many purposes.

With special adaptors, 5 or 10 µl capillaries can be centrifuged at up to 70000 rpm. Drummond microcaps are heat sealed at one end to a uniform round bottom, and are filled with the aid of a capillary pipette. The capillary is then put into a polyamide (nylon) or polyoxymethylene (Dynal) capillary adaptor. PTFE (Teflon) is unsuitable for adaptors, as it would be so deformed on high-speed centrifugation that breakage of the capillary would be inevitable. The adaptor (Fig. 4) has a central pocket 2 mm in diameter and 32 mm in depth (the length of the 5 µl capillaries). This pocket receives a polythene tube, of 2 mm o.d. and 1 mm i.d., which is open at both ends and is filled with the solvent used for the preparation of the suspension. The filled capillary is then pushed into the polythene tube and the amount of solvent displaced is blotted up with filter paper. This technique ensures that the capillary floats in a medium of the same density as that which it contains, and thus can be centrifuged for any length of time without breaking (NEUHOFF, 1968a, 1969).

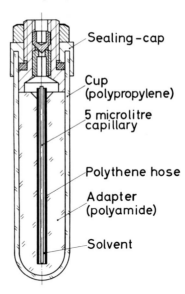

Sealing-cap

Cup
(polypropylene)

5 microlitre
capillary

Polythene hose

Adapter
(polyamide)

Solvent

Fig. 4. Adaptor for the high speed centrifugation of 5 μl capillaries in angle rotors

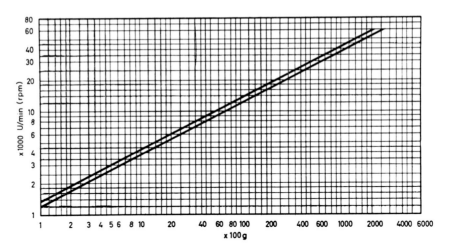

Fig. 5. Graph of speed vs. force for 5 μl capillaries in 60 000 rpm angle rotor

The milled recess in the top of the adaptors ensures easy removal of the capillary after centrifugation. These adaptors are put in normal centrifuge tubes and closed with vacuum tight sealing caps (see Fig. 4). The graph of speed vs. force for the angle rotor No. 9 730 (60 000 rpm) and the Omega II preparative ultracentrifuge of Heraeus-Christ[2] for a 5 μl capillary is shown in Fig. 5. At 60 000 rpm the angle rotor attains 254 000 g at the outermost point of the capillary and 213 000 g at the capillary top.

---

[2] See footnote 1, p. 206.

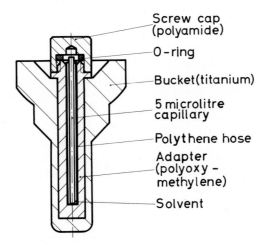

Fig. 6. Titanium bucket with Dynal adaptor for the high-speed centrifugation of 5 μl capillaries in the
70000 rpm swingout rotor

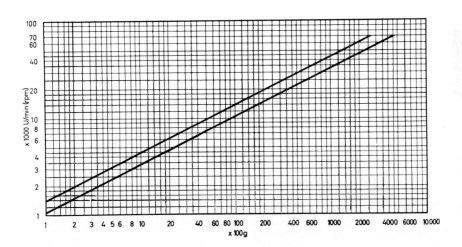

Fig. 7. Graph of speed vs. force for 5 μl capillaries in 70000 rpm swing-out rotor. (Figs. 4–7 from
NEUHOFF, 1968 a, 1969)

The special titanium adaptor shown in Fig. 6 was designed for still higher
forces in order to use shorter periods of centrifugation. The buckets are so constructed that the 6-place titanium swing-out rotor, designed for 5 ml tubes at
50000 rpm, can also be used at a speed of 70000 rpm ($\pm 0.5\%$). The 5 μl capillary
is put in a bucket with a vacuum-tight screw cap which is provided as before with
a polythene tube and solvent to support the filled capillary. The graph of speed vs.
force for this rotor (No. 9 792) is shown in Fig. 7. This includes the special buckets
for high-speed centrifugation of the 5 μl capillaries. The centrifugal force obtained

at the outermost point of the capillary at 70000 rpm is not less than 449000 g. The capillaries can withstand even this enormous force (which is equivalent to 1070 atm) for any length of time.

It is imperative, however, to ensure that the pocket in the Dynal adaptor, the polythene tube, and the capillary, are perfectly smooth and clean, since even minute irregularities of their surfaces would inevitably result in breakage of the capillary. Disregard of this precaution may not lead to the complete disintegration of the capillary but will result in a straight crack. However, should the capillary be filled with a solution of higher density than the solvent around it, it will burst; the cracks run parallel to the axis of the capillary, resulting in long glass splinters.

This method of high speed capillary centrifugation was originally developed in the hope that it might be useful for micro density gradient centrifugation. Centrifugation of this type could be carried out with normal capillaries containing a gradient of polyacrylamide. If the solution of polyacrylamide and the temperature during centrifugation are chosen correctly, the sedimentation bands in the polyacrylamide gel can be found by suitable staining after expulsion of the gel. Quartz capillaries can also be used for centrifugation, and the positions of bands recorded by means of a UV photo, or by UV-scanning.

However, for density-gradient centrifugation in capillaries, the optimal conditions must be worked out first. Systems which yield good separations in normal gradient centrifugation on the macro-scale cannot be simply transferred to capillary centrifugation. In capillary centrifugation, the molecules are sedimented appreciably faster than would be expected. For example, after 2 hrs. of centrifugation at 60000 rpm in a swing-out rotor (330000 g), human albumin (MW 64000) in water in a 5 µl capillary, is sedimented completely. Accordingly, high speed capillary centrifugation is very useful for the rapid concentration of proteins even with low molecular weights.

Fig. 8. Analytical cells reconstructed for receiving microcapillaries. Both forms are suitable for analytical capillary centrifugation

WORONOVA *et al.* (1972) used capillaries of 0.3–0.5 mm diameter, length
20–40 mm, for the high speed capillary centrifugation (35000 rpm for 24–72 hrs.)
of DNA in a two-step cesium chloride gradient. They used the techniques described here, with capillaries filled with acrylamide. After centrifugation, the acrylamide was subjected to photo-polymerisation and the gels were evaluated on a
specially designed micro-photometer (KUSMIN, MATWIEIEV, MIKITSCHUR, 1971).

Analytical Capillary Centrifugation

Capillaries can be centrifuged at high speed if they are embedded in a medium
which corresponds to the density of their contents. Thus, NEUHOFF and RÖDEL

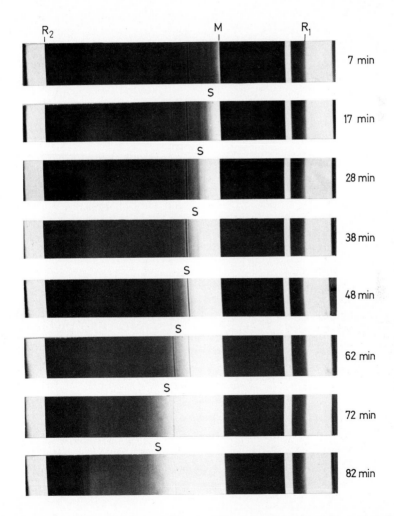

Fig. 9. Absorption photographs taken at 407 nm at different time intervals during analytical centrifugation of a 1% haemoglobin solution in 0.9% NaCl. Capillary volume 1.1 µl. R_1 and R_2 are the
reference edges, M = miniscus, S = moving boundary

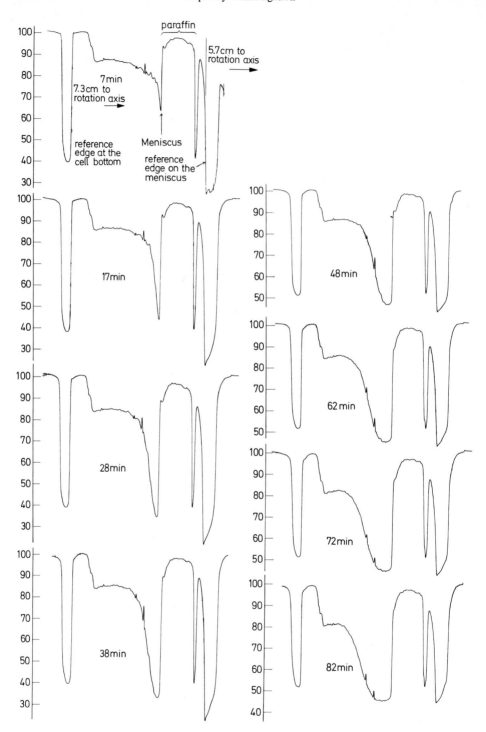

Fig. 10. Microdensitometer tracing from the film shown in Fig. 9

(1973) have developed a procedure for analytical capillary centrifugation. Theoretically, all problems of analytical centrifugation can be transposed to the capillary scale and a volume of approximately 1 μl used per analysis. The basic principles of analytical centrifugation on the macro scale (for details, see ELIAS, 1969; BOWEN, 1970) appear, according to experience so far, to be equally valid for the micro scale. Special rotors are not necessary, but perhaps simplified or even improved rotors having an improved optical system combined with optimal focussing can be built for capillary centrifugation.

Fig. 8 shows a normal analytical cell reconstructed to receive a capillary. The 5 μl Drummond microcaps (or, for particular problems, the corresponding quartz capillaries) are carefully and evenly flame-sealed at one end and, after scratching with a diamond glass writer, a length of 11 mm is broken off. This capillary is filled with the sample up to about $\frac{2}{3}$ of its length, using a fine capillary pipette as usual, made airtight with a layer of paraffin, and introduced into the analytical cell, which is filled with a medium of the same density. The further mounting of the cell is carried out exactly as instructed for the particular type. The only special feature is that a slit aperture corresponding to the diameter of the capillary is introduced to cut out stray light during observation and measurement of the centrifuging.

As an example of analytical capillary centrifugation, Fig. 9 shows absorption photographs of the progressive sedimentation of 1.1 μl of a 1% haemoglobin solution in 0.9% NaCl. Centrifugation was carried out for 82 min at 60000 rpm and 20° C in a Heraeus Christ Omega ultracentrifuge with supplementary analytical unit. Fig. 10 shows the microdensitometer curves corresponding to Fig. 9 from which the values for determination of the S-value may be derived as for analytical ultracentrifugation on the macro scale.

The appreciable advantage of this analytical ultracentrifugation depends less on improved sensitivity than on the considerably smaller sample volumes (by a factor of almost 10^3) necessary for analysis compared with the macro scale.

Literature

BOWEN, T.J.: An introduction to ultracentrifugation. London-New York-Sydney-Toronto: Wiley-Interscience 1970.

ELIAS, H.-G.: Ultrazentrifugen-Methoden. 3. Aufl. München: Beckmann Instruments GmbH 1969.

KUSMIN, S.W., MATWIEIEV, W.W., MIKITSCHUR, N.I.: Kolitschestwienny analis resultatow rasdelienia makromolekul mikroelektroforesom I ultrazentrifugirowaniem. Schkola Seminar Mikrometody Analisa Nukleinowych Kislot. Inst. Org. Chem. Novosibirsk/UDSSR 30–34 (1971).

NEUHOFF, V.: Simplified technique of high-speed capillary centrifugation. Analyt. Biochem. **23**, 359–363 (1968 a).

NEUHOFF, V.: Einfaches Verfahren zur Zentrifugation kleinster Volumina. Arzneimittel-Forsch. (Drug. Res.) **18**, 629 (1968 b).

NEUHOFF, V.: Einfaches Verfahren zur hochtourigen Kapillarzentrifugation. G-I-T Fachz. Laboratorium **13**, 86–87 (1969).

NEUHOFF, V., RÖDEL, E.: 1973 in preparation.

WORONOWA, T. G., KUSMIN, S. W., MIKITSCHUR, N. I., SANDACHTSCHIEV, L. S., NSCHUMILOW, J.: Analis Plawutschei Plotnosti DNK w Masschabie 10^{-8}–10^{-9} g. Dokl. Akad. Nauk SSSR **203**, 477–479 (1972).

Chapter 6

Micro-Electrophoresis for RNA and DNA Base Analysis

Electrophoresis in a microscopic fiber (microphoresis) separates nucleic acid components in amounts corresponding to 500–1000 pg RNA (EDSTRÖM 1960a, 1964a). This sensitivity permits RNA base analysis at the cellular level. In a scaled-up, somewhat simpler version, cellophane strips are used as the supporting medium (KOENIG and BRATTGÅRD, 1963; RÜCHEL, 1971) for the analysis of 3000–5000 pg RNA. Microphoresis in cellophane strips was introduced by KOENIG and BRATTGÅRD for the quantitative determination of radioactively labelled RNA, which, with the technique employed, required more RNA than is normally used for fibers. These techniques also lend themselves to base analysis of DNA and, with a simple modification, to the simultaneous determination of DNA content and base composition (EDSTRÖM, 1964b). Microelectrophoresis can also be used for enzyme determinations, in which substrate and product are separated after incubation with micro-isolated components, as used for intracellular localization of nucleases in starfish oocytes (SIERAKOWSKA, EDSTRÖM, and SHUGAR, 1964).

Compared to paper chromatography or electrophoresis, the increase in sensitivity on microphoresis is of the order of 10^5–10^6. The overall reduction in the dimensions of the supporting medium is responsible for this increase. This apparently simple modification necessitates, however, several modifications of the procedure.

Biological Material

Microscopic tissue constituents, such as single cells, cell nuclei, chromosomes, etc. can be isolated by a variety of methods. Components isolated from fresh tissue are placed on a coverslip ($12 \times 32 \times 0.17$ mm) and dried in air. The glass is then submerged into cold 1 N perchloric acid (0–4° C) for 5 min to remove free nucleotides, then rinsed three times for 5 min in 0.01 N acetic acid to remove the perchloric acid. If determinations are to be carried out immediately, it is rinsed in 96% ethanol for 5 min; otherwise it is stored in 96% ethanol in a test tube, and can be kept for several months. The ethanol treatment has been introduced in order to reduce the contamination of RNA extracts, and as an intermediate solvent between the dilute acetic acid and the 5 min treatment with chloroform which follows. The surface of the coverslip and the sample are then covered with liquid paraffin, inverted, and placed on top of an oil chamber slide (a $6 \times 40 \times 70$ mm glass with a central groove, 3×24 mm, parallel to the short sides, see Fig. 1), and the resulting space is filled with liquid paraffin (E. Merck A.G., Darmstadt).

Fresh tissue components are usually best isolated freehand under a dissecting microscope. Most useful are stereo microscopes which permit stepless variation of the magnification and are equipped with adjustable transmitted and incident illumination. If microdissection is carried out for prolonged periods, the stereo microscope must be equipped with a cooling stage to give temperatures between 0° C and 10° C, which permits work at room temperature. Freehand dissections are normally performed on slides covered with a layer of paraffin about 1 mm thick (histological quality), in a drop of a suitable buffer medium. The surface tension of a water drop on a paraffin layer is high enough to prevent spreading of the drop, and the evaporation is therefore low.

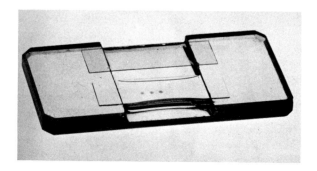

Fig. 1. Oil chamber with two halved cover glasses. The front part of the chamber is filled with liquid paraffin and three small drops of buffer are hanging in the chamber on the cover glass. The rear cover glass is free to receive extracts which are dried onto the surface

For the freehand isolation of single nerve cells, small pieces of tissue approximately 1 mm square are transferred into the buffer drop, and are stained weakly with methylene blue. Toluidine blue gives better contrast but stains irreversibly, whereas methylene blue does not. The nerve cells are released form the surrounding tissue either by agitating the piece of tissue with a fine steel wire (diameter 100–150 μm) in the buffer, or by reducing the tissue to even smaller pieces with two steel wires held crosswise. Single cells can then be lifted out of the drop (HYDÉN, 1959, 1961) with a fine steel wire (diam. 25–50 μm, 5–10 mm long) and fixed with Harvard glass wax[1] or paraffin onto a glass rod as shown in Fig. 2). The free nerve cells floating in the drop are clearly visible under the stereo microscope with either incident or transmitted illumination. The dendrites of the nerve cells become entangled in the wire, and the isolated cell has therefore to be removed from the wire by shaking in a drop of a suitable buffer. A trained worker should be able to isolate 2–5 single nerve cells within a minute, provide the tissue is rich in nerve cells having many dendrites. Cells without dendrites cannot be isolated by this technique. Liberated cells without dendrites can be sucked up in a suitable capillary under the stereo microscope. For extraction of RNA or DNA,

[1] RICHIER u. HOFFMANN, Harvard Dental Ges., Berlin.

isolated cells are dried on a coverslip and treated as described above. Fig. 3 shows some isolated neurons from human brain tissue.

Groups of ganglion cells, made visible by weak methylene blue staining, can be dissected out with tiny scalpels from various regions, such as the hippocampus, under a stereo microscope; this work is performed on paraffin-covered slides in a drop of a buffer solution. To make scalpels fine enough for microdissection, halved razor blades of untempered steel are broken transversely without damage to the edge with a pair of good-quality pliers (see Fig. 4). The resulting fragments have a cutting edge on the short side and can be fixed to handles, using De Kho-tinsky cement[2] with the aid of a heated platinum wire.

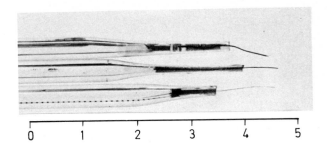

Fig. 2. Steel wires of different diameters (0.05, 0.10 and 0.15 mm, Nikrothal 1 from AB Kanthal, Halls-tammar, Sweden) fitted with glass wax in glass handpieces

Ribonuclease is used to extract RNA fragments from isolated tissue components. If their content of RNA is low, comparatively large amounts of tissue have to be extracted to obtain a given amount of RNA. For example, whole nerve tissue contains only one tenth as much RNA as do neurons (HYDÉN, 1963). There are thus more contaminants which can interfere with the micro-electrophoretic separations, so that special treatment of the extract is required prior to electro-phoresis. The procedure is given in Table 1 and compared with the standard procedure for RNA extraction of isolated cells. A cover with celloidin is used to hold the tissue pieces on the coverslip during the different incubation baths. The incubation with methanol/cyclohexane/carbontetrachloride removes mainly lipids. After treatment with perchloric acid, and incubation with ribonuclease as will be described, the dry extract contains contaminants of a crystalline nature. These are removed by incubation in N,N-dimethylformamide/n-butanol at $0°$ C, the incubation time depending on the amount of crystalline contaminants to be dissolved in the incubation mixture (NEUHOFF, 1966). After washing briefly in ethanol and chloroform, these extracts are treated further according to the standard procedure.

Micromanipulator isolations from sections of fixed and embedded tissue are often more suitable, mainly because the work can be performed on a much

---

[2] CENCO, BREDA, Netherland.

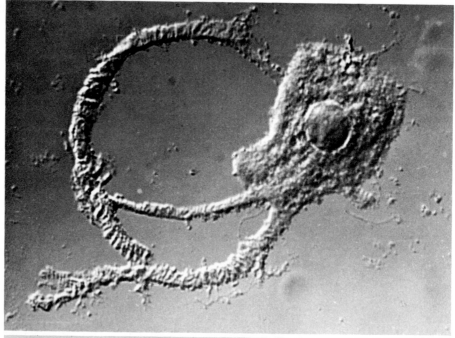

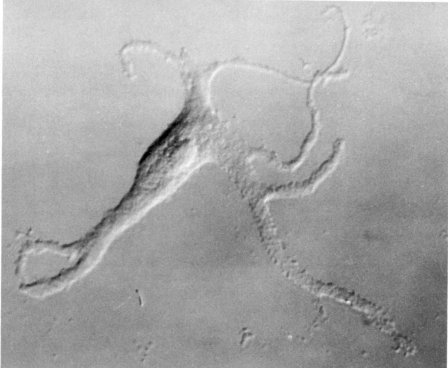

Fig. 3. Freehand-dissected human nerve cells from the nucleus spinalis nervus trigemini. Magnification 240 fold, Nomarski interference contrast

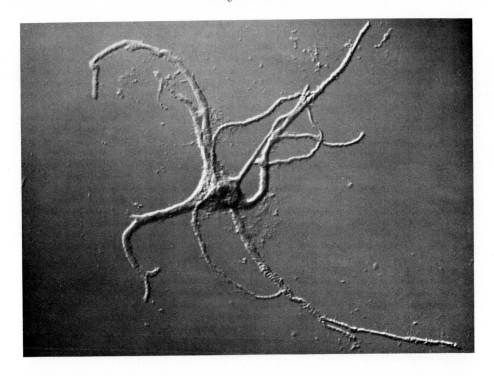

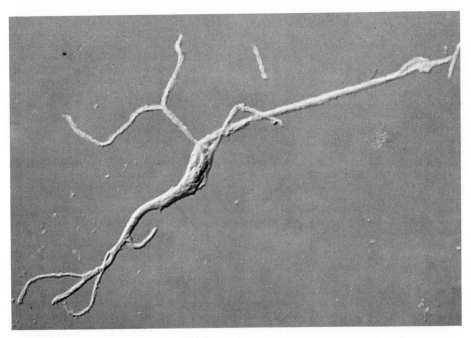

Fig. 3. (Continued)

Table 1. Preparation of isolated nerve cells and glia tissue for microphoresis of RNA bases

Isolated nerve cells	Glia tissue
air drying	Dispense in 0.01 N CH$_3$COOH
↓	↓
HClO$_4$/CH$_3$COOH/C$_2$H$_5$OH/CHCl$_3$	air drying
↓	↓
RNase 3 × 30 min, 37° C (0.4 mg/ml)	Celloidin (0.1 % in ethanol/ether 1:1)
↓	↓
combine extracts	methanol/cyclohexane/carbontetrachloride (1:1:1), 0° C, 2–15 hrs. dry from chloroform
↓	↓
hydrolysis (4 n HCl, 30 min)	HClO$_4$/CH$_3$COOH/C$_2$H$_5$OH/CHCl$_3$
↓	↓
dry	RNase, 3 × 45 min, 37° C (0.1 mg/ml)
↓	↓
take up in 4 N HCl	combine extracts
↓	↓
microphoresis	N,N-dimethylformamide/n-butanol (1:2) 0° C, 5–20 min
	↓
	wash briefly in abs-ethanol
	↓
	dry from chloroform
	↓
	hydrolysis (4 N HCl, 30 min)
	↓
	dry
	↓
	take up in 4 N HCl
	↓
	microphoresis

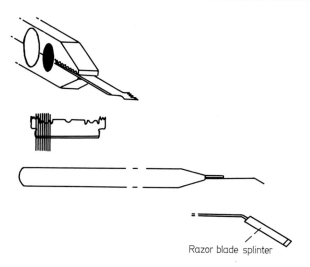

Razor blade splinter

Fig. 4. Production of very fine scalpels from razor blades for microdissection

smaller scale in the oil chamber under high magnification in a phase-contrast
microscope. A suitable fixative is CARNOY's solution. Tissue pieces, 2–3 mm thick,
are placed in a freshly prepared mixture of abs. ethanol, chloroform and conc. acetic
acid, 6:3:1 (v/v/v), for 90 min at room temperature, followed by abs. ethanol, then
benzene, 90 min each, and infiltration with paraffin for 4–18 hrs.

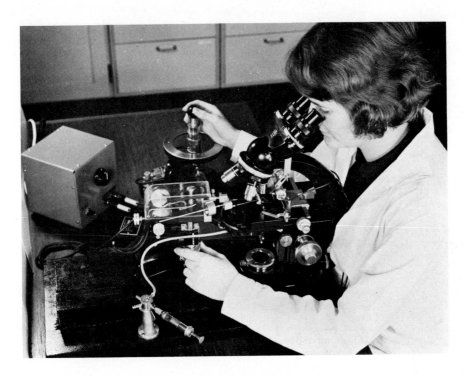

Fig. 5. Work bench for microdissection, consisting of a De Fonbrune micromanipulator and a Zeiss
phase-contrast microscope. The phase-contrast microscope and the receptor of the micromanipulator
stand on a wooden plate to which the syringe for introducing solutions into the capillaries is also fixed.
The lever control and the power supply for the platinum wire heater are fixed to the microscope stand
near the base plate. The investigator has her right hand on the lever control and her left hand on the
screw for making fine vertical adjustments to the receptor

For microdissection, embedded material is sectioned at a thickness adapted
to the size of the components to be isolated, and spread on coverslips. Paraffin is
removed from the sections by treating them with chloroform for 5 min, then
transferring them to abs. ethanol for 5 min and finally rinsing them in 0.01 N acetic acid. Excess acetic acid is blotted off, but the section should remain well
moistened. The coverslip is then placed over an oil chamber with the section
facing downwards, covered with liquid paraffin, so that it will remain moist. For
microdissection, it is better to have the tissue in a polar medium such as 0.01 N
acetic acid rather than an apolar one such as liquid paraffin.

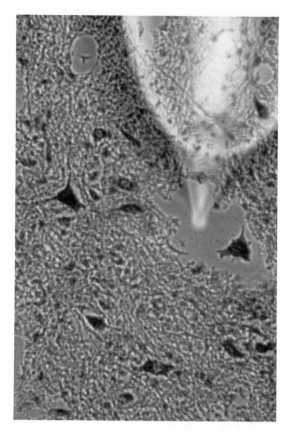

Fig. 6. Histological section of the upper cervical medulla, 20 μ thick, stained with methylene blue. The section is lying on an oil chamber. The cells were dissected free from the surrounding glia using the dissecting needle which is also visible in the figure. Magnification 180 fold; phase contrast

Tests have been carried out to check that CARNOY'S solution and the other media used do not dissolve RNA, and that the acetic acid treatment removes free nucleotides (EDSTRÖM, 1953; EDSTRÖM et al., 1961). The De Fonbrune micromani-pulator[3], equipped with two glass needles, is used to isolate microscopic objects under a microscope with phase-contrast optics. It is essential to use a microscope where focussing is performed with the microscope tube rather than with the stage; fine focussing can, however, be done with the stage. The De Fonbrune manipulator consists of two parts (see Fig. 5): the lever control, and the receptor. The lever is moved horizontally in a manner corresponding to the scaled-down movements of an instrument as seen in the microscope. It can be rotated simultane-ously around its axis to give vertical movements, so that movements in three dimensions can be carried out. The lever controls the receptor pneumatically; the two parts are connected to each other with rubber tubing, so that they can be arranged independently of each other.

[3] Etablissements Beaudouine, 1 et 3 rue Rataud, Paris 5ᵉ.

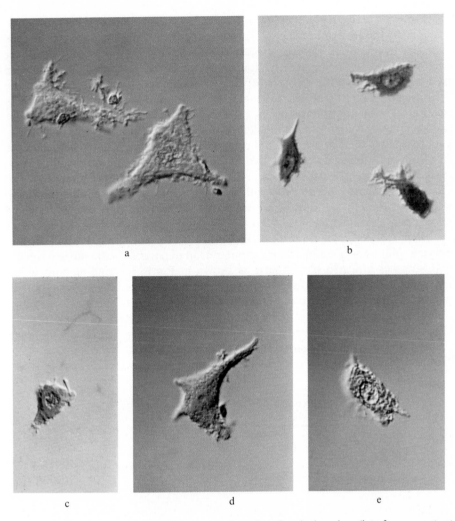

Fig. 7 *a–e*. Isolated fixed and stained nerve cells from the trigeminal nucleus (interference contrast [NOMARSKI]). *a* Feulgen staining 500 ×, *b* Methylene blue (M.B.) staining 400 ×, *c* M.B. staining after RNase treatment 400 ×, *d* M.B. staining after HCl hydrolysis 500 ×, *e* Ninhydrin Schiff staining 400 ×

Isolation of single nerve cells from stained or unstained histological sections, for direct cytophotometric analysis or for RNA extraction, can be performed with 20 µm sections (NEUHOFF, MÜLLER, and TER MEULEN, 1968). The coverslips are inverted over the oil chamber, and the resulting space is filled with liquid paraffin. Cells to be observed under UV light can be taken from embedded slices without first removing the paraffin, which does not interfere with UV spectrophotometric analyses. For the isolation of stained cells, paraffin must be removed from the slices before staining and dehydration. The slices are kept in xylene until they are placed on the oil chamber, and liquid paraffin is added before the xylene has evaporated from the section. The ganglion cells are isolated under a phase-contrast microscope (magnification 120 ×) with a round-tip needle bent upwards, operated by the De Fonbrune micromanipulator. The tissue surrounding the cells is pushed away from the nerve cells with the needle (Fig. 6). When 20–30 cells have been isolated and remain loosely attached to the coverslip, the needle is replaced by a micropipette with a tip of slightly bigger diameter than the isolated cells. The micropipette has a tip which is bent upward, smooth edges, and a constriction (see Figs. 13 and 14) to improve the regulation of pressure. The isolated cells are sucked up by the micropipette together with the paraffin oil. The cells then aggregate in front of the constriction and can be ejected onto a coverslip in a chamber free from oil, where they will lie floating in a small drop of liquid paraffin. More than 30 ganglion cells per hour can be isolated in this way. For cytophotometric measurements, the paraffin drop with the cells is covered with another coverslip and sealed in with Harvard glass wax. Fig. 7 shows isolated, fixed, and stained nerve cells as an example.

When isolated cells are to be treated with RNase or DNase, a small drop of enzyme solution is introduced into another drop of paraffin on the same coverslip and becomes attached to the glass as shown in Fig. 8. A small amount of the enzyme solution is transferred with a micropipette steered by the micromanipulator to the paraffin drop containing the isolated cells. These cells are then pushed with the capillary from the paraffin phase into the enzyme drop, in which they become hydrated and will remain. The oil chamber is then filled with liquid paraffin and new enzyme solution is added to the enzyme drop containing the cells until it is increased to at least ten cell volumes and the cells are floating freely.

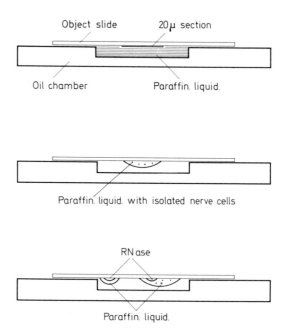

Fig. 8. Schematic drawing of the preparation procedure for the treatment of cells isolated from histological sections with DNase or RNase. Details see text

The incubation and repeated treatment with RNase is performed as described later; during this procedure the cells keep their shape and remain suitable for further cytophotometric analyses (cf. Fig. 7). After the last incubation the enzyme solution is completely removed. After removal of excess liquid paraffin the slide is placed on another slide and sealed with wax.

Preparation of Microinstruments and Their Use

Dissection needles: The needles are prepared from 3 mm dia. soft glass rod, pulled to give a piece 1 mm thick and 5 cm long, connected to 3 cm of the 3 mm thick piece. The end of the thin part is placed in contact with the heated platinum wire of the de Fonbrune microforge[4] (Fig. 9) and the molten glass is pulled at an angle of 50–60° from the axis to form a cone, the tip of which is relatively blunt, but which is sharpened by a renewed pull after a very light contact, during which the air blower of the microforge is used to obtain more localized heating. The tip should be still sharp, but not much extended. The needles are fastened onto adapters which are attached to the micromanipulator. One of the adapters fits onto the axis of the receptor part of the manipulator, and it is the movements of this axis that can be controlled in all dimensions by the lever control. It is not enough, however, to use only one needle for micro dissection because the object has to be held securely while it is cut or torn, etc. For this purpose the receptor is provided with a special holder for the other adapter. Both needles are thus fitted onto the receptor, one being under lever control, and the other not. The receptor itself is provided with controls for horizontal and vertical movements which, of course, move both needles at the same time. This is an advantage of the De Fonbrune micromanipulator and is preferable to arrangements where two needles are each, independently, under lever control. For micro dissection the two needles are placed so that their tips meet and they are held in the center of the viewingfield. If the microscope has been focused on the objects to be dissected in the oil chamber, the needles can be safely introduced after they have been lowered with the aid of the common vertical control screw at the base of the receptor. This screw is vital for the proper functioning of the manipulator during microdissection and gives enough fine control of the movements. There are De Fonbrune manipulators where, unfortunately, a much coarser control has been substituted for the screw in the original French-made manipulator, resulting in loss of an essential function for micro dissections. For dissection, the holder needle is placed against an object with the aid of the microscope controls and the vertical control screw on the receptor. The movable needle, being controlled independently, can be moved out of the way during this step and is subsequently used for dissection, cleaning, etc. During dissection the operator's right hand controls the lever (see Fig. 5), and the left hand the vertical control screw on the receptor, and the hands are used alternately for focussing and moving the microscope stage. The micromanipulator and microscope should be adjusted so that the work is always performed with the lever in a central position and the instruments in the center of the viewfield.

---

[4] See footnote 3, p. 222.

Fig. 9. De Fonbrune microforge with a capillary ready for working

Micropipettes: For all micropipettes, 8–9 mm dia. Pyrex glass tubing with a maximal wall thickness of 1 mm is used. It is pulled to give 0.5–1 mm capillaries; any increase in relative wall thickness by excessive heating must be avoided. The burner for pulling capillaries must allow fine regulation of the flame, which is achieved by mixing oxygen, air and gas to produce the right temperature. Pyrex glass tubing of approximately 10 cm length is heated in the flame so that a 1 cm section in the centre is brought to red heat. The tube is then pulled out by full extension of the arms, *outside the flame.* Even heating and rotation of the glass tube will produce a circular lumen and a constant wall thickness of the capillary; only such material is usable. The middle section, approximately 100 cm long, of the pulled capillary is cut into 15 cm lengths. A 10 cm capillary is heated and rotated close to an alcohol flame (70–80% ethanol), 2 cm from one end and pulled to give a short local thinning about 0.5 mm thick, and is then bent to form a hook distal to the thinning (see Fig. 10 b). This part of the capillary production is difficult but important for the quality of the micropipette. The straight end of the capillary is inserted into a 3–4 mm dia. glass tube bent to a right angle with arm lengths of 3 and 6 cm and the capillary is sealed into the long arm with paraffin, picein or pitch (see Fig. 10 c); this serves as a holder which can be adapted to the micromanipulator. The hook should be turned in the direction of the short arm of the holder.

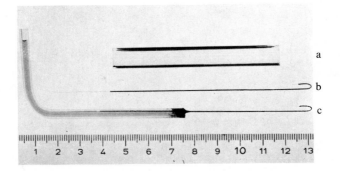

Fig. 10. Fine capillaries are pulled out from the 9 mm Pyrex glass tube *a*. One end of each capillary is tapered in the spirit burner and bent to a hook *b*. This capillary is fixed into a glass holder with pitch, and further worked on the microforge *c*

Fig. 11. Weights of rolled lead are hung on the hooks of the glass capillaries (Fig. 10, *b*) for working on the microforge (compare also Fig. 9)

A micropipette of microscopic dimensions is obtained as follows: a 0.05–3 g weight is hung on the hook, the capillary held at 35° to the vertical axis and the capillary thinning is moved close to the heated platinum wire for 20–30 sec (moderate heat to give a tip which is not too extended). Heating causes the capillary to soften and stretch until it breaks. The weights can easily be produced from rolled lead as shown in Fig. 11. The following relations exist between the weight, the diameter of the platinum wire (average diam. 300 μm), the heating temperature, the diameter of the capillary tip, and the length of the thin part of the pipette: a) the inside diameter of the capillary tip is proportional to the weight at a given temperature of the heating wire; b) with a given weight, a higher temperature gives a longer and thinner pipette tip; c) with a given weight and heat, the length of the thin part of the micropipette increases with increasing diameter of the platinum wire. A short pipette is produced by a heavy weight and a thin heating wire, and a long pipette by a light weight and a thick wire. The hot platinum wire of the microforge must *never* touch the glass capillary. Fig. 12 demonstrates the pro-

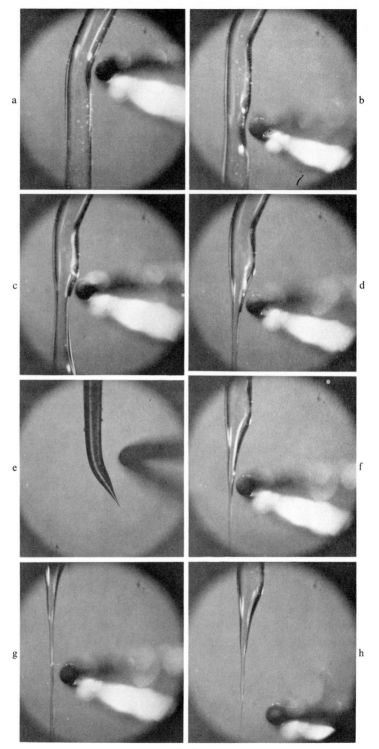

Fig. 12

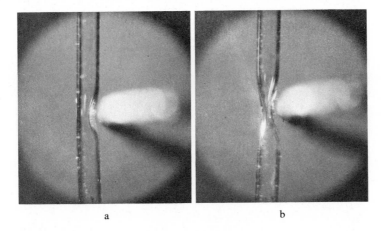

a   b

Fig. 13 *a* and *b*. Production of a micro-constriction pipette as seen by the operator through the microscope of the microforge. For details see text

duction of microcapillaries as seen by the operator through the microscope of the
microforge. After some experience a capillary can be produced with the required
shape. The standard pipette (Fig. 12, *a–e*) should have a continuous taper towards
the tip to give good elasticity. Fig. 12, *f–h* shows the production of a capillary
pipette with stepwise thinning of the capillary tip (see also Fig. 14). Capillaries
made in this way are more stable. The stepped shape of the capillary tip makes it
possible to shorten the tip slightly, on the edge of a coverslip, if it has become
blocked, without danger to the contents of the capillary.

Micro-constriction pipettes: These pipettes are prepared in much the same
way as described above. The preformed capillary is placed vertically in the holder
of the microforge and a weight of about 50 mg is attached to the hook to prevent
the capillary from bending towards the heated wire during the subsequent treatment. The very hot wire is brought towards the vertically positioned capillary
until the wall of the latter constricts (Fig. 13, *a*). The capillary is then rotated
through 180° on its axis and the procedure is repeated until most of the lumen is
obliterated (Fig. 13, *b*). The heater is then switched off and a heavier weight
(2–3 g) is attached to the hook. The capillary is next bent to an angle of 35° about
1–2 mm below the stricture, and then pulled out as described above. If constriction
pipettes are to be used for single cell preparations from histological sections, the

←————

Fig. 12 *a–h*. Production of microcapillaries as seen by the operator through the microscope of the microforge. In *a* the capillary is first bent. For this the instrument holder of the microforge is turned somewhat
to one side, and the hot platinum wire approached to the glass. The lead weight hanging on the capillary
causes it to bend. This procedure is repeated until the required angle is obtained. In *b* to *e* the capillary
tip is pulled out as required. For particulars, see text. *f* to *h* illustrate the production of a capillary
pipette with stepwise thinning of the capillary tip (compare also Fig. 14)

tip diameter should be more than 50 μm. The tip should not be too long and is pulled at the lowest possible temperature to give a distinct step (cf. Fig. 12, g). The tip of the pipette is broken off with the cold platinum wire (Fig. 14) to give an aperture of the required diameter. The edge can be smoothed off by moving the platinum wire, kept at low temperature, close to the tip from below. The diameter of the aperture can be measured with a measuring eyepiece in the microscope of the forge.

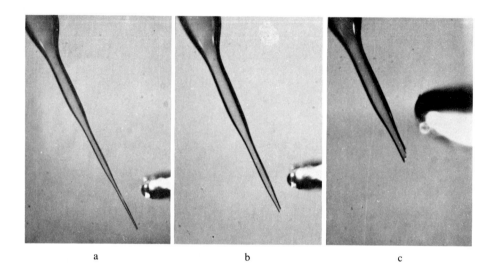

a   b   c

Fig. 14. *a* Stepwise thinning at the tip of a constriction pipette; *b* and *c* show how the tip is broken off with the cold platinum wire so as to give the required aperture diameter

The pipettes are filled with liquid paraffin in their distal parts and used only after they have been stored with liquid paraffin in the tip for at least a day. This gives the glass surface suitable hydrophobic properties. Although the micro-pipettes are maneuvered pneumatically, smooth operation can be achieved provided that the glass thickness is small, and that the pipettes are kept clean and hydrophobic with paraffin. After use, they are rinsed in 4 N hydrochloric acid and all aqueous solutions are expelled. Rapid passage of liquid into or out of the pipette should be avoided so as not to break the meniscus between the hydrophobic and the aqueous phases.

Volumetric pipettes: Various procedures are available for the production of volumetric pipettes. For one procedure (EDSTRÖM, 1964 b), one needs pipettes which are long, non-tapering, and have less inclination (about 10°) (Fig. 15) than the standard pipette. Such pipettes are obtained if the heating is carried on as late as possible before the expected breaking, and is subsequently applied to a region about 2 mm towards the hook. Volumetric pipettes are provided with graduations in the following way: a little India ink is shaken with liquid paraffin to give micro-

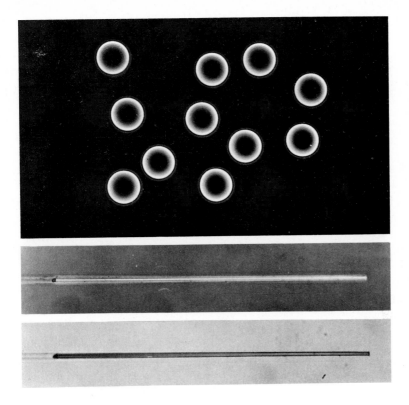

Fig. 15. A calibrated micropipette filled with aqueous liquid to the calibration mark, bright field and phase contrast, magnification 132×, and eleven 54 100 μ^3 spheres of oil in glycerol delivered by the pipette. Phase contrast, magnification 235×. [From EDSTRÖM, Biochim. biophys. Acta (Amst.) **80**, 399 (1964)]

scopic drops; the oil chamber is filled with this mixture and one or more drops of ink are allowed to fall in suitable positions on the measuring pipette. After drying in air the resulting spots form useful markings on the pipette. To calibrate a volumetric pipette, a column of liquid paraffin is introduced into the pipette, separated at the rear from other paraffin in the pipette by a moderate volume of water at the position of a marking, and followed at the tip a minute water volume. The paraffin trapped between the two lots of water is expelled into a droplet of glycerol under a coverslip in the oil chamber (see Fig. 15) and the diameter of the resulting sphere is measured. Predetermined volumes cannot be obtained by this technique but this is usually not essential.

In an alternative procedure, micropipettes with a stepped tip are used for the production of volumetric pipettes. If the step segments are relatively long, they normally have an inside diameter that is similar in different parts of the capillary. Parts with an even diameter can be produced by moving the heating wire continuously upwards towards the thicker part of the capillary while the softening capillary is pulled downwards by the weight in the hook. Calibration markings

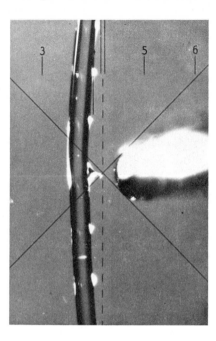

Fig. 16. Production of calibrations on a measuring pipette. A measuring eyepiece is fitted on the microforge. The bright spots on the capillary wall are fine calibration marks made with the warm platinum wire. In the cross-hairs of the measuring eyepiece a somewhat larger calibration mark can be seen

can be applied with the heating wire at regular intervals with the aid of a measuring eyepiece (see Fig. 16). A very fine platinum wire is heated to a temperature at which it will adhere slightly to the glass wall on contact. When the wire is removed, a tiny projection of glass is left on the capillary wall. The diameter of the capillary, and the distance between two or more glass tips are measured with the measuring eyepiece; the volume can be calculated, or determined as described above.

The determination of micropipette volumes in the nano-litre and even pico-litre range can also be performed with standard pipettes using the principle used for wet-weight determination of small tissue samples (cf. p. 403) (NEUHOFF, 1971; RÜCHEL, REHMANN, and NEUHOFF, 1973). Very small quantities of fluid can be transported in microcapillaries closed at one or both ends with liquid paraffin and silhouetted onto paper by means of a microscope using a suitable magnification and high-power illumination (see Fig. 17). If the volumes to be measured can be contained entirely within the region between the capillary tip and the inflection point P of the contour curve (Fig. 18) the problem reduces to the calculation of a hyperboloid of revolution. Nevertheless, the use of a liquid paraffin seal as described above necessitates corrections for one or two menisci (Fig. 18).

Calculations: If the contour curve of the capillary tip in the measurement region approximates to a hyperbola given by the equation $y = \dfrac{a}{x} + h$ $(a \geqq 0, h \geqq 0)$.

Fig. 17. Apparatus for projection of a capillary volume. On the right, the high-power illumination. The microscope tube has an adjustable prism at the top. The silhouette of the capillary is projected onto a piece of paper on the wall and the contours of the filled piece are traced

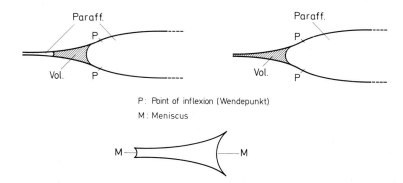

Fig. 18. Schematic drawing showing how a microvolume is arranged in a capillary tip for determination of the volume

Asymptotes of the hyperbola are $x=0$ and $y=h$, a is the area of a square specific for each hyperbola of the family, and h is the distance from the asymptote to the x-axis.

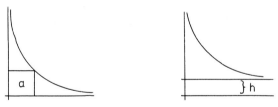

The volume of the body produced by rotating a curve $y=F(x)$ $(F(x)\geq 0)$ about the axis can be calculated between x_1 and x_2 $(x_1\leq x_2)$ by

$$V_{rot}=\pi\int_{x_1}^{x_2}(F(x))^2\,dx.$$

For the family of hyperbolae $y=\dfrac{a}{x}+h$ this yields:

$$V_1=\pi\int_{x_1}^{x_2}\left(\frac{a}{x}+h\right)^2 dx$$

$$=\pi\int_{x_1}^{x_2}\left(\frac{a^2}{x^2}+2ah\cdot\frac{1}{x}+h^2\right)dx$$

$$=\pi\left[a^2\left(\frac{1}{x_1}-\frac{1}{x_2}\right)+2ah\log\frac{x_2}{x_1}+h^2(x_2-x_1)\right],$$

since $y_i=\dfrac{a}{x_i}+h$ $(i=1,2,3,\dots)$ and $d=x_2-x_1$, then

$$V_1=\pi\left[a(y_1-y_2)+2ah\log\frac{y_1-h}{y_2-h}+h^2 d\right].$$

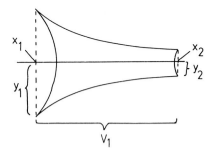

The volume correction for the larger meniscus, V_2, is obtained by rotation of a circular arc $y=\sqrt{r^2-(x-x_0)^2}$. The appropriate circle lies symmetrically on the x-axis, r is the radius and x_0 determines the position of the center along the x-axis. The integral from x_1 to x_0+r gives the volume V_2 corresponding to the large meniscus.

$$V_2 = \int\limits_{x_1}^{x_0+r} \left(r^2 - (x-x_0)^2\right) dx = \pi \left[r^2 (x_0 + r - x_1) - \tfrac{1}{3}\left(r^3 - (x_1 - x_0)^3\right)\right].$$

Under the restriction that the circle and hyperbola touch at x_1, one can eliminate x_0 and r because the functions are equal at the point of contact:

$$y_1 = \sqrt{r^2 - (x_0 - x_1)^2} = \frac{a}{x_1} + h.$$

The values of the derivatives at x_1 are equal.
    Let the slope of the arc at x_1 be

$$-\gamma = \frac{x_0 - x_1}{\sqrt{r^2 - (x_0 - x_1)^2}} = \frac{x_0 - x_1}{y_1}.$$

Therefore
$$-\gamma = -\frac{a}{x_1^2} = -\frac{(y_1 - h)^2}{a}$$

(slope of the hyperbola at x_1).
    Then, with the formula for V_2, one obtains:

$$V_2 = \frac{\pi}{3} \cdot y_1^3 \left[2(1 + \gamma^2)^{3/2} - 2\gamma^3 - 3\gamma\right]; \qquad \gamma = \frac{(y_1 - h)^2}{a}.$$

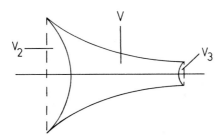

The volume of the small meniscus V_3 can be simplified to the formula for a hemisphere of radius y_2:

$$V_3 = \frac{2\pi y_2^3}{3}.$$

The slope γ of the hyperbola can be regarded as being zero here.
    The formulae for V_1 and V_2 are also applicable for $h < 0$ within the region $\frac{a}{x} + h \geq 0$. This restriction is automatically fulfilled for h in practical applications. By subtraction $V_1 - V_2 - V_3 = V$ one obtains the volume of the filled capillary tip segment.
    Templates can be made (Fig. 19) for a family of hyperbolae of the formula $y = \frac{a}{x}$ for the identification of the capillary contour curve. "a" will be set at $1/1.5/2/2.5/\ldots/11$.

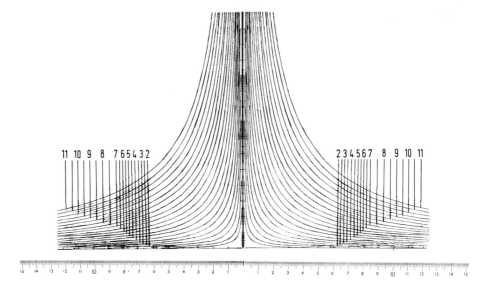

Fig. 19. Template of hyperbolae for the calculation of capillary volumes

Each hyperbola is computed at a few points, reflected across the x-axis, and plotted (length unit 1 cm). The construction is reduced photographically to length units of for example, 0.8/0.7/0.6/0.5/0.4/0.3 cm, and transferred to flat film as a positive. Using these templates the variables a, y_1 and y_2 can be obtained. For the case when the symmetry axis of the template and the axis of the capillary tip are parallel,—translated with respect to each other, the distance h (see above) can be determined.

The capillary contour curve can be generated more elegantly by the use of an enlarger, instead of by templates with fixed steps of length, and the family of hyperbolae on a slide film can be fitted exactly and continuously to the capillary contour.

Procedure for the Measurement of a Volume

The quantity of aqueous solution to be measured is sucked up into the capillary containing liquid paraffin, in such a way that the meniscus does not pass the inflection point of the contour curve (see Fig. 18). If possible, the capillary tip is closed with more liquid paraffin to bring the sample into the most favourable region of the capillary (as in "c").

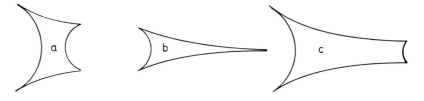

In example "a" the contour curve is so short that identification of the value of "a" with the template is difficult. In example "b" the value of y_2 cannot be measured with certainty.

The filled capillary is moved with a micromanipulator into a chamber containing liquid paraffin, and placed in the optical path of a microscope. With the help of a high-intensity lamp the silhouette of the capillary tip is projected through a prism onto a sheet of paper (see Fig. 17), so that the outline of the sample volume of aqueous solution comes optimally into view and can be traced. By the use of one of the templates described above, being careful to align it with the symmetry axis of the capillary, the hyperbola approximating to the capillary contour and hence the value of a are found. In every case the values of y_1 and y_2 are also measured in the units of length of the particular template. If a hyperbola approximating sufficiently closely to the contour can only be obtained by a parallel translation of the x-axis (symmetry axis of the template), the distance between these axes must be substituted for the value of h in the calculation, and the distance between x_1 and x_2 must be used for the value of d. In this calculation the optical enlargement factors must be taken into consideration; for example when a $40 \times$ objective is used with a $25 \times$ ocular at a projection distance yielding a factor $2.24 \times$ (total magnification $2240 \times$), a volume of 10^{-13} liter ($100 \, \mu^3$) is still measurable.

To check the method, small volumes of a aqueous solution of ^{14}C-glucose of known specific activity were taken up into capillary tips, projected, and evaluated. The glucose solution was blown out onto a cover slip, dried, and dissolved in 1 ml of ethanol in a counting vial; for scintillation measurements a suitable quantity of scintillation fluid was added. Comparison of the volumes obtained from the scintillation count rate with those calculated by capillary projection yielded, for a volume of 1.43×10^{-7} ml, an average error of about 5%. Smaller volumes, down to 10^{-12} ml, are measurable, but the obtainable counting rates are too small for comparison and have high deviations.

For example: The same volume (aqueous solution of ^{14}C-glucose) was drawn up into the tip of a microcapillary and projected at various levels in the capillary:

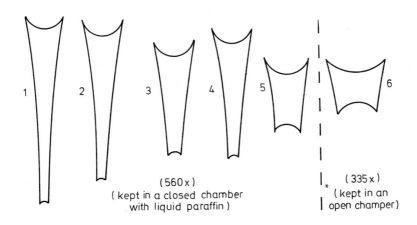

(560×)
(kept in a closed chamber
with liquid paraffin)

(335×)
(kept in an
open champer)

The following volumes were obtained by use of the formula derived above:

1   1.404×10^{-7} ml 97.6%   of 1.44×10^{-7} ml (100%) volume determinated
2   1.483×10^{-7} ml 103 %   by scintillation counts)
3   1.505×10^{-7} ml 104.6%
4   1.501×10^{-7} ml 104.2%
5   1.411×10^{-7} ml   98 %
6   1.586×10^{-7} ml 110* %

Extraction of Nucleic Acids

After isolation, the biological samples are transferred by means of the micro-needles or the constriction capillary to a second coverslip on the oil chamber, and their RNA and/or DNA is extracted. A standard micropipette, about 10 μm wide at the tip, is used to distribute and remove incubation volumes (Fig. 20).

The micropipette is connected by rubber tubing to one of the outlets of a 2 ml "Inaltera" syringe provided with a three-way (or two-way) stopcock[5]. The

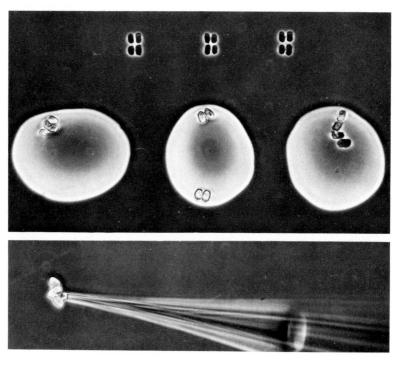

Fig. 20. Isolated nuclei before and during incubation in drops of aqueous solution, as seen under phase contrast in the oil chamber. The lower figure shows how an extract is taken up with a standard micro-pipette. Magnification 228 ×. [From EDSTRÖM, Biochim. biophys. Acta (Amst.) **80**, 399 (1964)]

* Overestimated because of the magnifying effect of the glass walls of the tube.
[5] Georg A. Henke, Tuttlingen, Germany.

stopcock itself is fastened to a cone-shapened brass stand fixed to a wooden plat-
form common for the microscope and the micromanipulator, and the stopcock
body with the syringe can be moved around this axis (see Fig. 5). Air can be
aspirated or ejected through the free outlet(s) and the micropipette is operated
by changes in air pressure.

For the extraction of RNA, a solution of ribonuclease, containing volatile
buffer electrolytes, is employed. Protease-free pancreatic ribonuclease (A) at a
concentration of 0.4 mg/ml is dissolved in 0.2 M ammonium bicarbonate + acetate
buffer, pH 7.6 (1.58 g ammonium bicarbonate in 100 ml dist. water, adjusted with
0.2 M acetic acid to pH 7.6, with a few drops of chloroform added). The solution
is stored in the refrigerator; the pH is checked before use and adjusted with acetic
acid if necessary. Freshly dissolved enzyme is always used.

The isolated tissue components are digested for 3×30 min at $37°$ C. A volume
of enzyme solution is placed with a standard type micropipette on each sample.
The incubation volumes are of the order of at least ten times the volume of the
sample to be extracted. After 30 min at $37°$ C, the incubating fluid is aspirated off
and removed (see Fig. 20). Volumes prepared in advance are then used for renewed
incubations, so that the pipette will be rinsed with the volumes for the second and
third extractions after it has been used to remove the first and second incubation
mixtures, respectively. A coverslip is placed on the oil chamber slide close to the
incubation coverslip, but is left untouched by liquid paraffin (a distance of 1–2 mm
is necessary to prevent the paraffin spreading over to the dry glass (see Fig. 1)).
Extracts are evaporated to dryness on this glass and extracts from the same sample
can be placed close together or combined.

Tissue for DNA analysis must be freed from RNA by prior treatment with
ribonuclease. The procedure is satisfactory for isolated chromosomes or free
nuclei; whole tissue can also be used, or cells with a relatively high DNA content
(e.g. material from thymus, testis, and cerebellar granular cortex). The samples are
incubated with 1 N hydrochloric acid at $37°$ C for 18 hrs. Liquid paraffin which
has been kept over hydrochloric acid is used in the chamber. In order to minimize
evaporation from the microscopic incubation volumes, the oil chamber is placed
in a small petri dish, which is itself placed in a larger one. Both petri dishes contain
filter paper soaked with 1 N hydrochloric acid, and the bottom of the oil chamber
is covered with filter paper soaked in hydrochloric acid. The samples are extracted
only once, since it is not necessary to collect the liberated purines quantitatively,
and the extracts are evaporated to dryness on a coverslip. For base analysis the
extracts are redissolved in 1 N hydrochloric acid in an oil chamber. Contact of
the dried extracts with liquid paraffin for more than a few minutes must be avoided
because of the solubility of the purines in oil.

Medium for Electrophoresis

The supporting medium for the electrophoretic separations of purines and
pyrimidines consists of cellulose regenerated by the cuprammonium method. It is
available as sheets (cellophane) or fibers (rayon silk).

The fibers were obtained from Farbenfabriken Bayer AG., Dormagen, Germany under the designation *Cupresa HW, Naturglanz, ungedreht, im Strang* with denier values for the individual fibers (filaments) of 0.81 or 0.44. (The latter was prepared specially for microphoresis.) Rayon silk threads containing several dozens of fibers are cut into 20–30 pieces, 2 cm long, which are moistened in distilled water, transferred to 1.5 N sodium hydroxide, and put in 2.25 N sodium hydroxide after 2 min, where they are kept for 5 min at 10° C with gentle stirring by means of a glass rod. The alkali solution containing the fibers is poured into a larger vessel containing distilled water; the fibers can then be removed by means of a glass rod and rinsed in three changes of distilled water, 2 min each. Excess moisture is blotted off and they are then put in the buffer for electrophoresis. The preparation is shaken for an hour and then placed in the refrigerator. Fibers can be stored for a couple of months after their preparation. The thicker fibers (0.81) are suitable for analysis of RNA in the range 500–1 000 $\mu\mu$g, the thinner ones (0.44) for 250–500 $\mu\mu$g.

The alkali treated fibers are stored in the viscous buffer solution as bundles of entangled fibers. A single fiber can be obtained by pulling a bundle apart. The free end of a fiber is caught with the tip of a needle, and the rest of the fiber is untangled slowly. Excess moisture is removed by allowing the fiber to come into contact with a clean glass surface several times, to leave wet tracks. When only thin and even tracks can be seen (after 6–7 contacts) the fiber is placed, slightly stretched, parallel to the long sides of a clean quartz slide, 24 × 30 × 0.5 mm.

Alternatively, microphoresis can be carried out in cellophane strips. The starting material is "Cuprophan" foil, Cuprophan-Normalware "M"/150 PT = 10 μ from Bemberg[6], or a cellulose-hydrate foil free from wetting agent from the same suppliers ("Probefolie ohne Weichmacher") 19 μ thick.

Sections 25 × 10 mm are cut out of the Cuprophan foil after its main direction of swelling in alkali has been determined. The sections are cut so that the swelling is parallel to the short side. The cellulose-hydrate foil has no pronounced swelling asymmetry and can therefore be used directly. In all cases untreated 25 × 10 mm pieces are wetted briefly, dipped in glycerol, and sandwiched between membrane filters[7]. Care has to be taken that no air bubbles are trapped between the layers, as such regions will take up paraffin during the embedding, which must be avoided. The cellophane piece (25 × 10 mm) is spread onto the membrane filter without creasing, a second membrane filter is spread on top, and the sandwich is pressed together with a glass rod. Excess glycerol is removed from the cellophane and absorbed by the filter by this procedure. The piece of cellophane with the filter is subsequently embedded vertically in paraffin (histological quality, solidification point 56–58° C) as shown in Fig. 21. During this procedure the temperature of the paraffin must not exceed 70° C. The temperature is optimal when the molten paraffin is about to solidify. Small plastic boxes with thin walls are used for the embedding, while the strip, coated with membrane filter, is held vertically with forceps. The box is immediately transferred to ice-cold water. The paraffin blocks containing the strips are easy to remove from the boxes after some time in a deep freezer. They are sectioned on a microtome at 40–80 μm, so that each section

---

[6] I. P. Bemberg AG, Wuppertal-Barmen, Germany.
[7] Sartorius Membranfilter GmbH, SM 11 200, Göttingen, Germany.

contains a strip from the long side of the foil. These sections are relatively thick and a heavy microtome is needed to produce sections with straight edges; if the blocks are sectioned on too light a microtome the vibration of the cutting knife produces a jagged edge. The strips, kept in a water bath at room temperature, are removed from the paraffin sections with fine steel wires, under a stereomicroscope. A simpler procedure is to place paraffin sections in a large beaker of water and stir them slowly with a glass rod. The strips will sink to the bottom of the beaker. The recovery of the strips is less in this procedure. The strips may be stored for several months in glycerol at $+4°$ C for future use.

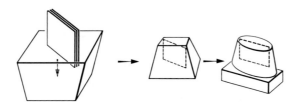

Fig. 21. Embedding of the membrane-coated cellophane strips in paraffin, and the preparation of the paraffin block for sectioning

For swelling, the strips are placed on a fine sieve, which must be alkali-resistant (e.g. a fine copper sieve), and treated as follows: 3 min in distilled water, 5 min in 1 N sodium hydroxide, 5 min in 2 N sodium hydroxide, and 2×3 min in distilled water. They are subsequently incubated for 20 min at $37°$ C in the buffer for electrophoresis, and should be stored for several days at $4°$ C in the buffer before use. They can be kept for several months before use.

For electrophoresis, the strips are removed singly from the buffer with a fine soft forceps (see Fig. 10 on p. 16), and once again placed on a membrane filter to remove buffer adhering to the strip. The strip is dipped briefly into 4 N hydrochloric acid/glycerol (5 ml + 4 drops) to remove buffer completely. It is then spread on a quartz slide ($25 \times 30 \times 0.5$ mm) and pushed with a razor blade so that it lies parallel to the long edges of the quartz slide at a distance of 4–5 mm from one edge.

Buffers for Electrophoresis

The buffers used for electrophoresis differ from the traditional ones in several respects. Diffusion of the migrating bands has to be reduced to an extent proportional to the reduction in the dimensions; this is obtained mainly by increasing the viscosity of the buffer. A high content of glycerol and glucose is used for this purpose. The increased resistance to electrophoretic migration is overcome by increasing the voltage.

Sharp separations are obtained only when the concentrations of buffer electrolytes are adjusted to the concentrations of the migrating compounds.

Because the latter are relatively high it is necessary to use high concentrations of electrolytes. This results in increased conductivity and heating, which is aggravated by the high voltage, but the temperature of the fiber or strip is little influenced because of its high surface to volume ratio. Two buffers are used, one for fibers, the other for strips. They are prepared as follows.

Buffer for fibers: 20 ml 8 N sulphuric acid, 8 ml water, 33 g glycerol (sp gr 1.26), and 72 g water-free D-glucose are mixed in a flask and heated with stirring on a water bath at 100° C. After 20 min the flask is removed, cooled, and stored at − 20° C.

Buffer for strips: 18 ml of conc. sulphuric acid (sp gr 1.84), 10 ml water, 66 g glycerol (sp gr 1.23) and 72 g of water-free D-glucose are mixed in a flask and heated with stirring on a water bath. After 20 min the flask is removed, cooled, and stored at − 20° C.

Micro-Electrophoresis (Microphoresis)

The quartz slide is placed on top of a chamber without oil, with the fiber or strip facing downwards behind a coverslip with hydrolysates kept over liquid paraffin (see Fig. 22). For the application of the hydrolysates, the fiber or strip is dried by a small electrical heater fastened to the microscope stand (cf. Fig. 5). This is also used to concentrate the hydrolysates by evaporation during their application. The heater consists of a 0.6 × 60 mm platinum wire connected to a variable transformer 220 V/3 V, maximum output 45 W.

Hydrolysates, taken with a standard micropipette, are ejected slowly, with continuous evaporation of the volatile constituents (water, hydrochloric acid) onto suitable points along the fiber or strip. Several hydrolysates can be run at the same time. To get a small area of application it is advisable to apply the hydro-

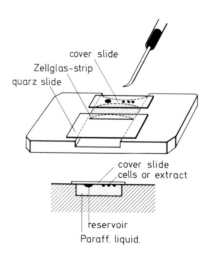

Fig. 22. Arrangement of an extract and of a microphoresis strip on the oil chamber

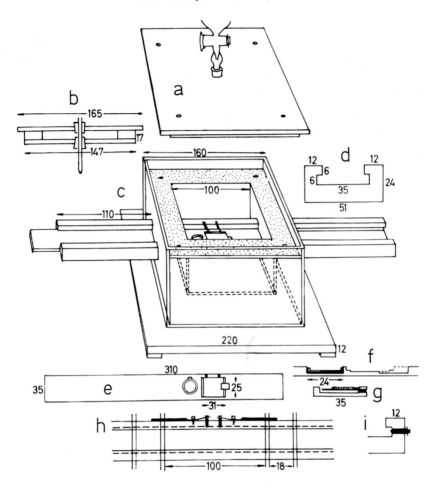

Fig. 23 *a–i*. Constant-humidity chamber for microphoresis: *a* the lid in perspective; *b* cross section through the lid at the center; *c* the chamber in perspective with guide and slide in position for microphoresis; *d* cross-section of the guide; *e* the slide with trough and cup for collecting liquid paraffin; *f* longitudinal section through the slide along the axis. The profile of the trough at an out-of-axis section is indicated by a broken line; *g* cross-section of the slide through the trough and one of the brass nails. A quartz slide in position, with a buffer bridge, is also shown; *h* view from above of the central part of the guide and the electrical connections; *l* cross-section of the guide through one of the brass nails. *d*, *f*, *g* and *i* have been drawn twice the size of *b*, *e* and *h*. *a* and *c* are at approximately the same scale as the latter three. [From EDSTRÖM and PILHAGE, J. biophys. biochem. Cytol. **8**, 44 (1960)]

lysate to the strip in several small portions. For the localization of the application points, a tiny drop of liquid paraffin can be placed on the empty quartz slide at suitable points close to the strip before the application of the hydrolysate. The fiber always has minor variations in width, but usually has many sections of uniform diameter. The application points are placed so that the distances travelled will be over uniform parts of the fiber. The strip has the advantage of being of uniform width.

During the isolation of individual fibers, and during the application of hydro-lysates, the fibers are in direct contact with the air. The relative water content of the buffer (i.e. the mole fraction of water) is a function of the relative humidity of the surrounding air, since an equilibration will soon be reached because of the small fiber dimensions. Since the humidity affects the water content of the buffer, it will also influence its viscosity, which is strongly dependent on water content. It is essential that the water contents and viscosities are reproducible during the runs. This is achieved by using a constant humidity chamber in which the fiber or strip is placed after it has been charged with hydrolysate. A saturated solution of zinc nitrate on the bottom of the chamber (Fig. 23) gives a constant humidity of 42%, so that the amount of water originally present in the buffer has no direct influence on the final viscosity.

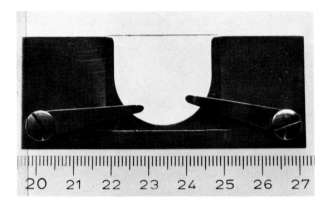

Fig. 24. Metal holder for the quartz slides for microphoresis

The metal electrodes of the chamber (Fig. 23, *h*) must make contact with the quartz slide close to the two ends of one of its long sides, so buffer bridges are placed to connect these points to the ends of the fiber or strip. The buffer bridges are made from a paste obtained by mixing finely-powdered silica oxide (SiO_2) with the buffer. After 5 min or 3 min of equilibration for thick or for thin fibers respectively, the slide is covered with liquid paraffin from a funnel mounted in the lid of the chamber (Fig. 23, *a*), and 1 500 V/cm D.C. are applied, for about 6 or 9 min respectively from a variable 500–4 000 V D.C. supply, with built in galvano-meter, connected to the chamber. After the run, the quartz slide is placed in a metal holder (Fig. 24) for observation under the ultraviolet microscope. If desired, electrophoresis can be continued after a preliminary inspection under the ultra-violet microscope.

Photographic and Photometric Measurements

The separations are inspected and photographed in an ultraviolet microscope at a suitable wavelength in the 260 nm region, and the proportions of the separated

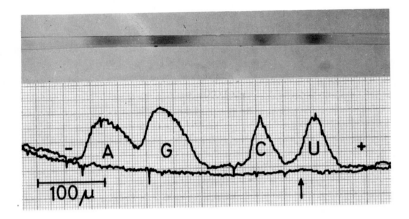

Fig. 25. A microphoretic separation scanned at 257 nm, of a hydrolysate containing about 350 μμg of RNA products from isolated sensory ganglion cells. A, G, C and U stand for adenine, guanine, cytidylic acid and uridylic acid, respectively. The arrow shows the origin

components are determined by microdensitometry of the photographic plates (Fig. 25), or by direct scanning photometry. UV-sensitive photographic plates (9 × 12 cm, Agfa Gaevert Scientia 23 D 50, Kodak, 0–250) are used, with a magnification of about 70 ×, so that the whole separation is included in one plate. The inclusion of a reference system for calibration of the blackening is unnecessary, since only the relative blackening of the bands are determined. An occasional test should be made, however, to check that the linear part of the photographic blackening curve is used. The optical apparatus should also be investigated regularly for stray light; a small amount (less than five percent) will not invalidate the measurements, since determinations are performed on moderately absorbant objects. A clean front lens surface is important in order to minimize stray light. In addition, other factors of importance in general microphotometric work (WALKER, 1958), should be checked carefully (spectral purity of the light, adjustment according to KÖHLER's rule, etc.; see also Chapter 9).

A source of error of particular importance for microphoresis is the instability of uridylic acid during ultraviolet illumination. Permissible exposure times will depend on the equipment, and must be established by measuring separations after exposure to ultraviolet light for varying periods of time. For the measurement of optical density of the plates an instrument that is self-recording and which responds linearly to *optical density* (rather than transmission) should be used. For determination of the relative amounts of the separated components, a base line is drawn along the tracing corresponding to the nonabsorbing parts of the fiber or strip, and parallel to the background which has been recorded along the fiber or strip (see Fig. 25). The areas under the peaks are copied onto transparent paper, cut out, and weighed or determined by planimetry. If many separations have to be evaluated the use of a densitometer equipped with an integrator is recommended. The weight or area is divided by the millimolar optical density coefficients (E_{mM}) at 257 nm, after which base quotients can be calculated.

Incompletely separated bands pose a general problem in every technique of electrophoresis. If the bands approximate to a symmetrical Gaussain distribution curve, the separation of individual bands can be accomplished simply by dropping a perpendicular at the minimum. However, the partitioning becomes difficult if the incompletely separated bands do not follow the normal distribution curve. A simple method will be described for the special case of RNA-base analysis, which permits an approximate solution for the general problem (RÜCHEL, 1971).

With some strip samples the CMP-band tends to develop a long tail which would be assigned to the steep UMP-band in the conventional band division. This is due to the changed structure of swollen cell glass strips. Therefore, if separation is incomplete, the content of CMP, which also has the highest correction factor at a measurement wavelength of 257 nm, will be depressed in favour of UMP.

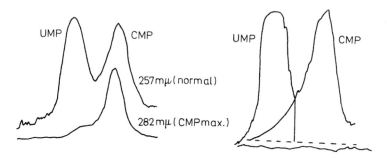

If the time of electrophoresis is increased so that CMP and UMP are separated completely, the bands of adenine and guanine become diffuse.

Making a simple mathematical assumption, a nomogram can be constructed for applying corrections to the usual band profiles (given by a microdensitometer).

The leading edge of the UMP band is drawn as a straight line with slope -10:

$$y = d(x - b) + c \qquad d = -10.$$

The trailing edge of the CMP-band will be represented as an exponential function given by the equation $y = e^{ax} - 1$, where a is the slope of the curve at $x = 0$ ($a > 0$).

At the intersection of the bands,

$$y = d(x - b) + c, \qquad c = e^{ab} - 1, \qquad a = \frac{\log(c + 1)}{b} \qquad (b > 0).$$

The area u under the curve is given by the integral:

$$u = \int_0^b (e^{ax} - 1)\, dx.$$

Hence,

$$u = \left[\frac{1}{a} e^{ax} - 1 \right]_0^b = \frac{1}{a} c - b.$$

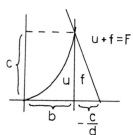

The area f under the straight line is given by

$$f = \left| \frac{c}{2} \cdot \frac{c}{d} \right| = -\frac{c^2}{2d}.$$

The total area $F = u + f$, hence:

$$F = \frac{c}{a} - b - \frac{c^2}{2d}.$$

Calculation of the nomogram: The slope d of the straight line $= -10$. For a, about twenty values lying between 0.1 and 1.0 are used. Each of these values defines a curve. For each of these curves, computations are carried out for a series of values of b.

$$(b > 0) \quad (b_1 = 30, b_2 = 40, b_3 = 50 \ldots \text{ to } 300).$$

By choosing suitable values, every area F, bounded by a straight line and an exponential curve, can be calculated.

In the usual method for the partition of overlapping UMP- and CMP-bands by a perpendicular through the minimum, the integral for UMP is decreased by the area f, and that for CMP by the area u:

$$F - 2f = u - f = F'.$$

$$F' = F - 2f$$

The area F' should be subtracted from the UMP area and added to the CMP area.

F' is calculated for all points of intersection for the families of straight lines and curves.

If the calculated lines and curves are plotted on millimeter graph paper, a network of intersections over a square $10\,\text{cm} \times 10\,\text{cm}$ is obtained. The network can be transferred to Plexiglas and the points of intersection drilled out. A plate of the same size is covered with millimeter graph paper, on which the points of intersection are plotted, labeled with the corresponding values of the area F', and the family of curves is sketched (Fig. 26).

The values of F' in the example given are expressed in the units of the integrator used.

The length units can be varied to take into account the varying heights of the band maxima, which depend on the quantity of extract, the time of exposure of the photographic plate, and the enlargement on the densitometer. In practice, the use of an average value ($E = 5\,\text{mm}$) has proved satisfactory.

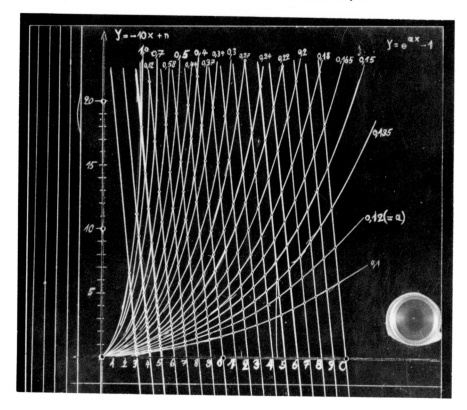

Fig. 26. Template and nomogram for the correction of incompletely separated bands

Application of the nomogram: The Plexiglas template is placed on the phero-grams keeping edges parallel whose UMP-CMP minimum lies above a third of the peak height. The origin of coordinates is moved, following the height of the band profile and keeping the edges parallel, until the trailing edge of the CMP-band approximates to a curve on the template. The nearest intersection point on this curve is marked with a pin. The template is then placed on the nomogram, and the corresponding value of F' is noted. For one series, Rüchel (1971) find for CMP:

Uncorrected:		Corrected:	
36.8 %		40.27%	
37.38%		40.72%	
32.66%		38.89%	
34.89%		39.68%	
34.39%		35.66%	
40.72%		37.37%	
40.27%		40.38%	
Standard deviation:	3.09	Standard deviation:	1.86
Mean error of the average:	1.17	Mean error of the average:	0.7

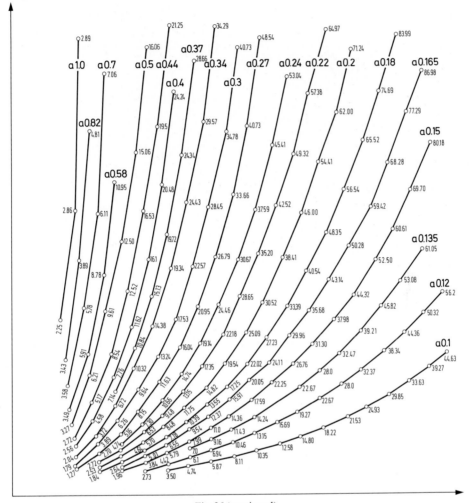

Fig. 26 (continued)

The separations can be evaluated much more quickly by direct scanning photometry in a UV scanning microspectrophotometer (Carl Zeiss) connected to a computer (PDP 12 Digital Equipment). Fig. 27 demonstrates the measurement of the separation on a strip. After adjustment of the optical system (see Chapter 9), the strip is scanned in 10 μm steps at 260 nm with a 0.25 mm monochromator slit. After adjusting the intensity range (Fig. 27, *a*) each 10 μm measuring step is displayed as a transmission value on the viewing screen of the computer (*b*). The computer then calculates and displays (*c*), the basis line as determined on parts of the strip ahead of the region of separation. The operator can adjust this line according to his observations, and adapt it optimally to the separation. Afterwards the transmission curve is transformed into the optical density curve (*d*),

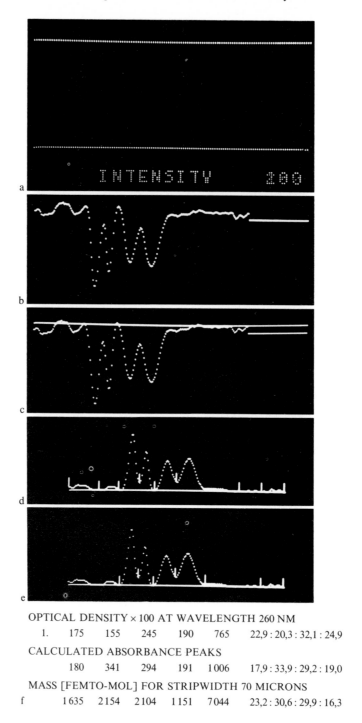

OPTICAL DENSITY × 100 AT WAVELENGTH 260 NM

1.	175	155	245	190	765	22,9 : 20,3 : 32,1 : 24,9

CALCULATED ABSORBANCE PEAKS

	180	341	294	191	1006	17,9 : 33,9 : 29,2 : 19,0

MASS [FEMTO-MOL] FOR STRIPWIDTH 70 MICRONS

f	1635	2154	2104	1151	7044	23,2 : 30,6 : 29,9 : 16,3

Fig. 27 a–f. Computer evaluation of RNA base microphoresis. For details see text

and the minima between the separated peaks as calculated by the computer are shown with pointers. The operator can eliminate pointers or change the position of a pointer along the curve (e). When the corrected curve has been accepted by the operator, the total optical density values for each band, the sum, and the percent distribution, are printed out (f). The next step includes corrections for the different extinction maxima of the four bases, the molar extinction coefficients of each component, and the strip width, are taken into account, and the values are printed out as 10^{-15} mole per RNA base. The extinction coefficients used for the computer program are measured photometrically in 6 N sulphuric acid, at the same pH as the buffer in the strips. The molar extinction coefficients are found to be, for CMP at 282 nm, 12.50×10^3; for UMP at 264 nm, 9.58×10^3; for guanine at 247 nm, 10.17×10^3; and for adenine at 263 nm, 11.60×10^3. On completing an experimental series the statistical evaluation is also printed out.

RNA Analysis

RNA is analysed after acid hydrolysis to a mixture of purine bases and pyrimidine mononucleotides. Acid hydrolysis is used because the excess hydrochloric acid can be evaporated easily, and because the products of acid hydrolysis give separations which are sharper than those of the products of alkaline hydrolysis. The ribonuclease extract is dissolved in 4 N hydrochloric acid (sufficient to give a concentration of less than one percent of RNA products) in a standard micropipette, into which 4 N hydrochloric acid has been aspirated previously to a length of several millimeters. Liquid paraffin is taken up between the acid and the hydrolysate, which is followed by a series of alternating volumes of hydrochloric acid and liquid paraffin towards the tip (Fig. 28). This arrangement prevents loss

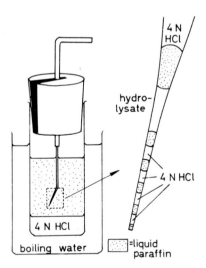

Fig. 28. Illustration of the procedure for quantitative hydrolysis of RNA in microamounts. [From EDSTRÖM, J. biophys. biochem. Cytol. **8**, 39 (1960)]

of the hydrolysate during the incubation at 100° C. The pipette is placed in an oil bath, which was previously saturated with 4 N hydrochloric acid at 100° C overnight, and is kept at 100° C for 30 min. The pipette is then removed, and the hydrolysate allowed to evaporate to dryness on a coverslip left untouched by the liquid paraffin in the oil chamber. The dried hydrolysate is dissolved under liquid paraffin in 1 N hydrochloric acid (about twice as much as is required for dissolving the hydrolysate) and is then ready for electrophoresis. This method of hydrolysis compares well with procedures used on a large scale (EDSTRÖM, 1960a). The following E_{mM} are used at 257 nm: adenine 11.2, guanine 9.6, cytidylic acid 5.15, and uridylic acid 9.5. Other wavelengths will give other specific extinction values. Fig. 25 shows the separation of the products of acid hydrolysis of the RNA from a nerve cell.

DNA Analysis

The extracted purines are evaporated to dryness, redissolved, and applied as for RNA hydrolysates, using the same conditions for separation. Assuming that the base composition of DNA reflects complementarity, the ratio of adenine to guanine gives sufficient information. Comparison between results obtained by this method with those of macrochemical analyses give good agreement on the whole (EDSTRÖM, 1964b).

Determination of Total Amounts of Nucleic Acids

If the total amount of RNA cannot be determined after microphoresis and measurement with a UV scanning microscope the technique of EDSTRÖM (1964a) can be used. For this, the ultraviolet optical density of extracts dissolved in lens-shaped drops on a quartz slide is determined. DNA can also be determined in this manner (EDSTRÖM and KAIWAK, 1961), but the electrophoretic method described below is simpler.

The total amount of DNA in an acid extract can be determined by adding a known amount of a reference compound included in the incubation fluid. Cytidine has been used since it is stable, absorbs ultraviolet light, and migrates as a separate band (slower than adenine and guanine, but fast enough not to overlap with contaminating material at the application point, Fig. 29). Since it is added to the sample with the acid for extraction, aliquots of the extract can be used for one or more measurements, and it is not necessary to transfer volumes quantitatively from the tissue to the fiber (which might be difficult). A calibrated pipette is used to transfer cytidine solutions.

The total amount of DNA (x) is given by the formula:

$$x = 2 \cdot \frac{A+G}{C} \cdot c \cdot 308.9$$

where A, G, and C are the integrated absorption values for adenine, guanine, and cytidine, divided by the corresponding E_{mM} (11.2, 9.6, and 4.3, respectively at

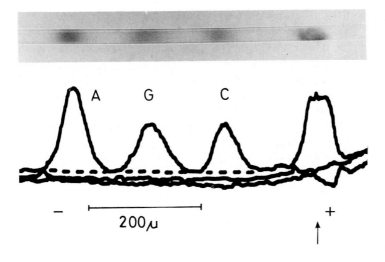

Fig. 29. A separation by microphoresis on a thin fiber. Scan at 257 nm of DNA purines from an extract of 8 axolotl spermatids with 0.34 μμmole of cytidine added to the incubation fluid. *A*, *G*, and *C* stand for adenine, guanine and cytidine. The arrow denotes the application point

257 mμ), c is the amount of added cytidine expressed in μμmoles, and 308.9 the average nucleotide mole weight for DNA containing 29% adenine and thymine and 21% guanine and cytosine.

Biological Applications

The composition of RNA from different components of large cells such as oocytes and salivary gland cells, has been investigated. The base composition of RNA in the nucleolus is similar to that of the bulk of cytoplasmic RNA (ribosomal RNA), in contrast to nucleoplasmic RNA, sap RNA, or chromosomal RNA, which have their own characteristic compositions (EDSTRÖM, 1960b; EDSTRÖM *et al.*, 1961; EDSTRÖM and BEERMANN, 1962; EDSTRÖM and GALL, 1963).

Chromosomal RNA in lampbrush chromosomes shows a general resemblance to DNA, but deviates significantly from a complementary composition (EDSTRÖM and GALL, 1963). RNA derived from Balbiani rings derived from single bands, is also asymmetric, with high adenine and low uracil values (EDSTRÖM and BEERMANN, 1962).

For the determination of DNA in mammalian cells by microphoresis a relatively large number of cells is required for an analysis: 30–40 diploid cells for thin fibers. However, certain urodeles, *Necturus* and *Amphiuma*, contain enough DNA for analysis of single diploid cells. Giant chromosomes offer particularly suitable material, not only for RNA analyses, by microphoresis but also for the determination and analysis of DNA (DANEHOLT and EDSTRÖM, 1967 and 1969).

Analysis of RNA by microphoresis has been used extensively by HYDÉN and colleagues in studies of nerve and glia cells during different functional and pathological conditions, and in pharmacological treatments (for review see HYDÉN, 1967). The conversion relations between the RNA purine-precursor pools and the pyrimidine-pools were the same in nerve cells and glia, although the labeled RNA in the two types of cells differed in composition. The overall rate of synthesis of RNA was twice as rapid in the glia as in the nerve cells (DANEHOLT and BRATT-GÅRD, 1966, EGYHÁZI and HYDÉN, 1966). The total RNA content in motor nerve cells from man (spinal cord, C_5, *Nucleus ventralis lateralis*) increases up to the age of 40; after the age of 60 the RNA content falls rapidly (HYDÉN, 1960). RINGBORG (1966) has observed differences with age in the amount of RNA per pyramidal cell in the CA_3 zone of the hippocampus of rats, and also differences in the RNA base compositions. The $G+C/A+U$ quotient increased with increasing age of the rats. Changes in the RNA base composition in the CA_3 layer of the hippocampal neurons occur after chronic treatment of rabbits with LSD 25 (NEUHOFF, 1967; RÜCHEL, 1971). Intraperitoneal injection of Reserpine into rats results in an increased RNA content in the ganglion cells of the lumbar sympathetic ganglia, but the RNA base composition remains constant (JARLSTEDT and MYTILINEO, 1971). ANDERSON, EDSTRÖM and JARLSTEDT (1970) have demonstrated the presence of axonal RNA in giant nerve fibers of the crayfish *Procambarus clarkii*, and have evidence that the axonal RNA is synthesized independently of the cell body; its base composition, which is DNA-like, differs distinctly from the RNA of nerve cell bodies.

According to ECCLES and MCINTYRE (1953) a possible model for the learning process is offered by post-tetanic potentiation. While previous work by HYDÉN and EGHYHÁZI (1964) indicated changes in the RNA base composition in the cortical neurons of rats in learning experiments, the post-tetanic potentiation of monosynaptic reflexes in cat's spinal cord produced no significant changes in the RNA base composition of α-motoneurons (ALTHAUS *et al.*, 1972).

A cytophotometric analysis was carried out on 238 nerve cells isolated from histological sections from the trigeminal nucleus of a subacute sclerosing panencephalitis (SSPE) brain; they were then compared with 236 control cells from the same region. The measurements were carried out in the ultraviolet region on fixed, unstained cells, and in the visible light region after staining the sections with methylene blue, and according to the Feulgen, ninhydrin Schiff, and Millon methods. The SSPE cells showed an increase of 40% in proteins, an increase of 34% in RNA and a decrease of 29% in DNA as compared to control cells. Computer analysis of the data following the ninhydrin Schiff staining revealed increased protein, not only within the cytoplasm, but also within the nucleus and nucleolus (TER MEULEN *et al.*, 1969).

Literature

ALTHAUS, H.-H., BRIEL, G., DAMES, W., NEUHOFF, V.: Zelluläre und moleculare Grundlagen der nervösen Erregungsspeicherung. 2. Neurochemische Mikroanalysen des Rückenmarks der Katze nach posttetanischer Potenzierung monosynaptischer Reflexe. In: Sonderforschungsbereich 33, Nervensystem und biologische Information, Göttingen 1969–1972, p. 107–121.
ANDERSSON, E., EDSTRÖM, A., JARLSTEDT, J.: Properties of RNA from giant axons of the crayfish. Acta physiol. scand. **78**, 491–502 (1970).

DANEHOLT, B., BRATTGÅRD, S.-O.: A comparison between RNA metabolism of nerve cells and glia in the hypoglossal nucleus of the rabbit. J. Neurochem. **13**, 913–921 (1966).

DANEHOLT, B., EDSTRÖM, J.-E.: The content of deoxyribonucleic acid in individual polytene chromosomes of *Chironomus tentans*. Cytogenetics **6**, 350–356 (1967).

DANEHOLT, B., EDSTRÖM, J.-E.: The DNA base composition of individual chromosomes and chromosome segments from *Chironomus tentans*. J. Cell Biol. **41**, 620–624 (1969).

ECCLES, J.-C., MCINTYRE, A. K.: The effects of disuse and of activity on mamalian spinal reflexes. J. Physiol. (Lond.) **121**, 492–516 (1953).

EDSTRÖM, J.-E.: Ribonucleic acid mass and concentration in individual nerve cells. A new method for quantitative determinations. Biochim. biophys. Acta (Amst.) **12**, 361–386 (1953).

EDSTRÖM, J.-E.: Extraction, hydrolysis, and electrophoretic analysis of ribonucleic acid from microscopic tissue units (microphoresis). J. biophys. biochem. Cyt. **8**, 39–43 (1960a).

EDSTRÖM, J.-E.: Composition of ribonucleic acid from various parts of spider oocytes. J. biophys. biochem. Cyt. **8**, 47–51 (1960b).

EDSTRÖM, J.-E.: Microextraction and microelectrophoresis for determination and analysis of nucleic acids in isolated cellular units. In: Methods in cell physiology I, ed. by D. M. PRESCOTT, p. 417–447. New York: Academic Press 1964a.

EDSTRÖM, J.-E.: Microelectrophoretic determination of deoxyribonucleic acid content and base composition in microscopic tissue samples. Biochim. biophys. Acta (Amst.) **80**, 399–410 (1964b).

EDSTRÖM, J.-E., BEERMANN, W.: The base composition of nucleic acids in chromosomes, puffs, nucleoli and cytoplasm of Chironomus salivary gland cells. J. Cell Biol. **14**, 371–380 (1962).

EDSTRÖM, J.-E., GALL, J. G.: The base composition of ribonucleic acid in lampbrush chromosomes, nucleoli, nuclear sap and cytoplasm of *Triturus oocytes*. J. Cell Biol. **19**, 279–284 (1963).

EDSTRÖM, J.-E., GRAMPP, W., SCHOR, N.: The intracellular distribution and heterogeneity of ribonucleic acid in starfish oocytes. J. biophys. biochem. Cytol. **11**, 549–557 (1961).

EDSTRÖM, J.-E., KAWIAK, J.: Microchemical deoxyribonucleic acid determination in individual cells. J. biophys. biochem. Cytol. **9**, 619–626 (1961).

EDSTRÖM, J.-E., PILHAGE, L.: Extraction, hydrolysis, and electrophoretic analysis of ribonucleic acid from microscopic tissue units (microphoresis). J. biophys. biochem. Cytol. **8**, 44–46 (1960).

EGYHÁZI, E., HYDÉN, H.: RNA with high specific activity in neurons and glia. Brain Res. **2**, 197–200 (1966).

HYDÉN, H.: Quantitative assay of compounds in isolated fresh, nerve cells and glial cells from control and stimulated animals. Nature (Lond.) **184**, 433–435 (1959).

HYDÉN, H.: The neuron. In: The cell, ed. by J. BRACHET and A. MIRSKY, vol. IV, p. 215. New York: Academic Press 1960.

HYDÉN, H.: Satellite cells in the nervous system. Sci. Amer. **205**, 62–70 (1961).

HYDÉN, H.: Biochemical and functional interplay between neuron and glia. In: Recent advances in biological psychiatry, ed. by J. WORTIS, vol. VI, p. 31–54. New York, London: Plenum Publ. Corp. 1963.

HYDÉN, H.: Behavior, neural function and RNA. In: Progress in nucleic acid research and molecular biology, ed. by J. N. DAVISON and W. E. COHN, vol. 6, p. 187–218. New York, London: Academic Press 1967.

HYDÉN, H., EGYHÁZI, E.: Changes in RNA content and base composition in cortical neurons of rats in a learning experiment involving transfer of handedness. Proc. nat. Acad. Sci. (Wash.) **52**, 1030–1035 (1964).

JARLSTEDT, J., MYTILINEO, C.: Effect of reserpine on sympathetic neuronal RNA. Brain Res. **28**, 355–356 (1971).

KOENIG, E., BRATTGÅRD, S.-O.: A quantitative micromethod for determination of specific radioactivity of H³-purines and H³-pyrimides. Analyt. Biochem. **6**, 424 (1963).

NEUHOFF, V.: Präparation der Neuroglia zur Mikroelektrophorese der Ribonucleinsäure-Basen. Arzneimittel-Forsch. **16**, 779–781 (1966).

NEUHOFF, V.: Die Wirkung von Lysergsäurediäthylamid auf Ganglienzellen. 1. Mitt. Arbeitshypothese und Ribonucleinsäure-Basen im Hippocampus. Arzneimittel-Forsch. **17**, 176–181 (1967).

NEUHOFF, V., MÜLLER, D., TER MEULEN, V.: Präparation von Ganglienzellen für cytophotometrische Untersuchungen. Z. wiss. Mikr. **69**, 65–72 (1968).

NEUHOFF, V.: Wet weight determination in the lower milligram range. Analyt. Biochem. **41**, 270–271 (1971).

RINGBORG, U.: Composition and content of RNA in neurons of rat hippocampus at different ages. Brain Res. **2**, 296–298 (1966).

RÜCHEL, R.: Mikroelektrophoresen von RNS-Basen. Anwendung zur Untersuchung bestimmter Hirnregionen und kritische Analyse der Methode. Inaugural-Dissertation Göttingen, 1971.

RÜCHEL, R., REHMANN, U., NEUHOFF, V.: In preparation.

SIERAKOWSKA, H., EDSTRÖM, J.-E., SHUGAR, D.: Intracellular localization of nuclease enzymes by a microdissection-microelectrophoretic technique. Acta biochim. pol. **2**, 497–507 (1964).

TER MEULEN, V., MÜLLER, D., NEUHOFF, V., JOPPICH, G.: Immunhistological, microscopical and neurochemical studies on encephalitides. V. Subacute sclerosing panencephalitis. Cytophotometric studies on isolated nerve cells. Acta neuropath. (Berl.) **15**, 128–141 (1970).

Chapter 7

Determination of the Dry Mass of Small Biological Objects by Quantitative Electron Microscopy

Prerequisites

Before initiating quantitative mass determination of biological objects in one's own laboratory, the following prerequisites must be available:

Electron Microscope

Any electron microscope is suitable for quantitative work.

Accelerating Voltage

The higher the accelerating voltage at which the instrument can be operated, the better, because then one can work with thicker objects. Lower accelerating voltages, however, are often satisfactory for most requirements.

Contrast Aperture

A relatively large contrast aperture such as $100\,\mu$, is chosen to begin with. If the operational data of the microscope (i.e., focal length of the objective lens, and the position of the aperture in the back focal plane of the lens) is known, the angular-subtense in steradians for the aperture opening can be calculated. In our laboratories a value of 2×10^{-2} is used.

On removing the contrast aperture from the beam, an increase in diffusely-scattered electrons will usually be noticed. The emulsion under the shadow of the object support grid should normally show no more density than that in an unexposed area (edge). This may be used as a rough indicator for the suitability of an aperture; a "large" aperture should not increase diffuse scattering.

Exposure System

There must be means for controlling the exposure, preferably by a combination of electron sensor with the shutter mechanism. The sensor should sample a small area in the magnified electron beam, somewhere between the projector lens and the fluorescent screen. The sensor should be small, as is the case in most microscopes, because the method requires that the beam intensity be sampled in an object-free area of the field to be photographed, so that exposures with closely comparable background densities will be produced. The sensor must be small in order to fit between densely packed objects. Sensors (photometers) covering the entire field, or multiple sensors which have to be illuminated simultaneously, are less desirable because they do not allow sufficient control of exposure when preparations have particles of varying density and size, and because they cannot allow for the shadow of a grid bar when this is part of the field.

Fig. 1. Simple electron sensor. Fluorescent screen of the Siemens 1A electron microscope to the right. A small brass block holds the plastic insulator, through which the arm of a small disk-sensor is anchored to the edge of the frame. A flexible copper braid connects the sensor arm to an electrical feed-through in a brass plate (left) replacing a window in the microscope viewing chamber. The feed-through is soldered to an amphenol connector held by a bracket; from here a coaxial cable (not shown) connects to a micro-ammeter

With some electron microscopes the operator must rely on visual judgment of intensity, and must time the exposure himself. If the instrument lacks an automatically-timed shutter device there is little one can do, other than obtain one from the manufacturer. Most makers of electron microscope have constructed automatic shutter systems as accessories for their older models. If it is not possible to fit such a shutter, one can construct some simple device for interrupting the beam somewhere in its path. If a magnetic shutter is used, it must be designed so that the coil is not activated during exposure, because its field would distort the electron beam (BAHR, SACKERLOTZKY, and ZEITLER, 1963).

For a sensor, a small metal disk of about 1 cm diameter such as a coin is chosen. It is soldered to a conducting metal arm so that it is either permanently or adjustably in the beam. Fig. 1 illustrates a sensor for a Siemens 1A microscope. The arm is held by an insulating piece of polystyrene plastic, from which a flexible copper braid extends to a glass insulated wire glued to the oval brass disk with epoxy polymer. The brass plate replaces one window of the microscope's viewing chamber. Outside the brass plate an amphenol connector is mounted on a metal bracket and soldered to the feedthrough wire. A coaxial cable connects the sensor to a microammeter for the direct measurement of beam current density. Arrangements similar to the one described can be assembled easily for other makes of microscopes.

The method of reading current density depends on available resources. We obtained satisfactory measurements for several years with a set of variable resistors and a light-spot galvanometer. When electronic microammeters became cheaper, we changed over to them, and obtained excellent performance; they are faster to read, and cover a wider range of currents than galvanometers.

Some of the most modern microscopes have automatic, coupled sensor-exposure systems with sensitivities comparable to separate microammeters, and, if the sensor is reasonably small, may be used without alteration.

Darkroom

Darkroom requirements for quantitative electron microscopy are different from those for tissue work. Complicated printing work is not required, and one can do without an enlarger. Few electron microscopes are installed without access to complete facilities for enlarging and printing. For quantitative work one needs processing tanks in which *sets* comprising one microscope-load of plates or sheet film can be developed. Batch processing may be done on individual hangers handled simultaneously, but it is much safer and more convenient to have simple holders for plates or cut film made from acrylic plastic. Two types of holders which we find most satisfactory are shown in Fig. 2.

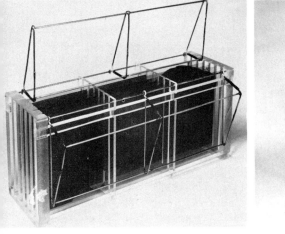

a   b

Fig. 2 *a* and *b*. Plate holders of plexiglass for batchwise development of negatives. *b* arrangement for glass slides. *a* holder for triple exposed sheet film; welded wire holders (one shown opened) serve to retain film and to carry assemply

The developing bath must be kept constant at the temperature recommended for best results by the manufacturer of the photographic emulsion. A thermometer should be available for checking.

We found a satisfactory arrangement to be a set of stainless steel tanks, measuring 10 cm wide, 28 cm high and 12.5 cm deep, standing in a thermostatically-controlled water bath. In many darkrooms a mixing valve has been installed to adjust the flow of cold and warm water to give a selected temperature; such a valve is easily installed and would also be useful for other photographic work in the darkroom. Four tanks are needed, one for developer, which must always be covered by a floating lid when not in use, one rinse bath, one bath of hypo fixer, after which the emulsions should be rinsed in plenty of circulating water. A final quick rinse in a fourth tank with a dilute solution of a surface active agent such as Photo-Flo 200 from Kodak is useful, because it prevents the formation of "water spots" on the photographic material, which may interfere with subsequent densitometry.

Each laboratory must establish suitable techniques for development in accordance with the specific photographic material and the properties of developer used.

Densitometry

The Principle of the Method of Dry Mass Determination with the Electron Microscope

The basis for the photometric determination of the dry mass of biological objects with aid of an electron microscope has been developed in collaboration with Dr. ELMAR ZEITLER. It is briefly as follows:

Loss of electrons from the beam passing through the object is caused by electron scattering, and can be approximated closely by a formula analogous to BEER's law.

$$I = I_0 \cdot e^{-\alpha S w} \tag{1}$$

where I_0 is the unscattered beam intensity, I the beam intensity after passing through the object, and α an instrumental factor determined by the operating parameters and the design of the microscope. S is analogous to the extinction coefficient of absorption measurements for light, and is the scattering cross section per gram of substance (cm^2/gm). The exponent w is the transradiated mass per area.

The form of Eq. (1) indicates that the object's mass cannot be derived by integrating the current density of the transmitted beam I over the object area directly. It would be necessary to measure I point by point in the electron image and sum the logarithms of the individual measurements.

We use the photographic emulsion as an analog converter of I to photographic density D. (1) This conversion is linear for density values below 1.2 and is given by

$$D = E \cdot b + \delta, \tag{2}$$

where E is exposure of the emulsion to the electron beam ($I \times$ time in amps), and b is a constant governed by the response of the emulsion to electrons and by the conditions of development; δ is the fog produced in the emulsion by diffusely scattered electrons and by the chemical development procedure. LIPPERT (1969) has recently investigated Eq. (2) in some detail and concludes that curves of density D should be drawn for varying I rather than time, because for certain types of emulsions there is a noticeable increase in density after exposure.

Scientists experienced in quantitative measurements on photographic emulsions often have the more complex relationship of density and exposure to light in mind. Fig. 3 illustrates schematically the fundamentally different effects of light and electrons for a low density range.

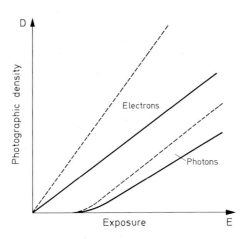

Fig. 3. Different effects of electrons and photons at low photographic density. For electrons, density is proportional to exposure (disregarding fog); for exposure to photons there is a lag (toe) before density develops. Broken line shows effects of longer development than that used for solid line, i.e., the photographic contrast increases with prolonged development

The linear curve for electrons intersects the origin, i.e., exposure is proportional to density from the start. Density caused by exposure to photons increases only after an emulsion-specific lag period, and starts slowly with a toe. Linearity for electrons therefore makes it possible to convert electron intensity I [Eq. (1)] directly to photographic density D [Eq. (2)] and to use the relationship of photographic density to transmission.

$$D = -\log_{10} T \tag{3}$$

in combining Eqs. (1) to (3) with the simplified result:

$$\frac{T - T_0}{T_\infty} \approx \alpha \cdot S \cdot w, \tag{4}$$

where T is the transmission of an image point in the electron micrograph of the object plus its underlying supporting film, T_0 is the corresponding transmission through the supporting film (background), T_∞ is the transmission of the un-exposed photographic material, which is chiefly determined by the fog level δ [Eq. (2)]. We find that the transmission difference becomes proportional to mass:

$$\frac{\overline{T} - \overline{T_0}}{T_\infty} \approx W. \tag{5}$$

It has been found experimentally by HALL (1966) and REIMER (1967) that Eq. (1) and hence Eq. (4) are largely independent of the elemental composition of the object.

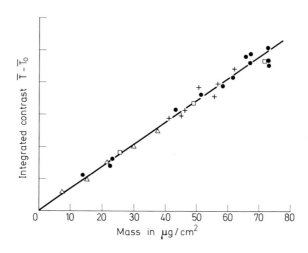

Fig. 4. The mass of steps in a Nylon crystal △, evaporated Carbon ●, Silver □, and Beryllium +, has been determined by independent methods, such as multiple interference measurements, or by comparison with objects of known mass, and is plotted against integrated transmissions measured with apertures in the Zeiss IPM-2

In Fig. 4 the increase of contrast in the negative was compared with increase of mass for such diverse objects as nylon crystals, carbon layers, and silver and beryllium layers.

Estimation of the Useful Mass Range of an Electron Microscope

For biological objects (2) S is in the order of $5 \cdot 10^4 \text{ cm}^2/\text{gm}$. It has been stated by ZEITLER and BAHR (1965) that $S \cdot w$ should be approximately 1, i.e., the mass limit would be $2 \times 10^{-5} \text{ gm/cm}^2$ or 2000 Å for an object of specific gravity 1 gm/cm^3. More recently the limit has been found by HALLIDAY and QUINN (1960) to extend to $S \cdot w = 3$, i.e., to about 60 μg/cm^2 or to a layer of 0.6 μ thickness. Also recent measurements in our laboratory indicate a critical thickness, W_{lim}, of 80 μg/cm^2, up to which BEER's law can be applied safely. A biological object containing 20–30 percent solids of specific densities close to 1.3 gm/cm^3

would be up to 3.3 μ thick before dehydration. Experience, however, shows that the hydrated biological specimen may often be almost twice as thick (e.g., up to 6 μ) as the suggested limits, which were extrapolated from measurements on solid films of evaporated carbon. An object loses all its free water in the vacuum of the electron microscope, furthermore most preparation procedures usually lead to flattening of the object.

A pragmatic procedure for estimating the range over which a microscope at a given, preferably high, accelerating voltage can operate is simply to determine this range. Although layered films of carbon or of another material are suitable,

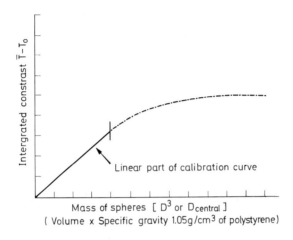

Fig. 5. A curve like this can be established easily in any laboratory, in order to estimate the range of mass for which a given electron microscope will give acceptable data

the most convenient, and with respect to irradiation damage the most object-like material for range determinations are polystyrene latex spheres which can be obtained in a so-called monodisperse form, as well as emulsions with great variations in sphere size. A curve of integrated contrast against diameter cubed is easily constructed. If only point measurements on an existing microdensitometer are possible, T is measured in the center of a sphere image $D_{central}$ for which the diameter is also measured. The latter is then plotted on the abscissa, and $T - T_0$ for the very same small measuring spot is on the ordinate. The principle is illustrated in Fig. 5.

Images of biological objects should be checked at their most dense regions to assure that the technique is applicable, because any contrast exceeding the straight portion of the curve in Fig. 5 will produce falsely low mass values.

For very thin objects, one should thus choose a small aperture fulfilling the requirements for a minimum density difference of 0.02 between object and background (see below). Generally, one chooses a large aperture, as mentioned above, and determines the limits of mass for which the relation of contrast to mass is linear (Fig. 5), if it is intended to cover a relatively large mass range, as illustrated for an aperture of $2 \cdot 10^{-2}$ in Fig. 6. But if greater sensitivity towards small differences of mass is required, one uses a smaller aperture and gains a larger range of contrast over a narrower range of mass.

Returning to Eq. (4) and Fig. 5 we realize that the difference $T - T_0$ will be zero in unexposed as well as in grossly overexposed electron micrographs. Between these two extremes there will be a range in which the slope of the proportionality

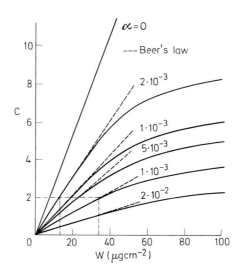

Fig. 6. Variations of contrast, C, with mass per cm², (w), of carbon at 75 kV for various aperture angles (radians). The broken lines indicate plural scattering theory. (After SMITH and BURGE, 1963.) Range of adherence to Eq. (1) is clearly seen for various aperture sizes

curve is at a maximum. In theory and practice the range of the relatively broad maximum lies between 0.17 to 0.35 for T_0/T_∞. Electron microscopic exposures and photographic development have to be monitored appropriately to assure background transmissions (T_0) in this range. All micrographs for which the background transmission falls within this range can be compared quantitatively, as the error due to variations in T_0 is at a minimum.

In the work of LIPPERT (1969) the point is made that for certain emulsions the density D increases after exposure. It is therefore advisable that the photographic material chosen should be checked for this effect by developing equally exposed plates at intervals of, say, 2 min. A plot of D against time will reveal quickly if, and when, this process levels off. Exposure will then have to be adjusted to take this storage effect into account.

Measurements of the differences of relative mass *within* one plate are permissible over a wider range of T_0. Such determinations can in fact be made on slightly over or under-exposed plates.

Upper and Lower-Limits

Small objects should be magnified in the electron microscope to give a diameter of at least 2 mm on the negative, because of the limitations set by the photographic grain and the interference of phase effects at the limits of resolution of the instrument.

To reduce the statistical effects of photographic graininess, the minimum density difference of the object over the background should be 0.02, which, for 100 kV, corresponds to a mass of about $2 \cdot 10^{-7}$ gm/cm^2, or a layer 20 Å thick. Taking into account the lower limits for size and mass/cm^2, the smallest mass that can be determined by the weighing procedure with the electron microscope is 10^{-18} gm for the 100 kV used in our experiments.

The upper limit for determination of mass is given by the limit for object thickness, i.e., the limits of validity for Eqs. (1), (5), and by the size of the object in the photographic plate in the electron microscope, which must be commensurable with the restrictions of the photometric evaluation. In the IPM-2 densitometer to be described below, a circular area of 18 mm diameter sets the limit for the size of the object's image. For most electron microscopes it is easy to adjust the operating conditions to these various limits by changing the accelerating voltage, aperture size, and magnification.

Consideration of the Photographic Fog

In Eqs. (4) and (5), T_∞, the diffuse density of emulsions, which is the product of storage and chemical processing, called fog, is the devisor for both T and T_0. The level of fog can be measured at the unexposed edge of the emulsion, which is often very narrow.

By definition, transmission is expressed as fractions of 100, i.e., from zero transmission to 100 %. Allowance for fog can be made simply by defining transmission in an unexposed area as 100 %. Whatever the level of fog, it is now subtracted from every measurement of transmission in that emulsion.

In the photometric procedure to be described one measures the transmission of relatively large areas. Consequently levels of fog have also to be measured over large areas. We have solved this problem by mounting a grounded metal disk of about 15 mm diameter to a suitable part of the microscope camera, as closely as possible to the plane of emulsion, in order to shield a circular area from exposure. Fig. 7 illustrates this for a single exposure. In some instances it is possible to increase the distance between successive exposures (such as on film), leaving space for measurements of fog between exposed areas.

Measurement of Transmission

While the exponential Eq. (1) requires that every point of the object be measured, followed by transformation of the exponential values and addition to give the total mass, no such complications are encountered when the measured value is related linearly to the mass. This is the case for the transmission over the background to at least a useful approximation.

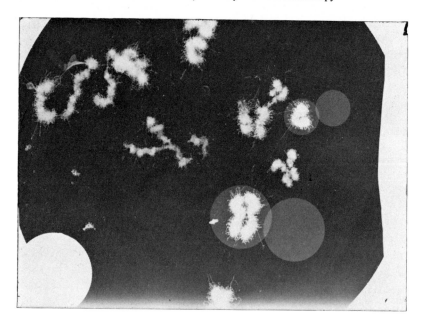

Fig. 7. Typical electron micrograph (negative) showing chromosomes (from collaborative work with
E.J. DUPRAW). Edge of glass can be recognized, as well as circular limitation of field by projector
lens. A circular metal disk has covered the lower left corner of the slide for the determination of the
100% transmission value (making correction for fog unnecessary). Lighter disks, over and next to,
two chromosomes simulate measurements of T and T_0. The unevenness at lower edge of slide is due
to storage. No measurements should be made in this area

All values of contrast, $T - T_0$, for all image points can thus be determined
by only two measurements, one over the object area, the other in its immediate
vicinity. The difference of these two is proportional to the total mass.

For such integrating measurements, Dr. ELMAR ZEITLER and the author
developed a photometric procedure, for which the author and Dr. LEON CARLSSON
had previously developed a prototype photometer while working at the Karolinska
Institute, in Stockholm (BAHR, CARLSSON, and ZEITLER, 1961).

ZEITLER and BAHR (1962, 1965), and BAHR and ZEITLER (1965) obtained
experimental and theoretical evidence for the validity of the approach, and
defined its limitations.

A telecentric lens system, Fig. 8, projects the image of a light source Q onto
a detector DE, usually a photomultiplier. Into the parallel portion of the light path.
The micrograph, OS, is inserted between L_1 and L_4. Since T is determined as
the integrated transmission over the area of the image, a physical aperture is
introduced into the light path at A, and its *image* is projected into the plane
of the micrograph, using two auxiliary lenses, L_2 and L_3. The detector then
sees only the average transmission over the area A, which will be exactly the
same for measurements of both object and background. Circular apertures are
convenient to manufacture and their area can be determined easily and precisely

from the diameter of the hole. The fact that the detector senses average transmission only as an intensity is the reason why any shape of aperture may be used.

Lenses L_5 and L_6 form a telescope, through which one can look via mirror M
at the object-plane and judge the position and size of the aperture image relative
to the object image.

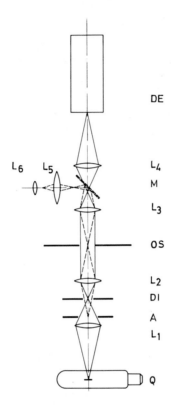

Fig. 8. A simple photometer, designed by BAHR and CARLSSON, for integration of transmission
measurements. Q is a tungsten ribbon lamp, DE a photomultiplier P 28 with opal glass in front of the
multiplier entrance. The other optical elements are described in the text

One of the stringent requirements which must be fulfilled by the optical
system is evenness of illumination, differing by not more than $\pm 1\%$ from one
part to another for the largest measuring area given by the largest aperture used.
This condition is checked routinely by moving a small aperture on a plain glass
plate around in the object plane.

Other more sophisticated, partially automatic versions of the apparatus
shown in Fig. 8 were built in this laboratory before Carl Zeiss, Oberkochen,
West Germany, designed the final version, the Integrating Photometer, IPM-2.

The IPM-2 is illustrated in Figs. 9 and 10. The optical path is partially analogous to the light path for Köhler illumination in a microscope, Fig. 9. In one

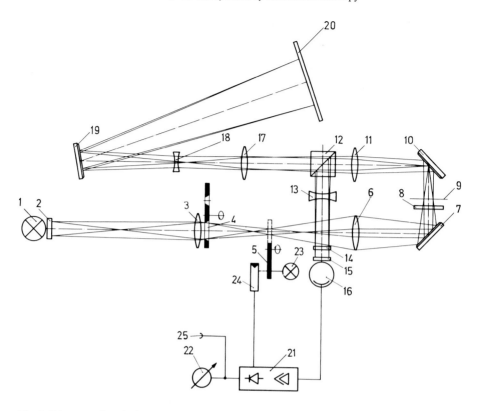

Fig. 9. Diagram of optical components and light path in integrating photometer by Carl Zeiss, Oberkochen.

1 Light source, *2* Opal glass screen, *3* Light collector, *4* Measuring aperture, *5* Sector wheel, *6* Condenser, *7* Deflecting mirror, *8* Plane cover plate, *9* Electron micrograph negative, *10* Deflecting mirror, *11* Objective lens, *12* Beam splitter, *13* Adjusting lens, *14* Neutral filter, *15* Opal glass screen, *16* Photomultiplier, *17, 18* Projecting lenses, *19* Deflecting mirror, *20* Projection screen, *21* Amplifier and linear rectifier, *22* Mirror galvanometer, *23* Modulator lamp, *24* Photo-cell, *25* Analog output

case, the measuring aperture (*4*) is illuminated by the light source (*1*) through a collecting lens (*3*) which forms a 1:1 image of the aperture through the condensor (*6*) in the plane of the micrograph. Light passing through the aperture, area *A*, in the micrograph, and not deflected by the beam splitter (*12*), produces a 5× magnified image of the measuring area on a ground-glass screen (*20*). In the other case, the light source (*1*) is imaged by the collecting lens (*3*) into the plane of selector wheel (*5*), and from there, via the beam splitter and lens (*13*), into the photomultiplier (*16*). The electrical signal generated is amplified, rectified (*21*), and fed to a galvanometer (*22*) and simultaneously to an analogue output (*25*).

The light source is a 50 W photometer lamp, supplied by a separate, precisely stabilized, DC power supply. Uniform illumination is achieved by having opal glass in front of the lamp, (*2*), and in front of the photomultiplier, (*15*).

Ten circular apertures (4), are arranged on a turret so that sizes of 1.5 to 18 mm diameter can be selected, and brought individually, precentered, into the beam. In order to compensate for the considerable difference in luminous flux between the smallest and the largest apertures, a set of neutral density filters is changed automatically in front of the photomultiplier. Exact compensation is attained by automatic slight modification of the amplification factor. If, for example, 100% transmission has been adopted for unexposed emulsion with one aperture, change to another size of aperture will have little or no effect on this setting.

As already discussed, the integrating procedure may also be used with non-circular apertures. For this purpose a second aperture turret is employed, into which free-form apertures can be inserted; the turret with the neutral density filters is disengaged from the set of circular apertures and must be operated by hand.

Fig. 10 shows the front view of the IPM-2. The carrier for the electron microscope plates or films rests on four Teflon feet, so that it can be shifted easily and quickly for centering and orientation of the object. Plates are inserted directly

Fig. 10. Front view of integrating photometer.

1 Projection screen, *2* Mechanical zero control, *3* Galvanometer scale, *4* Signal lamp, indicating power on/off, *5* Short-circuiting switch for galvanometer, *6* Electric zero control, *7* 100 point control for analog output (not shown), *8* Power switch, *9* Background illumination control, *10* Push-buttons for printer control, *11* Electron micrograph negative, *12* Aperture number indication, *13* Four-digit switch, *14* Change-over between circular and non-circular apertures, *15* Selector switch for circular apertures, *16* Selector switch for non-circular apertures, *17* 100-point control for galvanometer readout, *18* Holder for plates or cut films, *19* Stage for negative, *20* Desk

into the top plate of the carrier (exchangeable for various plate sizes). Cut film is sandwiched between two glass slides. Roll film (e.g. 70 mm) is fed from one spool to another at tension, past two roll bars to keep the film parallel to the plane of measurement, Fig. 11.

Fig. 11. Device for handling 70 mm roll film for integrating photometry. The spools are directly compatible with camera, and the developing rig

For orientation, the entire negative is illuminated by means of variable background illumination. To prevent this background illumination and stray room light from affecting the measurements, the measuring beam is chopped at three times the line frequency, by a rotating three-segment wheel. This light modulator controls photoelectrically, and at the same frequency, the auxiliary voltage required for linear rectification of the signal voltage, Fig. 9.

Most functions of the integrating photometer shown in Fig. 10 should be familiar from the foregoing discussion. A few features require comment. Two nixi-tubes indicate the number of the aperture in the measuring position. This value is also available at the output connector for digital processing. Four digital switches (thumb-wheels) can be set to identify the plate or the series of measurements when these are electronically recorded. This value is also available at the output connector.

Measures for keeping luminous flux at the photomultiplier nearly constant when apertures are changed have been discussed. Small adjustments may nevertheless be required, using the same control (17) as is used to set the 100% value initially. The IPM-2 is designed so that one can choose whether to read all values into a calculator-printer or into computer-compatible data-storage, such as paper- or magnetic-tape. Each transfer of data may be triggered by pushing the buttons (10), on the left side of the desk.

The IPM-2 provides the fastest, most accurate, most convenient way of making mass determination from electron micrographs. When measurements are recorded for computer processing one can trigger them in rapid sequence with a foot switch. We have routinely made 150–250 measurements per hour, suggesting that the procedure is not limited by the time required for photometric evaluation, but that electron microscopy, and photoprocessing, set the pace. When the masses of objects on the same negative are compared, the maximum relative error is only $\mp 2\%$; however, when calibration for absolute mass is involved, errors will be nearly $\mp 9\%$ (BAHR and ZEITLER, 1965).

There is another means of evaluating electron micrographs for quantitative purposes, namely by scanning densitometry. If one has an instrument with the capacity for automatic x and y scanning, and a suitable off- or on-line digital-recording system of evaluation, the computer algorithms can be written to convert density to transmission, for sensing and deducting the background transmission and for calculating a total mass value. Since areas of measurement in scanning microdensitometers are usually difficult to change, there will be non-relevent information in each field scanned. The image can be cleaned either by corrections entered into the computer on punch-cards, or by "cleaning" the image with an interactive process with a light pen on a cathode ray tube (WIED et al., 1970).

In the author's laboratory, both integrating photometry with the IPM-2, and scanning densitometry with a modified Joyce-Loebel microdensitometer, digitally interfaced to an incremental magnetic tape recorder, have been used. The IPM-2 is superior in every respect for the mass determination, notwithstanding the usefulness of scanning microdensitometry for other purposes such as pattern recognition (BAHR, 1968).

Alternatively, a simple integrating densitometer, using a polarizing microscope, can be built as described by SILVESTER and BURGE (1962). This apparatus is primarily suitable for smaller objects.

From Sample to Result

To illustrate the *second step*, of establishing quantitative electron microscopy in one's own laboratory, we will follow a preparation through the various procedures.

Electron Microscopy

Work with one type of microscope the Siemens 1 A, will be considered, however, it is easy to adapt the technique for any other make. We use photographic plates, which are put into the microscope fresh from the package, without pre-evacuation, and the twelve plates are then pumped on the forepump for 15 min. The diffusion pump is switched on, and the first exposure is made at a vacuum of 5×10^{-3} torr. We expose all plates within 15 min because of the gradual release of moisture from the photographic material after its introduction into the electron microscope. That fraction of the emulsion which is effectively penetrated by the beam will contain increasing concentrations of silver halide, and consequently will produce increasing photographic density for the same electron exposure as the emulsion shrinks. Pre-evacuation of the plates is of little use, because significant amounts of moisture are usually picked up by the gelatin in the short period when the plates are transfered through the ambient laboratory air. Estar-based emulsions in a non-gelatinous carrier are much less hygroscopic and contain little initial moisture; they are finding increased use in electron microscopy.

We proceed as follows: A selected object area is centered on the fluorescent screen, with the magnification set so that the largest object fits into the largest aperture of the integrating photometer, which is 18 mm in diameter. Some fine marks with a lead-pencil on the fluorescent screen are helpful. In some microscopes, magnification on the fluorescent screen is significantly smaller than on the emulsion proper because of increased distance from the projector lens. The image is moved so that the sensor for the beam intensity is covered only by background. After focussing, the microscope illumination is adjusted to the correct current reading, and the emulsion is exposed for a brief, timed period, e.g., 2 sec.

Exposure systems in which the time of exposure is adjusted automatically to the beam intensity are a great convenience for quantitative mass determinations. Since our instrument does not have such a system, we adjust the condenser current and/or the bias until the micro-ammeter indicates the correct value. When manipulating these two intensity controls even illumination over the image area must be maintained. Generally, low beam intensities are also recommended in order to keep contamination of the object at the lowest possible level.

The object is selected and centered, preferably without increasing the beam intensity. Refocussing is only carried out if objects are far apart on the grid. Focus, if fairly close, is not critical; see the discussion by Bahr and Zeitler (1965). In this way, we expose 9 of the 12 plates, then remove the specimen and insert a standard preparation of polystyrene spheres. Two more exposures, without alteration of the magnification and background beam current, are obtained. Next, we insert a magnification standard, again without touching the magnification controls, and expose at any suitable intensity. All preparations are put into the microscope with the object facing the beam, so as to lie in the same optical plane in the microscope. If refocussing for appreciable distances from preparation to preparation is necessary, the magnification will not remain constant, because the objective lens has a fixed focal length.

When using a roll of 48 frames of film, one starts with the magnification standard and two mass standards, and may conclude the series in the same manner in order to keep a close check on the conditions.

Next, all 12 plates are placed in the holder (Fig. 2), and developed for 5 min at 20° C (70° F) with Dektol diluted 1:1 with deionized water. At intervals of one minute the holder is lifted out of the tank, letting the developer drain off briefly while holding it tilted at about 45° parallel to the plane of the emulsions. This is done, tilting to alternate sides, to avoid the uneven development known as bromide streaking. After fixing the plates in acid fixer for twice as long as it took to clear them, but for not longer than 10 min, they are washed in running tap water for $\frac{1}{2}$ hour, then dipped briefly into a surface-active rinse such as Photo-Flo 200 from Kodak, and dried at room temperature in a dust-free place.

Measurements with the IPM-2

The instrument should be warmed up for about 15 min. A plate holder is selected to suit the format of plates used (cut film is inserted between two special glass plates which are provided for the purpose) and mounted on the carriage (Fig. 10). The emulsion should be face up and precisely in focus on the ground glass observation screen, from both left to right and top to bottom; this is done by lifting, lowering, and tilting the negative. If necessary, the stage level is corrected, using the adjustment screws at the corners of the stage.

The galvanometer is short circuited and mechanically set to Null with the knob on the left upper side of the photometer housing. Electrical zero is adjusted with the beam blocked by an opaque portion of the stage, and the object is centered on the ground-glass observation screen. The aperture big enough to cover the whole of the object's image is selected; it should not fit too closely, so that it is outside the fringe of intensity modulations produced on slight under- or over-focus. Neither should the fit be too wide, as this not only decreases the numerical difference between the transmission of the object T and the background T_0, but also makes it often more difficult to find a suitable free area for measuring the background transmission. The size (area or code number) of the selected aperture should be recorded in the log sheet.

The carriage is then moved so that the bright disc of the aperture image falls clearly in the unexposed area provided for this purpose, and the transmission T_∞ is adjusted to 100 on the galvanometer using the helipot control on the right lower side of the photometer table.

Next the image of the object is brought into the aperture and the total transmission $\overline{T}$ over the object area is measured. The aperture image is then moved to a free background near the object, and T_0 is recorded as the third and last entry in the log sheet. Should there be considerable variations in background, due perhaps to debris, several measurements of the background are obtained, and an average value is derived. This should also be done if the exposure shows a "slant", due to uneven illumination in the electron microscope.

The measuring process with the IPM-2 is often improved by non-glossy positive prints from the negatives, because the objects to be measured can be marked, as well as the directions of "cuts" (see description of cuts later). Orientation of the negative with the carriage is much facilitated by a print, but is also helped by the variable understage illumination.

If the objects have shapes that differ significantly from a circle, e.g., short segments of fibers, a suitable aperture is selected from among the free-form apertures, or one is made and inserted into the free-form holder (Fig. 9); this can be activated by turning the change-over switch between the two aperture selector switches. For this to be done both selector switches have to be in the filled white-spot position first.

Free-form or non-circular apertures must be calibrated with respect to area. The areas of all circular apertures, inserted by the right-hand selector switch, are precisely known. One of these, A_{cir}, is first set to 100% transmission using a clear glass plate, then the free-form aperture is brought into view and its transmission, T_{ff} is recorded. From the transmission reading T_{ff}, which should be smaller than the 100% of the selected circular aperture, the area A_{ff} of the unknown can be calculated easily.

$$A_{ff} = \frac{T_{ff} \times A_{cir}}{100}. \tag{6}$$

It sometimes happens that the transmission of the object can be measured with an aperture, but that no equivalent free background area can be found. A smaller aperture must then be selected for the background measurement and its T_∞ set to 100% separately. Only the aperture size for T is used in subsequent calculations.

Some short cuts in the procedure may be justified under certain conditions. If, for example, a large series of particles is to be measured from negatives with even and consistent backgrounds, only one or a few background transmissions need be noted. For the quick assessment of the distribution of mass in a sample it may be unnecessary to adjust the transmission of each of the various apertures used to 100% precisely, as the error may be only a few percent. Only minor adjustments of the 100% T_∞ value are usually necessary to obtain 100 for each aperture, which in such a situation may be insignificant and can therefore be left out.

In practical applications of the method it is often desirable to weigh only a part of the object, e.g., only the head of an intact spermatozoon. This is easily achieved by "cutting" the image of the object with the aperture edge at the desired place, i.e., the aperture contains only that part of the object to be measured. The measurement of the total fiber length per chromosome is another good example of the measurement of partial transmission. The total weight of the chromosome is determined with a large aperture first. Then pieces of chromosomal fibers are measured with a small rectangular aperture of known length. Division of the total chromosome weight by the average weight of a section of fiber, and multiplying it by the length of the section gives the total fiber length per chromosome.

Calculation of Mass

The formula

$$W = R \frac{0.5495 : 10^{-3}}{M^3 \cdot \beta} \text{ [gm]} \tag{7}$$

has been found useful in the method of mass determination. W is the weight of the object in gram. R, the reading is

$$R = \frac{\overline{T} - T_0}{T_\infty} \cdot A \tag{8}$$

where $\overline{T}$ is the transmission over the object averaged photometrically, and T_0 is the background transmission, both corrected for fog. These two values are obtained from the entries on the log sheet. Their difference is multiplied by the area, A, of the aperture used, which is also recorded on the log sheet. All readings are thus normalized for aperture size and are freely comparable, regardless of the aperture used. At this point statistical evaluation on a relative basis can be performed. The portion

$$\frac{0.5497 \cdot 10^{-3}}{M^3 \cdot \beta}.$$

of Eq. (7) represents the values which convert R to an absolute weight in grams. Its components are derived as follows:

$$\tfrac{4}{3}\pi r^3 \cdot \rho \cdot D^3 = 0.5497\, D^3.$$

We used a specific gravity ρ of 1.05 g/cm^3 for the polystyrene spheres most often used for calibration (see next section "Standardization").

M is the magnification, and β a factor relating the integrated transmission measurement R_s on a sphere, to its volume

$$\beta = \frac{R_s}{D^3}.$$

The diameter of a sphere D is known from its image in the negative, and the magnification is known from the image of the magnification standard. Since D has been measured in millimeters, it is multiplied by 10^{-3}. We now write

$$W = \frac{\pi \cdot \rho \cdot 10^{-3}}{6} \cdot \frac{1}{M^3 \times \beta}.$$

The factor β is derived from a graph in which R_s is ordinate, and D^3 abscissa. The resulting distribution of entries is usually smoothed, and the straight line of best fit is drawn through the origin. At $D^3 = 10$ the slope of the curve is read off to give β, which is inserted in Eq. (7).

When large series of measurements are made routinely, much time and effort can be saved by using electronic data collection, and preferably some evaluative computer programs such as the system in use in Washington, D.C.

Standardization

The electron micrograph of the standards of mass (Fig. 12) shows an area of an air-dried suspension of spheres of polystyrene latex, in which some 3–10 spheres lying singly can be seen. These are numbered on a working print, and their R_s is measured as described above under "Measurements with the IPM-2".

Subsequently, the diameter of each sphere is measured twice as precisely as possible under a low-power microscope, and the average is obtained, because D is used to the third power. If the volume of each sphere is multiplied by the specific gravity of polystyrene (1.05 gm/cm^3), we obtain the absolute weight for each sphere measured, and can draw a graph of R_s (ordinate) against the true weight of a sphere. This is a most straightforward calibration curve; it should intersect at or very close to the origin and should be a straight line which fits

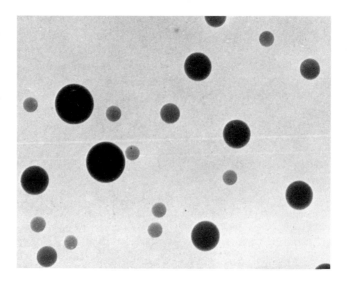

Fig. 12. Electron micrograph of a mixture of polystyrene latex spheres, for calibration of the determination of dry mass in absolute units (gram)

the data closely. Any reading R can now be read off the calibration curve as an absolute weight.

Single-Line Scanning Measurements

There are two cases when one may not be able to obtain integrated measurements with either circular or free-form apertures. In the first, the object's dimensions, or one of them, such as in fibers, may exceed the diameter of the largest circular aperture. In the second, a small rectangular aperture, scanning across the object's image, can only give the cross-sectional mass.

In the scanning mode of the IPM-2, the regular stage is replaced by a motor-driven stage providing movement of the upper half of the stage. Usually a narrow slot aperture is selected for scanning, with a length which exceeds the dimensions of the object slightly, i.e., it is slightly wider than the muscle fiber to be scanned lengthwise. The plate is oriented on the scanning stage in such a way that the object's long axis moves orthogonally to the long axis of the slot aperture. A strip-chart recorder is coupled to the output of the IPM-2. In scanning, the total cross-sectional mass is recorded per unit length. The topology of the mass distribution is preserved along one axis of the specimen.

Scans of objects are compared to scans of weight standards by using the identical settings for the recorder and preferably by having the same magnifications. The area under the scan for a whole sphere is equivalent to the total mass of the sphere and serves to interface other scan recordings. Scanning of a magnification grating provides an absolute measure of size for scanning.

Small, twisted fibers are dealt with in a similar manner. The fiber is scanned across at any short, reasonably straight, part, using a small slit just long enough to provide a useful record, but short enough to fit the few straight stretches of fiber. With the same setting, a sphere is scanned, not necessarily through its center. The area under the recording from the sphere is set equivalent to the slice of a sphere, of known specific gravity and thickness, given by the length of the scanning slot. The areas under the curves are evaluated either by planimetry or by weighing the cut-out pieces and comparing them with the weight of pieces of known area.

Concentration of Solids

All of the above techniques are applied to quantitative measurements on whole, isolated objects, where the total or partial mass is to be determined. For thin sections, another parameter is assessed, namely concentration. After the usual embedding and sectioning procedures, one obtains a solid layer of organic material of rather even thickness, i.e., the specimen in a hardened plastic. As a whole, few objects of biological interest ever occur as a thin section. Any variation in the contrast of the image of sectioned material is due to differences in the concentration of matter, commonly known as specific gravity (ρ). Only by comparing the transmission of a standard of known ρ [g/cm^3], embedded and sectioned together with the specimen, can ρ be calculated for the specimen. This approach is useful when assessing concentration gradients within a structure, or evaluating the increase of contrast after an enzyme reaction which produces a precipitate of high density material. To compare any two sections, one must know the thickness of each section. Incorporation of a standard makes it possible to calculate the thickness of the section. Techniques for thin sections have been described in a series of papers by SILVERMAN and GLICK (1969).

Preparing an Object for Quantitative Electron Microscopy

Thin sections are the standard objects of transmission electron microscopy, TEM, in the biological sciences. As discussed previously, a typical thin section of embedded tissue can only be analyzed quantitatively in terms of the concentration of biological substances, and its change, or the concentration of contrast-enhancing reagents if the contrast before enhancement is known. Stereology and autoradiography can also be considered as quantitative methods in electron microscopy, but will be disregarded for the purpose of this discussion since they do not evaluate contrast.

The first methods of preparation for electron microscopy consisted simply of putting objects on a thin membrane over holes in the specimen carrier. This approach is still used in principle for the preparation of objects for the quantitative evaluation of contrast. Some essential refinements were added to the procedure, however, long before scanning electron microscopy, SEM, forced existing techniques to be reviewed. Most prominent among the rediscovered methods is critical-point drying, decribed by T.F. ANDERSON in 1951, and revived by GALL (1966) and by others in the sixties. Surface spreading was described by FERNANDEZ-MORAN as early as 1948, but was fully developed by KLEINSCHMIDT and colleagues (1962).

All these technique are used frequently for quantitative measurements, and are essential for delicate objects. The major preparatory principles are listed below.

Simple Preparatory Steps

1. A droplet of the suspension of interest is placed onto a grid coated with Formvar. Too much should not be applied, because the drop will extend beyond the circle of the grid and thus decrease the thickness of the fluid layer on the film, so that fewer particles attach to it; also preparatory results become unpredictable. The fluid is left to stand for 5–10 min, the small quantities of fluid remaining are blotted off by touching the edge of a filter paper to the edge of the grid.

If the fluid contained salts, sucrose, or similar material, droplets of rinsing solution (most often distilled water) are allowed to roll down the grid while it is held obliquely; it is then allowed to dry in air.

The quality of the suspension is, of course, decisive for the success of this simple operation. Factors affecting this are the quantity of debris present, the concentration of objects (number/cm^3), the nature of other solutes, and fixation with formalin for 10 min by diluting the suspension with $\frac{1}{3}$ of 10% (4% absolute) neutral formaldehyde.

2. Extended objects such as spermatozoa tend to become entangled and to overlay each other. A microscope slide is covered with a film of Parlodion (a collodion dissolved in amyl acetate); this film can be relatively thick, e.g., 500 Å, because Parlodion looses 40–50% of its mass upon irradiation with beam electrons. The suspension is smeared onto the film very gently, taking care not to scratch it. While still wet the slides are rapidly placed in an oven at 85° C so as to heat fix them. The fixed objects are then washed with distilled water and air-dried again. In our experience whole cells, spermatozoa, or red blood cells are well preserved and are not lysed at all. The film is broken by scoring the glass edge with a needle or razor blade, and is floated off by immersing the slide very slowly and at a shallow angle into distilled water. A wood-handled preparation needle is kept ready to ease off the film if it does not come off by itself. If the object side of the film becomes wetted it must be allowed to dry before repeating the attempt. We place grids of 100 or 150 mesh on the floating film and remove the entire assembly from the water by overlaying it with a piece about twice its size of oil paper, parafilm (a thin, commercially-available sheet composed of long chained paraffins), or any plastic that is stiff enough to get the film-grid assembly from the surface of the water without wrinkling it.

Whatever means used to lift film and grids from the water, they are now air-dried, and are ready for use.

3. In both of the above methods the strong effects of surface tension cause flattening and often distortion of the objects. This is more noticeable for unfixed material than for objects fixed with formalin. Surface tension is eliminated entirely in the critical-point drying technique of ANDERSON (1951). Although it was not very much used for 15 years following its introduction, the technique is now essential for Scanning Electron Microscopy and for the preparation of many objects for Transmission Electron Microscopy. We have found ANDERSON'S

technique indispensable for the preparation of delicate chromosomes. It should be noted here that none of the variations of the original carbon dioxide technique produces acceptable results. We also tried Freon and Nitrous Oxide but abandoned them.

Preparing Human Chromosomes

There are of course many variations for preparing specimens for quantitative work when critical-point drying is the method chosen. We will describe the method of preparing whole mounted chromosomes for quantitative determinations of dry mass as an example, because many of the steps are useful for other objects. First of all, a list of the things needed:

Table of Required Items

1. Clean microscope slides
2. Coverslips
3. Acetic acid orcein stain in bottle with pipette
4. Phase microscope
5. Tapered centrifuge tubes
6. Swinging bucket or angle-head laboratory centrifuge, non-refrigerated
7. Stand for 10 centrifuge tubes
8. Hanks solution
9. Pasteur pipettes, disposable, glass
10. Bulbs (suction) for pipettes
11. Stainless steel spatula with 4 mm wide tounge
12. Distilled water
13. Diminac (Demineralizer)
14. Spreading trough, teflon coated
15. Towels aplenty
16. Formvar (polyvinylformal), powder
17. Ethylenedioxide, chemically pure
18. 200-mesh copper grids for electron microscopy
19. Stainless steel tweezers, No. 3, from Dumont, Switzerland
20. Holder for grids to be inserted into CPD apparatus
21. Ethyl alcohol, 100%
22. Amyl acetate, chemically pure
23. Ethylene glycol, chemically pure
24. Ethylene glycol monoethyl ether (Cellosolve)
25. Wide-mounthed glass jars, 25 cc
26. Critical Point Drying (CPD) apparatus suitable for use with liquid CO_2
27. Carbon dioxide bottle, with siphon tube, with transport cart, and safety mount
28. Crushed ice
29. Hot or heated water $\approx 55°$ C
30. Plastic storage boxes for grids with numbering system for grid retrieval. (Several companies supply these.)
31. Large wide-mouthed glass jar (e.g., desiccator), one-third filled with a granular drying agent, for storage of grid boxes.

Chromosome Culture

Commercial chromosome culture kits from Difco Company[1] are used for growing chromosomes.

Peripheral blood is obtained from a donor preferably several hours after a meal. Fluid from the separation vial (heparin) is drawn into a 10 cc syringe, and then 10 cc of blood is obtained by venipuncture of the donor. The heparinized blood is kept upright in the syringe for 30–60 min, until separation of the leucocyte-rich plasma is obvious. Approximately 1–2 ml of this plasma is injected directly into a TC chromosome culture bottle which has been mixed previously with reconstituting fluid.

The innoculated cultures are incubated at 37° for 66–72 hrs. They should be checked after 24 hrs. and 48 hrs. to see that the medium indicator is not too yellow (acidic, the cap must be loosened to allow CO_2 to escape) or too red (alkaline, tighten the cap, to prevent loss of CO_2). On the morning of the third day, a vial of arresting solution (colchicine) is added to each culture bottle and the mixture is agitated gently. Approximately 6 hrs. later the cells are harvested. The contents of each culture bottle are poured into a test tube and centrifuged for 10 min at 200 g. The supernatant is discarded and the pellet of cells is re-suspended in hypotonic Hanks solution diluted 1:1 with distilled water, to swell the cytoplasm and nucleus. Hypotonic treatment is allowed to proceed for 10 min before a second centrifugation for 5 min at 100 g. Some types of cells, often long-established tissue-culture lines, require longer exposure to diluted Hanks and possibly to distilled water alone. Swelling should be monitored by phase contrast microscopy. The supernatant is discarded and the remaining pellet is spread directly on a modified Langmuir trough.

The modified Langmuir trough (Fig. 13) consists of a teflon coated metal through, 50 mm by 125 mm by 7 mm deep. Teflon coated bars[2], 12 mm by 12 mm by 125 mm are used to clean the surface of the solution in the trough. Other suitably prepared troughs and bars, for example, glass coated with paraffin can be used. Distilled and deionized water is the preferred medium for chromosome spreading. After the bars are passed over the convex water surface several times to remove all contaminants, they are placed approximately 25 mm apart and thus define the spreading area. Part of the pellet of swollen cells is pipetted from the test tube onto the tip of a metal spatula, and applied to the surface of the spreading area. The already swollen cells swell more in the distilled water, burst, and release their chromosomes and nuclei which, because of their fibrous composition, become entangled with cellular proteins on the surface and so do not sink as do most of the other cellular components (BAHR, 1971).

Chromosomes and nuclei are removed from the trough by touching FORMVAR-coated grids to the surface. Six grids are usually applied before any are removed, so that the surface film of denatured cell proteins is disturbed as little as possible. The grids are then removed one by one from the surface and are placed in 30% ethanol for the start of dehydration.

[1] Detroit, Michigan.
[2] Available from C.W. French Division, 5 Shasheen Avenue, Bedford, Mass.

Fig. 13. Utensils for spreading biological objects on a surface. Teflon coated, anodized aluminium trough; 2 bars for limiting the surface of spreading; demineralized water; holder onto which rings receiving grids and spacer rings are stacked under 30% ethanol; spatula for applying pellet or droplet of concentrated object; little tube with talcum powder, only to be sprinkled onto the surface to visualize spreading the first few times

Preparation of FORMVAR-Coated Grids

FORMVAR-coated grids are prepared by dipping a clean 1×3 cm slide into a 0.35% solution of FORMVAR dissolved in ethylenedichloride. The slide is placed upright on filter paper, leaning against a glass beaker. When dry, the film is scored along its edges. The slide is then slid gently, at a shallow angle, into a dish filled with water, and the film separates from the glass and floats on the surface. Electron microscope grids are placed gently on the floating film. Film and grids are then picked up as described previously. The grids are now sandwiched between film and the material used, for pick-up. The paper is placed in a dust-free atmosphere and allowed to dry; on drying, the film becomes attached to the grids and becomes the specimen support.

Dehydration and Critical-Point Drying

In the critical-point method, dehydration is the first step for bringing the wet spread specimens to a dry condition without their suffering the disruptive and distorting effects of surface tension during drying. If a liquid in equilibrium with its own vapor is heated in a constant space, a critical temperature is reached at which liquid and gas are indistinguishable, and above this temperature no additional compression will force the gas to condense back to the liquid phase. As the pressure increases, the density of the vapor phase increases until at the critical pressure, it is equal to that of the liquid. Surface tension is reduced until it becomes zero, and the border between liquid and gas vanishes. The critical point is the particular combination of critical temperature and critical pressure. Particles immersed in the liquid and carried through the critical point are therefore brought to the dry condition (being surrounded by CO_2 gas) without having passed through the surface. The gas is then bled off slowly above the critical temperature, in order to avoid adiabatic cooling. The particles remain dry. ANDERSON (1952) calculated that the tensional stress exerted on a bacterial flagellum, 200 Å in diameter (the diameter of an average interphase chromatin fiber) in a drying water film may be as high as 325 tons/in². ANDERSON used carbon dioxide (critical temperature=31° C, critical pressure=72 atm) as the transitional fluid. For the intermediate fluid, miscible both with the dehydrating agent (i.e., ethanol) and the transitional (carbon dioxide), ANDERSON used amyl acetate.

The preparations, in 30% ethanol, are placed one by one in a brass grid-holder (each holds 6 grids), 5 of which can be stacked on a post to be processed at one time. The stack (with 30 grids) is passed through a series of wide-mouthed bottles containing ascending concentrations of ethanol (50, 70, 95%) for ap-

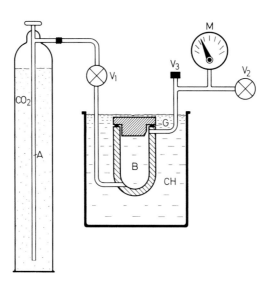

Fig. 14. Critical-point drying apparatus. At left a flask of carbon dioxide, with raiser tube to deliver liquid CO_2. V_1 is the inlet valve of normal design, V_2 is the outlet needle valve. Bomb B is filled from V_1 and must withstand 2400 pounds per square inch pressure (safety factor of one-third). It is a blow-out valve triggered at the maximum pressure the bomb will withstand. Pressure is measured by manometer M. The bomb is designed to have an internal volume of 50–75 ml, depending on quantities to be dried. It is sealed by some mechanical means, often employing a plastic or rubber gasket G. The bomb can be immersed in ice water or hot water, CH

proximately 3 min each, and then through two changes of truly absolute ethanol. This is followed by three passages through water-free amyl acetate. The specimens in their holder, wet with amyl acetate, are placed quickly into a bomb attached by high pressure stainless steel tubing to a tank containing liquid carbon dioxide (Fig. 14). The bomb is flushed, through valve V_1, with liquid carbon dioxide (tubing and bomb are immersed in an ice bath, B) to replace the amyl acetate. Flushing continues, with the bomb's escape valve (V_2) slightly open, until the characteristic smell of amyl acetate disappears from the exhaust gas. With the bomb filled with liquid carbon dioxide alone, and closed at both intake and escape valves $(V_1 + V_2)$, the temperature is raised by immersing tubing and bomb in hot (50° C) water, (B). During this warming process, the pressure increases and the volume remains constant. The escape valve (V_2) is opened gently to allow the gaseous carbon dioxide to blow off very slowly. During this process the pressure of CO_2 decreases to the ambient value. Critical-point drying is now complete; the bomb is opened, the brass grid holder removed, and the specimen is ready for examination in the electron microscope without having suffered the damaging passage through a phase boundary. Grids dried by the critical-point method are stored in boxes, which in turn are stored in a suitable jar or desiccator containing a solid desiccant.

Summary

A method is described in detail, in which electron microscopic contrast is used to determine relative and absolute dry masses of biological objects. The weight of an object can be ascertained, regardless of shape and composition. Photometric procedures and instrumentation are explained, by which large numbers of measurements in the range from 10^{-10} g to 10^{-18} g can be made rapidly. Photographic and preparatory procedures are described, and illustrated by examples. The error for relative mass is $\mp 2\%$ and that for absolute mass $\pm 9\%$.

Acknowledgements: This work has been supported by contracts from the Medical Research and Development Command, United States Army, Washington, D.C.; and grants from the American Cancer Society. The close cooperation with Carl Zeiss, Oberkochen, in the development of the IPM-2, is gratefully acknowledged.

Literature

ANDERSON, T.: Techniques for the preservation of three-dimensional structure in preparing specimens for the electron microscope. Trans. N.Y. Acad. Sci., Ser. II, **13**, 130–134 (1951).

BAHR, G.F.: Quantitative electron microscopy: The present state and a look into the future. Invited lecture, Fourth European Regional Conference on Electron Microscopy, Rome, Italy, 567–572, 1968.

BAHR, G.F.: Human chromosome fibers, considerations of DNA protein packing and of looping patterns. Exp. Cell Res. **62**, 39–49 (1970).

BAHR, G.F., CARLSSON, L., ZEITLER, E.: Determination of the dry weight in populations of submicroscopic particles by means of quantitative electron microscopy. First International Biophysics Congress, Stockholm, 1961, p. 327.

BAHR, G. F., SACKERLOTZKY, O., ZEITLER, E.: Electromagnetic exposure system for Siemens Electron Microscope. Rev. Sci. Instr. **34**, 1443–1444 (1963).

BAHR, G. F., ZEITLER, E.: Study of bull spermatozoa, quantitative electron microscopy. J. Cell Biol. **21**, 175–189 (1964).

BAHR, G. F., ZEITLER, E.: The determination of dry mass in populations of isolated particles. Symposium on Quantitative Electron Microscopy (editors: BAHR, G. F., and ZEITLER, E.), p. 217–239. Baltimore, Md.: Williams and Wilkins 1965.

FERNANDEZ-MORAN, H.: Examination of brain tumor tissue with the electron microscope. Ark. Zool. (Stockh.) **40**, 1–15 (1948).

GALL, J. G.: Chromosome fibers studied by a spreading technique. Chromosoma (Berl.) **20**, 221–233 (1966).

HALL, C. E.: Introduction to electron microscopy, 2nd ed. New York: McGraw-Hill 1966.

HALLIDAY, J. S., QUINN, T. F. J.: Contrast of electron micrographs. Brit. J. Appl. Phys. **11**, 486–491 (1960).

KLEINSCHMIDT, A. K., LANG, D., JACHERTS, D., ZAHN, R. K.: Preparation and length measurements of the total desoxyribonucleic contant of T_2 bacteriphages. Biochim. Biophys. Acta (Amst.) **61**, 857–864 (1962).

LIPPERT, W.: Erfahrungen mit der photographischen Methode bei der Massendickenbestimmung im Elektronenmikroskop. Optik **29**, 273–278 (1969).

REIMER, L.: Elektronenmikroskopische Untersuchungs- und Präparationsmethoden, 2nd ed. Berlin-Heidelberg-New York: Springer 1967.

SILVERMAN, L., GLICK, D.: Measurement of protein concentration by quantitative electron microscopy. J. Cell Biol. **40**, 773–778 (1969).

SILVERMAN, L., SCHREINER, B., GLICK, D.: Measurement of thickness within sections by quantitative electron microscopy. J. Cell Biol. **40**, 768–772 (1969).

SILVESTER, N. R., BURGE, R. E.: A method for the routine evaluation of dry mass from electron micrographs. J. Roy. Micr. Soc. **81**, 25 (1962).

SMITH, G. H., BURGE, R. E. A.: Theoretical investigation of plural and multiple scattering of electrons by amorphous films, with special reference to image contrast in the electron microscope. Proc. Phys. Soc. **81**, 612 (1963).

ZEITLER, E., BAHR, G. F.: A photometric procedure for weight determination of submicroscopic particle by quantitative electron microscopy. J. Appl. Phys. **33**, 847–853 (1962).

ZEITLER, E., BAHR, G. F.: Contrast and mass thickness. Symposium on quantitative electron microscopy (editors: BAHR, G. F., and ZEITLER, E.), p. 208–216. Baltimore, Md.: Williams and Wilkins 1965.

WIED, G. L., BIBBO, M., BAHR, G. F., BARTELS, P. H.: Computerized microdissection of cellular images. Acta Cytol. (Philad.) **14**, 418–433 (1970).

Chapter 8

The Construction and Use of Quartz Fiber
Fish Pole Balances

The quartz fiber balance is an extremely simple spring balance, in which a quartz fiber of suitable dimensions is mounted horizontally in a case to protect it from air currents and static charges. Its displacement with microgram or smaller loads is measured by an eyepiece micrometer in a horizontally mounted binocular microscope which is also used for loading and unloading the balance. The precision is approximately 1.0–2.0%; it has a relatively narrow range of 50–100 times its sensitivity; however, balances can be constructed with useful sensitivities which far surpass those of other mechanical balances at present available.

The subject of the construction and design of quartz fiber balances has been discussed in great detail recently by OLIVER H. LOWRY (O. H. LOWRY and JANET V. PASSONNEAU: A Flexible System of Enzymatic Analysis, p. 236–249. New York: Academic Press 1972). All serious students of micromethodology should read this chapter. The present chapter will summarize and augment some of the technical aspects of balance construction as given by LOWRY.

Optical Measurements

The displacement of the quartz fiber tip is generally measured on an eyepiece micrometer possessing a 5 mm scale. The displacement of the quartz fiber as measured by this scale will depend upon the objective magnification; what is actually being measured is the displacement of the real image produced by the objective lens. Thus, if the objective lens is 1× then the image is displaced 1 mm for every mm of object displacement; if the objective magnification is 2×, then the image will be displaced 2 mm for every 1 mm of object displacement. It should be noted that eyepiece micrometer scales are commonly placed in the ocular of the microscope at a convenient focal point for the viewer, depending upon the microscopy technique of the user; this may be a point determined by focussing the eyes at infinity, or at a near point. To be of use, the micrometer scale must be placed in the image plane of the objective. It is possible that users employing various techniques may, on refocussing the ocular, alter the plane of the micrometer disc, and so vary the effective magnification within a range of ±5%. It is therefore advisable, when calibrating a particular balance with a particular horizontal microscope, to focus the ocular micrometer scale on an absolute micrometer scale mounted conveniently in the plane of the balance.

Thus, the actual magnification of the ocular micrometer disc under the conditions of balance calibration can be determined and can be checked conveniently at any time.

The Choice and Manufacture of Suitable Quartz Fibers

Quartz fibers of various diameters down to approximately $3\,\mu$ are commercially available. However, with a little practice, fibers of almost any diameter can be made in the laboratory. The relationship between the diameter of a fiber and the sensitivity of the balance is expressed by the formula of LOWRY

$$S = \frac{L^3}{d^4}\ \text{mm}/\mu\text{g}$$

where S is the sensitivity, L is the length, d is the diameter, and mm/µg is the displacement in millimeters per microgram. This formula applies to displacements which are relatively small in relation to length, usually less than 10%. LOWRY has constructed a set of curves which relate the balance sensitivity to the diameter, length, and self displacement of the fiber. Since the diameter of the fiber determines its weight per unit length, the self displacement, or "droop", is a function of the length and diameter of the fiber. Thus, if the length and droop of the fiber are known, its diameter and the balance sensitivity are automatically determined.

In practice, the sensitivity of a particular fiber can be determined easily, by mounting it temporarily in a suitable container to reduce air currents, and estimating the displacement of the tip of the fiber when the fiber is rotated from a vertical to a horizontal position. The sensitivity, expressed in millimeters per microgram, can then be found by the following formula of LOWRY:

$$S = \frac{3 \times 10^6\, D^2}{L^5} = 3 \times 10^6 \left(\frac{D}{L}\right)^2 \left(\frac{1}{L^3}\right)\ \text{mm}/\mu\text{g}$$

where D is the droop in millimeters and L is the length in millimeters.

Since the sensitivity of a fiber is proportional to the cube of its length, a relatively long fiber may be shortened so as to give a lower sensitivity. In practice, fibers with a droop of greater than 15% of their length are not practicable, nor will the displacement be entirely proportional to the weight placed at the tip. To achieve higher sensitivities it is better to use thinner, shorter fibers (which will also have proportionately less self displacement in relation to sensitivity).

Although balances of a particular sensitivity can be accurately constructed by the method outlined, the sensitivity can usually be roughly estimated, especially for balances with sensitivities of 0.01 µg/mm or less (i.e. relatively large balances with relatively low sensitivities), by noting the droop of a fiber in relation to its length when held by one end in a relatively draft-free room. This experience is gained by constructing a set of balances with overlapping ranges of sensitivities.

Making the Fiber

The method described here is different from that of LOWRY, and probably allows greater control over the diameter of the fiber than obtained by any other method of fiber manufacture. In essence, it involves the use of a micro torch to spin an extremely thin fiber from a fiber or rod with a thickness of 50–500 times the required diameter. This technique requires only a few hours of practice in order to make fibers with diameters of from 50 μ to less than 1 μ. As a guide, a fiber of 1 μ, 6 mm long, will have an approximate sensitivity of 5 nanograms per millimeter of displacement. The art in the method to be described is to start with a suitably thin piece of quartz, use a suitably small flame of the proper temperature, and to heat the quartz at precisely the right point in relation to the place from which the fiber is drawn. The thinner the fiber required, the thinner must be the stock from which it is made and the cooler and smaller the flame. The most suitable torch for micro work is one manufactured by the Tescom Corporation, 315 14th Ave. Southeast, Minneapolis, Minnesota 55414, U.S.A., which uses methane or propane and oxygen, or standard "illuminating" gas (normally used for cooking) plus oxygen. Suitable quartz rods ranging from 0.05 to 2 mm in diameter may be made by gathering a bead of quartz heated to white heat *(welders goggles or their equivalent must be worn)* in a gas-oxygen blast-lamp or blow-torch and then pulling rapidly or slowly immediately after removing the quartz bead from the flame. The quartz bead may be made from 5–8 mm tubing or rod. Beads with a diameter of 5 to 8 mm will yield rods or fibers of 0.2 to 2 mm, depending upon the rate of pulling. Smaller beads will yield relatively thinner fibers. Having made a supply of stock fibers or rods with suitable diameters, one can then move to the micro torch. Conveniently, such a torch is set up with the nozzle pointing toward the operator, and manoeuvers are best observed through a long working distance stereo microscope under magnifications of 10–20 ×. The nozzle of the burner is placed close to the table so that the stock rod can be held by the operator between the thumb and forefinger of the favored hand (this hand will control the position of the stock in the flame) while resting the hand firmly on the table to allow fine control and positioning of the stock in the flame. The stock is then drawn to a fine taper in the flame. Almost immediately, the thicker portion of the taper is moved into the flame so that the heat is applied a short distance away from the thinnest part, i.e. that part from which the fiber is to be spun. At the same time the other hand, holding the other end of the tapered rod, is moved away from the flame at a steady rate, so spinning the quartz fiber off the end of the taper. As material is removed from the tapered end, the quartz stock is advanced into the flame to maintain the relative position between the heating point and the spinning point. If the spinning is too rapid the fiber may break and will often whip back and become ravelled and useless. If the operation is performed carefully, uniform fibers can be produced between two pieces of tapered stock, by removing the stock from the flame at the moment that pulling is stopped. If the quartz fibers are exceedingly thin they are best observed by the Tyndall effect, with oblique lighting against a black background. The fibers are best stored by laying them on a piece of black velvet and breaking the relatively thicker ends, near the taper of either side, against the velvet, much in the manner in which one would break the tip of a pencil by pressing on it too hard.

During the process of fiber spinning it is extremely important that the feeding hand should hold the stock as close to the flame as possible without burning the fingers. With tiny flames this might be as close as 1 cm or less. The remaining fingers of that hand are pressed firmly against the table, and all movements of the stock are accomplished by moving the entire hand against the resistance of the fingertips, in such a way that the fingertips roll over the table without actually sliding. Such a manoeuvre allows extremely fine repositioning of the tip of the stock, with the kind of control one would ordinarily expect only from a micro-manipulator. Although this is an extremely simple manoeuvre, most people not familiar with it are astonished at the accuracy obtainable for small displacements. When the process has been demonstrated by the author, his audience usually responds as to a virtuoso performance; however, within less than two hours of practice, even those with little skill or experience can make very adequate quartz fibers by this technique. Moreover, the technique can be used to manufacture quartz fibers with various controlled tapers, and it is relatively easy to make quartz points which are relatively rigid at one end and which taper to extremely small diameters very rapidly. Such quartz points, when glued to a hairpoint, provide extremely convenient tools for handling pieces of frozen dried tissue weighing less than 50 nanograms. Quartz fibers can be cut with scissors; portions selected for balance construction should be of uniform diameter and relatively straight. A slight uniform curvature is often useful if the fiber can be mounted with the concavity facing upwards so that the final, mounted fiber will be nearly straight, the curvature compensating for the self displacement. Such balances will have the most nearly linear relationships of weight to displacement.

Balance Cases and Final Assembly

Any convenient chamber of the appropriate size may be used for mounting the quartz fiber. LOWRY has devised a most convenient balance case for balances intended to weigh samples of 10 µg or less: a glass syringe barrel is cut off perpendicularly near its tip, and the plunger is inserted backwards into the barrel. The flared end of the barrel is ground flat against a glass plate with carborundum to give a relatively airtight seal with a glass plate. A spiral of copper wire is mounted to the end of the plunger, using a quick-setting epoxy resin, with the central portion of the spiral wire protruding at right angles to the plane of the spiral for approximately 1 cm. It is often convenient to sharpen this protruding end to a relatively fine point. The quartz fiber is then mounted to this end with a quick-setting epoxy resin. This type of case is extremely convenient, since the plunger may be advanced so as to protrude beyond the end of the barrel for mounting the quartz fiber or for adjusting its position by bending the free end of the copper wire behind the point of attachment of the quartz fiber using either forceps or artery forceps. The mounted fiber can then be drawn into the barrel by withdrawing the plunger, and the fiber can be positioned optimally by rotating the plunger. Preliminary testing for sensitivity can then

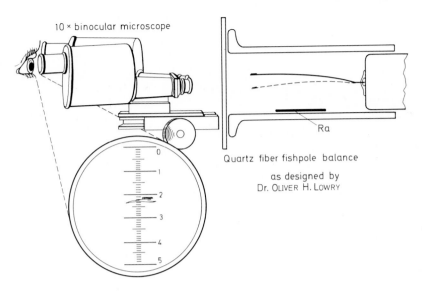

10 × binocular microscope

Ra

Quartz fiber fishpole balance

as designed by
Dr. OLIVER H. LOWRY

Fig. 1. Diagram of the balance arrangement. The circular insert shows the position of the balance pan with a load as seen through the microscope

be performed while the barrel and plunger are held in a temporary mounting. The free end of the fiber should be positioned approximately 1 mm behind the plane of a plate of glass which is brought flush against the ground end of the barrel.

A balance with a load limit of 0.05 µg or more usually requires a pan mounted on the end of the fiber to hold the sample. Since the pans must be extremely light in weight and should not ordinarily cause more than a 10% displacement of the end of the fiber, they are conveniently made from fragments of extremely thin bubbles of pyrex glass. Such fragments are easily made by sealing one end of a piece of pyrex tubing in a relatively hot flame and collecting a bead at least as large as the diameter of the tube. The tube is then removed from the flame and maximum pressure is applied at the open end by blowing. A large, thin bubble usually forms which breaks; the flakes are collected on a piece of black velvet and a conveniently thin portion is placed on the stage of a binocular microscope on a piece of plexiglass (Perspex). The thin glass, usually thin enough to show inter-ference colours, can be trimmed using micro knives (described elsewhere in this book) to a square or rectangle with dimensions of 1 to 4 mm on an edge, depend-ing upon the sensitivity of the balance and the size of sections to be weighed. The pan is then placed, with its convexity upwards, on the end of a teflon slide or object carrier.

Having determined the optimal position of the quartz fiber in its case, a mark is placed on the end of the plunger, denoting the top of the balance. The balance is then turned through 180° and mounted in a convenient, adjustable clamp which permits the positioning of the balance and the fiber under the stereo microscope. The plunger is then advanced, bringing the free end of the quartz fiber into position directly on the center of the pan.

The balance pan is then mounted to the end of the fiber in the following way: a piece of DE KOTINSKY or similar laboratory cement or sealing wax is pulled into fine fibers. A short piece of such a fiber with a diameter no greater than twice that of the quartz fiber is then placed between the quartz fiber and the balance pan. While observing the process through the stereo microscope, the heated tip of a micro soldering pencil (such as is used for small transistor work) is brought near enough to the cement to cause it to melt and is then immediately removed. The quartz fiber with the pan attached can then be lifted away from the teflon with a hairpoint, and the balance is reversed into its normal position. With the tip of the fiber and pan still supported by the hairpoint, the plunger is carefully withdrawn into the barrel until the free end of the quartz fiber is 1 mm to 3 mm behind the plane of the cover. In its working position the pan will be seated on top of the end of the fiber; the copper wire to which the quartz fiber is mounted may then have to be bent up a bit in order to compensate for the additional droop of the fiber caused by the weight of the pan. The fiber should be checked by placing an appropriate weight on the pan to ascertain that the weight will displace the fiber in a vertical direction. This should usually be performed both before and after mounting a pan, to ensure that the pan will remain horizontal and that the fiber will be displaced vertically by weights. If these conditions are not attained, the balance will not be accurate.

Mounting the Balance

Several balances may be mounted side by side on a single balance table. An arrangement which has been found useful in the author's laboratory is to use a heavy table of desk height (85–90 cm). The individual balances are mounted horizontally with heavy clamps, each clamp suspended from a horizontal bar which may be mounted either between two heavy ring stands or permanently fixed to the table by means of a chemical support rack. A balance should be mounted in such a way that the open end of the syringe barrel faces the operator at approximately eye level when the operator is seated at the table (see Figs. 2 and 3). A convenient hand support can be provided by a horizontal heavy wooden bar mounted separately to the table at a level approximately 1 cm below the barrels of the balance cases. This bar should be wide enough to allow space for specimen carriers close to the balances, and should be solid enough to give firm support for the hands of the operator while his elbows are firmly on the table so that the loading process can be controlled well. Usually the hairpoint is held in one hand like a pencil, guided by the forefinger while the remaining fingers rest firmly on the hand support. In our laboratory we have found that a wooden beam, 10 cm × 10 cm, running the length of the balance table, provides an excellent support (Fig. 3). A balance should not be moved once it has been mounted. Some workers have designed turntables on which several balances may be mounted like the spokes of a wheel, and with which the proper size of balance can be selected by rotating it into position. However, it must be emphasized

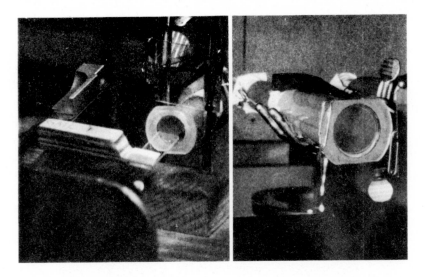

Fig. 2. Two quartz fiber balances are shown in operation. On the left is a small balance with a load limit of 0.05 µg. The cover is open and the tissue carrier in the loading device is shown in position for loading. A small tissue fragment can be seen on the carrier. The objective of the binocular microscope is at the lower left. The quartz fiber is too fine to be seen without magnification. On the right is a balance with a load limit of 100 µg. Here the pan can be seen approximately $\frac{1}{4}$ of the way down from the top of the balance case. The cover is closed

that absolute mechanical rigidity and reproducibility of balance position are imperative in the design of such turntables. Furthermore, it is important that the hand rest is not linked mechanically to the balance mounting. The relative position of the balance with respect to the table must be absolutely, mechanically stable. The horizontal stereo microscope which is used for viewing the balance while loading and also for taking balance readings should be provided with a heavy stand which has a 3-point contact with the table. It should also be provided with a focussing rack-and-pinion which works in the horizontal position, as well as with a rack-and-pinion which allows the vertical height of the microscope above the table to be adjusted accurately.

The final adjustement of the balance is usually made after the balance case has been permanently mounted. In order to minimize air currents it is imperative that the case be no larger than absolutely necessary. A balance for the load range of 25–50 nanograms, whose fiber is approximately 6–8 mm long and approximately 1 µ in diameter, is conveniently mounted in a syringe such as an insulin or tuberculin syringe, whereas larger balances require correspondingly larger cases. The final adjustment of the quartz fiber balance should be such that the tip of the fiber is located 2–3 mm above the point of support in order to give maximum linearity in the response. In order to prevent the build-up of static charges on the fiber or on hairpoints or tissue sections, a small piece of metal foil bearing a sealed source of alpha particles is placed in the bottom of the balance case. Although radium was used initially, it is also a gamma emitter and

Fig. 3. A typical arrangement of three balances mounted side by side in operative position. Reading from upper right to left we can see the light source, *L*; the balance, *B* with its cover in a closed position, the rubber bands extending backward from the cover; the supports for the cover when open can be seen on either side of the front of the balance; the tissue carrier, *C* is seen mounted in the loading device, *L*, which has been affixed to the wooden beam of the hand rest. The upper knob controls the horizontal movement of the carrier into and out of the balance, and the lower knob controls the vertical position of the carrier with respect to the balance. The measuring microscope, *M* is seen on the right

current radiation safety standards usually prevent its use. We have found americium 241, a pure alpha emitter, in sealed sources with a total radioactivity of 5–25 µC, to be extremely well suited for this purpose.

Since the calibration of the balance may change somewhat if the fiber should get dirty and droop to a point lower than its original position at the time of calibration, it is usually convenient to record the distance of the tip of the fiber from the top of the syringe barrel.

The closure of the balance is made from a thin flat glass plate such as a good-quality microscope slide. The ends of the slide should protrude approximately 1 cm on either side of the syringe barrel so that clips may be mounted to it, to take large rubber bands which may be mounted to separate clamps, in such a way that the cover plate is held firmly against the ground front end of the syringe barrel without being attached directly to the balance case (Figs. 2 and 3). Two large nails in the hand rest on either side of the balance case, approximately 2–3 mm behind the opening of the balance, make a convenient support for the balance cover when it is open. The nails should make no direct contact with the balance case and should be long enough to protrude at least twice the height of the balance case above it. The rubber bands, holding the balance cover, run along the far side of these supports in such a way that, when the balance cover is lifted up and away from the balance case, it will rest against the supporting nails. The rubber bands should be tense enough to ensure firm closure, but should not pull so tightly against the balance case as to displace it when it is closed. Adequate lighting for the balance is imperative for easy viewing and accurate weighing. A small, focussable, low-voltage light source, such as those provided by many microscope manufacturers, can be mounted either directly on the same bar as the balances or near the balances in some other way. Such light sources, because of their high intensity, usually generate radiant heat in addition to light; they can be provided with an adequate heat filter such as can be obtained from various glass manufacturers. Excessive heating of the balances will produce erratic behavior as well as altering the calibration of the balance. The smallest balances (those with load capacities of 0.05 µg or less) which have no pans, and the diameter of whose fiber is often near the limit of visibility, provide the most difficult lighting problem. These fibers can be visualized most easily when the light beam strikes the fiber at a rather acute angle, and if the background is kept relatively dark.

Loading is best accomplished by arranging a group of samples on a sample carrier (Fig. 3) in such a way that the carrier can be inserted into the balance case directly beneath the tip of the fiber or the pan. The process is considerably simplified if the sample carrier can be mounted on a simple manipulator capable of advancing the carrier for loading and unloading, and retracting it so that the balance cover can be closed for reading. Such a manipulator can be constructed easily from a detachable mechanical microscope stage, turned on its side and provided with a clip for holding the sample holder. This instrument can be mounted to the hand rest (see Fig. 3).

Weighing

Prior to adjusting the zero point, the verticality of the eyepiece micrometer is confirmed by focussing the microscope on one edge of the balance pan or on the tip of the fiber. The vertical rack-and-pinion on the binocular microscope is then racked up and down to ascertain that the image falls on the center of the scale at both ends. The zero point is then adjusted in such a way that the top edge of the balance pan or the fiber image just touches the zero line on the optical scale. Opening and closing the cover of the balance or other manipulations should not change the position of this image. If this occurs then the balance must be more rigidly mounted. Other sources of mechanical instability may be loose mounting of the microscope or an unstable base. The balance case is opened and with the rack-and-pinion micromanipulator the samples on the holder are moved into the balance just under the fiber tip or pan. A sample is loaded onto the balance pan or tip with a hairpoint, the holder is withdrawn, the glass cover is closed, and a reading is made. The scale can be read directly to two places and the third place can be accurately estimated. In making a reading, the same reference point on the balance must be used as for the zero point setting. Usually, ten to forty samples can be lined up on one holder and weighed in succession without touching the sample carrier. Repeated weighings of the same sample should be reproducible to within 0.02 mm on the optical scale. Unless the absolute stability of the balance has been ascertained, it is useful to check the zero point between every fourth and fifth weighing. While reading the zero point, or while weighing, it is important that no weight such as the operator's elbows or hands be placed on the table. Large specimens are best transferred with a hairpoint, but small specimens are often difficult to release from the hairpoint, and it is recommended to use a fine quartz tip, glued to the end of a hair, for making transfers of specimens below 0.05 μg in weight.

The technique of picking up a section with a hairpoint will be relatively easy to learn if one remembers that adherence of the hairpoint to the substrate on which the section is lying (whether this be the balance pan or the carrier), will cause the operator to overshoot, and when the adhesion between hairpoint and substrate is finally broken the hair may spring up and flick the section away. It is therefore important that only the section is touched with the hairpoint, usually under a protruding edge without touching the substrate. Thus, the section may be lifted easily with the very tip of the hairpoint or quartz point, and can then be released by reversing the operation of picking up.

Calibration

Balances with a capacity of 0.5 μg or more can be calibrated using any pure, anhydrous, highly coloured, crystalline substance; p-Nitrophenol is convenient. Each crystal is weighed and then dissolved in an exact volume of a buffer composed of equal amounts of 0.1 M sodium carbonate and 0.1 M sodium bicarbonate. These solutions may be read against appropriate gravimetric standards in a spectrophotometer at 400 nm. Smaller balances can be calibrated by using crystals of quinine hydrobromide, which are then dissolved in a measured volume of 0.1 N sulfuric acid and read in a fluorometer against appropriate standards.

New balances can be calibrated against reliable old ones by using weights made from pieces of tissue section. This is a rapid and convenient method. The linearity of the balance is checked by weighing samples which cause approximately 1 mm deflections of the image, first individually, and then in groups of two, three, or four. Perfect linearity implies that the total of four pieces weighed together equals the sum of their individual weights.

Once a well-standardized balance is available, standard weights may be prepared from gold or platinum leaf. These should be somewhat crumpled in order to make it easier to pick them up, and to prevent their adherence to a smooth surface. They may be stored conveniently in electron-microscope grid carriers. If gold leaf is unavailable, small pieces of gold EM grids can also be used. Such standard weights can be used to re-check balance calibration periodically, as well as for calibration of new balances.

Weight Correction for Adsorption of Gases and Moisture in Tissues

Freeze-dried tissue sections adsorb gases and moisture from the air. When a sample is removed from the vacuum chamber, equilibrium is reached within a few seconds. At 50% humidity, a tissue sample will weigh approximately 7% more than its dry weight. When calibrating a balance, it is often convenient to record both the absolute calibration as well as a calibration factor which incorporates a correction of 7%. LOWRY has determined that tissue rich in lipids will increase in weight if left in air for several days, probably as the result of lipid oxidation.

Chapter 9

Microphotometry

Preface

Microphotometry is a special branch of microscopy and photometry. It offers the microscopist the opportunity of measuring his specimens photometrically, and the biochemist that of extending photometry to minute samples.

The following chapter is intended to help workers in medicine and biology to understand microphotometry better. The principles of photometry and microscopy are not dealt with in detail because there is sufficient literature on this topic. Training courses and demonstrations have shown that the problem lies particularly in the acquisition of the techniques for microphotometric measurements. Although there is comprehensive information on instrumentation and results of measurements, the relevant literature only rarely gives hints on the practical applications. In this chapter the emphasis is, therefore, laid on practical examples and detailed descriptions of the manipulations essential for correct measurements.

Only absorption measurements on minute quantities are described, because fluorescence measurements in the micro range are treated by RUCH and LEEMANN in chapter 10 of this book. Although measurements of reflectivity and reflectance under the microscope are important for structure analyses of solid materials, they are not of primary interest in molecular biology.

Photometry

The principles and methods of photometry are described in many books, for instance, EWING (1954 and 1964), KORTÜM (1962), PESTEMER (1964), and WEISSBERGER and ROSSITER (1972). Those not familiar with photometry should read at least one of these books before starting work in microphotometry. The following introduction does not replace the afore-mentioned literature; it only repeats the basic definitions to ensure comprehensible and uniform terminology.

One of the basic photometric procedures is the determination of concentrations, for example of potassium chromate in 0.05 N KOH. Such a solution appears yellowish, due to the strong absorption of the violet portion of the visible spectrum, so displaying the complementary color of violet, i.e. yellow. The absorption maximum lies at 375 nm, outside the visible range which extends from 400 nm (violet) to 700 nm (red).

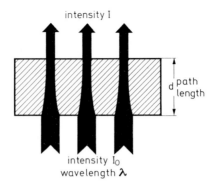

Fig. 1. Principles of measurement of transmission or absorbance

For the measurement of concentration a cuvette is filled with the solution. A standard cuvette has a path length, d, of 1 cm. Depending on the width and height of the cuvette, about 1 ml solution is required for filling. The cuvette is brought into the photometer beam path. The measuring principle is as follows (see Fig. 1): a light flux of intensity I_0 is measured in front of the cuvette. Part of the light is absorbed by the solution. The intensity I is measured behind the cell. The quotient

$$\tau = \frac{I}{I_0} \qquad (1)$$

is the *transmittance* of the sample. Transmittance is a non-dimensional figure between 0.00 (total absorption of light) and 1.00 (total transmission). For practical work it is more convenient to express the *transmission* in percent:

$$T = \frac{I}{I_0} \cdot 100\%. \qquad (2)$$

To explain the operating principle of a normal photometer, Fig. 1 should be turned through 90°, as the cuvettes are normally in an upright position and the light comes from the side. In microphotometry, however, the light passes in most cases vertically through the sample.

Actually, the term "radiant power" would be physically more correct than the term "light flux intensity". But as the term light is generally also applied to UV radiation, it is retained in this chapter.

I_0 is in fact not measured in front of, but behind a cuvette which is similar to the measuring cuvette, but which contains pure solvent. This eliminates measuring errors due to reflection at the walls of the cuvette and absorption by the cuvette and solvent. Furthermore, I_0 is not determined in absolute figures, but is used to set display of the instrument to 100 scale divisions. This is achieved either electrically, by changing the photocurrent or its amplification, or optically, by reducing the light intensity by filters or diaphragms. As a result, the transmission T is displayed directly on the instrument when the sample is measured. The transmission for the potassium chromate solution will, for instance, be $T = 38\%$.

Table 1. Table of absorbance values

$T\%$	0	1	2	3	4	5	6	7	8	9
0	∞	2.00	1.70	1.52	1.40	1.30	1.22	1.15	1.10	1.05
10	1.00	0.96	0.92	0.89	0.85	0.82	0.80	0.77	0.74	0.72
20	0.70	0.68	0.66	0.64	0.62	0.60	0.58	0.57	0.55	0.54
30	0.52	0.51	0.49	0.48	0.47	0.46	0.44	0.43	0.42	0.41
40	0.40	0.39	0.38	0.37	0.36	0.35	0.34	0.33	0.32	0.31
50	0.30	0.29	0.28	0.28	0.27	0.26	0.25	0.24	0.24	0.23
60	0.22	0.21	0.21	0.20	0.19	0.19	0.18	0.17	0.17	0.16
70	0.15	0.15	0.14	0.14	0.13	0.12	0.12	0.11	0.11	0.10
80	0.10	0.09	0.09	0.08	0.08	0.07	0.07	0.06	0.06	0.05
90	0.05	0.04	0.04	0.03	0.03	0.02	0.02	0.01	0.01	0.00

The *absorbance*, E, is defined in terms of the directly measurable transmittance:

$$E = -\lg \tau = \lg \frac{I_0}{I}. \qquad (3)$$

Extinction, optical density, abbreviated to OD, and absorbance are just different terms for the same property. lg is here the decadic or BRIGGS' logarithm.

A table such as Table 1 is the simplest means of converting transmission to absorbance values. In our example, according to Table 1, transmission $T = 38\%$ corresponds to an absorbance $E = 0.42$. Like transmittance absorbance is a non-dimensional quantity, and may lie between 0.00 (total transmission, corresponding to $T = 100\%$) and ∞ (total absorption, corresponding to $T = 0\%$). The limit $E = \infty$ is useless because of the limited accuracy of measurements. In practice, absorbance values exceeding 2.00 (corresponding to T less than 1%) will be used for evaluations only in rare cases.

Tables are not necessary for practical work, since most photometers possess scales of absorbance, or calculate and display the logarithm electronically.

The reason for the conversion of transmission into absorbance is the Lambert-Beer law. (The physicist J. LAMBERT established the principles of the theory of light measurements; in 1852 A. BEER investigated the absorption of solutions, and generalized the results found by P. BOUGUER in 1729 and J. LAMBERT in 1760.) The *Lambert-Beer law* is as follows:

$$E = c \cdot d \cdot \varepsilon \qquad (4)$$

where E is the absorbance of the sample; c the concentration of the absorbing substance, having the dimensions $\dfrac{mol}{liter}$ or $\dfrac{millimol}{cm^3}$, which gives the same numerical value; d is the path length in cm; ε the molar absorptivity, having the dimensions $\dfrac{cm^2}{millimol} = \dfrac{liter}{mol \cdot cm}$ and depending on the wavelength at which the absorbance is measured. Provided the substance under test follows the Lambert-Beer law strictly, ε is independent of concentration and path length. However, the Lambert-Beer law is not a natural law but a rule, applicable in many cases. In microphotometry in particular, reproducible measurements alone do not furnish quantitative results.

If the Lambert-Beer law is applicable, the concentration of a solution can be determined by measuring the absorbance, since Eq. (4) yields:

$$c = \frac{E}{d \cdot \varepsilon}. \tag{5}$$

The path length d is known or can be measured easily; for many substances the value of ε can be found in tables. The potassium chromate solution we took as an example has an absorption maximum at $\lambda = 375$ nm, and a molar absorptivity $\varepsilon = 4815 \dfrac{cm^2}{millimol}$ (PESTEMER, 1964, p. 57). Accordingly, a concentration $c = \dfrac{0.42}{1 \cdot 4815} \dfrac{mol}{liter} = 0.0087 \dfrac{mol}{liter}$ is found for $E = 0.42$ and $d = 1$ cm.

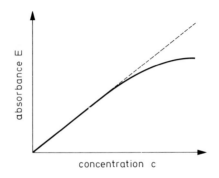

Fig. 2. Calibration curve for determination of concentration

If ε is not known, the absorbance for known concentrations of a series of dilutions can be measured at a constant path length, and the empirically determined ratio between the absorbance E and the concentration c is plotted to give a calibration curve (Fig. 2). The Lambert-Beer law applies to the range where this calibration curve is linear. The slope of the straight line is a measure of the molar absorptivity ε. A calibration curve allows concentration determinations to be performed in regions where the Lambert-Beer law is no longer valid.

According to Eq. (5), the accuracy of a photometric determination of concentration depends on the accuracy with which d, ε, and E can be measured. By repeated measurements, the path length d and the molar absorptivity ε can be determined carefully and with low errors. Under such conditions, the accuracy of the determination depends above all on the accuracy of the measurement of absorbance. Differentiation of the Lambert-Beer law yields the following expression for the relative error in the absorbance:

$$\frac{dE}{E} = 0.43 \cdot \frac{dI}{I_0} \cdot \frac{10^E}{E}. \tag{6}$$

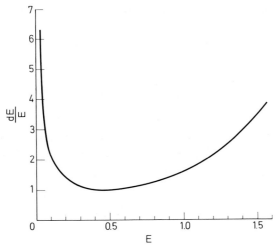

Fig. 3. Relative error, $\dfrac{dE}{E}$, of the absorbance, E, plotted as a function of E in arbitrary units of $\dfrac{dE}{E}$

The derivative of the equation is given by EWING (1954, p. 149), and KORTÜM (1962, p. 247). Fig. 3 shows the error function in arbitrary units of $\dfrac{dE}{E}$. The curve has a minimum at $E=0.43$, and is so flat that within the wide range from $E=0.1$ to $E=1.2$ the error rises to only twice its minimum value. Outside these boundaries the relative error increases considerably, below $E=0.03$ and above $E=1.8$ it exceeds five times the minimum value.

The minimum value itself is determined by $\dfrac{dI}{I_0}$, i.e. by the "noise" of the amplified photo-current. The most favorable conditions are given by low amplification and low high-voltage for the photomultiplier. This in turn calls for a strong light flux, to which limits are set by the instrument and sample. However, the light flux should not be attenuated unnecessarily by narrow stops or gray filters.

To calculate absorbance from transmission, the sample must be completely homogeneous. The errors in measurement caused by inhomogeneous samples are of decisive importance in microphotometry, and it is therefore illustrated here by a simple example: a solution of substance X has concentration c, with the values $T=10\%$, and thus $E=1.0$, when the measurement is correct. Now the cuvette is only half filled with this solution, and pure solvent is poured on top of it. The upper half of the cuvette then has the transmission $T=100\%$, the lower half $T=10\%$. The mean value measured by the photometer is $T=55\%$, with the corresponding absorbance $E=0.26$ (see Table 1). The concentration being proportional to the absorbance, a concentration $0.26\,c$ is thus "measured" instead of the correct value $0.5\,c$.

This *distribution error* with inhomogeneous samples is caused by the non-linear relationship between transmittance and absorbance: $E=-\lg\tau$. The mean value of two absorbances, E_1 and E_2, is $\frac{1}{2}(E_1+E_2)=-\frac{1}{2}(\lg \tau_1+\lg \tau_2)$. According to the rules for logarithms this equals $-\lg \sqrt{\tau_1 \cdot \tau_2}$ and differs from $-\lg \frac{\tau_1+\tau_2}{2}$, the logarithm of the mean value of the two transmittances.

Photometers

Photometers are instruments for measuring the transmission or absorbance of a sample. Of the many possible designs, a simple one will be explained. Technical details are omitted; the basic principle alone is described to illustrate the difference between micro-photometers and normal photometers.

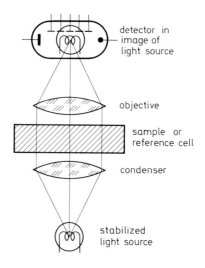

Fig. 4. Basic principles of a photometer

Fig. 4 illustrates the principles of the photometer. The condenser produces an image at infinity of a light source. The result is an almost parallel light bundle which passes through the sample or reference cuvette. Parallelism is necessary to make sure that all rays have the same path length within the sample. The objective behind the sample produces an image of the light source in the rear focal plane of the objective, where the light intensity can be measured by a photo-electric detector.

If a cuvette containing pure solvent is placed in the sample compartment, a photo-current which is proportional to I_0 is obtained at the detector output. When the cuvette containing the solvent is replaced by one containing the sample, I is obtained the same way. The calculation of the absorbance, $E=\lg \frac{I_0}{I}$, is true only if the change in the radiant power during measurement is produced merely by differences in the absorption of measuring and reference cuvettes. If this is not the case, false results will be obtained. The following are the most important sources of error: instability of the light source, contamination of the cuvette, changes in the beam path caused by cuvette exchange, and instabilities of detector or display.

Modern electronics keep the instabilities of lamp current, radiation detector, and amplifier small. They are inherent in the design and can be improved only by the photometer manufacturer. Faulty measurements due to contaminated or misplaced cuvettes, or inhomogeneous samples are caused by the operator alone. One must make sure that no reflection occurs at the side walls of the cuvette or on the sample surface, for these may impair the measurement. The vertical cross section of the cuvette should not restrict the light bundle used for the measurement.

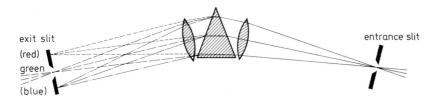

Fig. 5. Operating principles of a monochromator

The photometer shown in Fig. 4 is not complete for correct measurements because it has no capacity for limiting the light to a certain range of wavelengths; for this, filter glasses or interference filters can be brought into the beam path between condenser and sample. This is a cheap and convenient solution if only a few specific wavelengths are required for the measurement.

However, the recording of absorption spectra calls for continuous wavelength scanning and sometimes for adjustable bandwidth of the radiation. To this end, a monochromator is used instead of filters. In the monochromator, the light is dispersed by a prism or a diffraction grating. Fig. 5 shows the operating principles of a monochromator.

The entrance slit of the monochromator is illuminated with "white" light from a high-intensity light source. Lenses or mirrors produce an image of the entrance slit in the exit slit. The prism or the grating converts the "white" slit image into a continuous sequence of "colored" slit images in the order of the wavelengths. The desired wavelength is obtained by placing the exit slit within this sequence of slit images. The slit width determines the bandwidth of the monochromatic light.

The exit slit of a monochromator can be used as a light source for a photometer. Depending on the design, monochromators can be used for wavelengths between 200 nm and 2500 m, i.e. for UV, visible light, and IR.

For biological investigations in the visible spectral range (400 nm to 700 nm) the bandwidth control can often be abandoned. It will then be cheaper and more convenient to use a so-called continuous interference filter instead of the monochromator. Such filters have bandwidths between 13 nm and 20 nm.

Microphotometry

The boundary between normal photometry and microphotometry can be defined by the sample volume. Normal photometry ends with volumes of a few microliters $(1\ \mu l = 10^{-3}\ ml = 10^{-6}\ l = 1\ mm^3)$. This limit is set, above all, by problems inherent in the instrumentation: it is difficult to bring a volume of $1\ \mu l$ reproducibly into the sample compartment of a photometer. Physicians and biologists need to examine quantities in this range. Their demands go beyond $1\ \mu l$, down to $1\ \mu m^3$, i.e. $10^{-9}\ \mu l = 10^{-15}\ l$. Microphotometry may be defined as the photometry of volumes between $1\ \mu l$ and $10^{-9}\ \mu l$. These are, of course, not fixed boundaries, but good references.

There is ample literature on the instrumentation and methodology for normal photometry, but only little for microphotometry. In the relevant literature only short references are given, by e.g. KORTÜM (1962, p. 320), or WEISSBERGER and ROSSITER (1972, part IIIa, p. 266). Microphotometric equipment, which by now can be called "classical", is described by FRANÇON (1961). The most comprehensive descriptions of instrumentation, methodology, results, and problems can be found in SANDRITTER and KIEFER (1965), WIED (1966), and WIED and BAHR (1970a).

With reference to microphotometry, COCKS and JELLEY state in WEISSBERGER and ROSSITER (1972, part IIIa): "It is a technique that requires considerable knowledge and skill on the part of the operator." The following paragraphs are intended to help the reader acquire such knowledge and skill.

Fundamentally, the same laws that are valid for normal photometry apply to microphotometry. However, the minuteness of the samples makes it much more difficult to obey the theoretical requirements and to check the validity of the laws. The measurements are complicated by the fact that both a photometer and a microscope must be operated. The user is rewarded, however, by the high sensitivity of microphotometric measurements.

For an estimation of the *detection limit*, both terms of Eq. (5) are multiplied by the volume V (in cm^3), thus converting the concentration c into the mass m in millimoles:

$$m = c \cdot V = \frac{E \cdot V}{d \cdot \varepsilon} = \frac{E}{\varepsilon} \cdot a. \tag{7}$$

In Eq. (7) a is the measuring area, i.e. the cross-section of the light cone in the sample. The detection limit, i.e. the smallest mass measurable by microphotometry, is found by insertion of the limit values of E, a, and ε in Eq. (7). The smallest absorbance measurable is $E = 0.01$. The smallest measuring area obtainable in an optical microscope is $a = (0.25\ \mu m)^2$, i.e. $0.0625\ \mu m^2 = 6.25 \cdot 10^{-10}\ cm^2$. The maximum molar absorptivity possible is $\varepsilon = 5 \cdot 10^5 \frac{cm^2}{millimol}$ (PESTEMER, 1964, p. 9). These three values yield as the *theoretical limit of detection* the mass

$$m_{MIN} = \frac{10^{-2} \cdot 6.25 \cdot 10^{-10}}{5 \cdot 10^5}\ millimol$$

that is 10^{-17} millimol, or 10^{-20} mol, which means that, in this extreme case, 6000 molecules could be detected photometrically.

The above calculation gives the theoretical limit of a microphotometer. In practice, the limit of detection is much higher. Considering the accuracy of measurement, $E = 0.10$ will be inserted in (7) as the lowest absorbance value, $a = 1 \, \mu m^2 = 10^{-8} \, cm^2$ as the smallest measuring area, and $\varepsilon = 10^4 \dfrac{cm^2}{millimol}$ as the maximum useful absorptivity. These values yield the mass

$$m_{MIN} = 10^{-13} \, millimol = 10^{-16} \, mol$$

as the *practical limit of detection*. The main reason for the high sensitivity of microphotometry is the extremely small measuring area. When changing from the microscopic measuring area of $1 \, \mu m^2$ to the normal measuring area of $1 \, cm^2$, the practical detection limit will be $10^{-8} \, mol$, instead of $10^{-16} \, mol$.

In contrast to a normal photometer, the microphotometer is not suited to the determination of small concentrations. Eq. (5) for the concentration $c = \dfrac{E}{d \cdot \varepsilon}$ has the path-length d in the denominator. For a fixed minimum value of the absorbance E, determined by the accuracy of measurement, long path-lengths d are necessary for measuring small concentrations c. However, the beam path in the microphotometer allows only path-lengths from $1 \, \mu m$ to $1 \, mm$. The example of potassium chromate solution, discussed on p. 297 ff., gave a concentration of 0.0087 mol/liter at 1 cm path-length. The same absorbance requires a concentration of 0.087 mol/liter at 1 mm path-length. Mathematically, the result for $1 \, \mu m$ path-length would be 87 mol/liter, an amount which is impractical.

Supplementing the conditions $E = 0.10$ and $\varepsilon = 10^4 \dfrac{cm^2}{millimol}$, used for an estimation of the practical limit of detection of $10^{-16} \, mol$, by the assumption that the path-length $d = 1 \, \mu m = 10^{-5} \, cm$, it follows from Eq. (5) that be substance must have a concentration $c = 1 \, mol/liter$. A path-length $d = 10 \, \mu m$ still requires a concentration of 0.1 mol/liter.

As shown by these examples, the preferred field for the application of microphotometry is the determination of small amounts of substances, and the measurement of high concentrations at small path lengths. In this instance the Lambert-Beer law can only be used as a rough approximation. For absolute measurements the relationship between absorbance E and mass m or concentration c must be established or checked by means of calibration curves. Due to the extremely small dimensions and the difficulties involved in preparation, the outlay is considerable. With some pigments such determinations are even impossible. Relative results are, therefore, often considered adequate, and the product $E \cdot a$ is referred to as "mass in arbitrary units [A.U.]". Accordingly, the quotient E/d is referred to as "concentration in arbitrary units".

Microphotometers

Before starting microphotometry, the user should become familiar with the design of the instrument, because microphotometers are more difficult to operate than normal photometers, and yield correct measurements only when adjusted

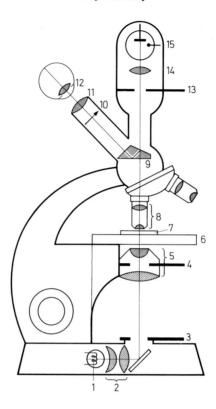

Fig. 6. Diagram of a microscope with photometer attachment: *1* light source, *2* collector, *3* luminous field stop, *4* aperture stop, *5* condenser, *6* stage, *7* specimen, *8* objective, *9* deflecting mirror, *10* plane of intermediate image, *11* eyepiece, *12* observer's pupil. With the mirror *9* removed, the light is measured in the photometer attachment: *13* measuring diaphragm, *14* projective, *15* multiplier

carefully. The reasons for this are the small size of the sample and the measuring area. Samples of a few μm must be positioned 10000 times more accurately than a cm-cuvette which can be brought easily into the beam path. Whereas, in a normal photometer the correct position of the sample can be checked without auxiliary means, a microphotometer needs a microscope for this purpose.

Whoever intends to work in microphotometry should become familiar with the microscope, not so much with its theoretical principles, as with its proper handling. From the ample selection of books MÖLLRING (1968), or MOELLRING (1973), convey the knowledge of the microscope in a condensed form, even for those who have never used a microscope before. Of more comprehensive publications MICHEL (1964), and NEEDHAM (1958) should be mentioned.

Fig. 6 shows a microscope schematically. The light source *(1)* is imaged by the collector *(2)* into the rear focal plane *(4)* of the condenser *(5)*, where a variable diaphragm (the aperture stop) is provided. This is used to adjust the illuminating aperture, i.e. the aperture of the cone of light which hits the specimen *(7)*. A diaphragm *(3)* (the luminous field stop) lies between collector *(2)* and condenser *(5)*.

An image of the luminous field stop is produced by the condenser in the plane of the specimen, which means that this stop *(3)* determines the size of the illuminated spot at the specimen. The aperture stop *(4)* determines the aperture angle under which this spot is illuminated (Fig. 7). The correct use of the stops *(3)* and *(4)* is the keystone of microscopy and of microphotometry in particular.

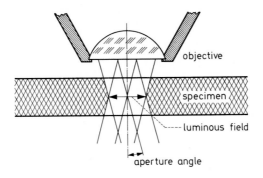

Fig. 7. Luminous field and aperture at the specimen

Microscopists are used to refer to the angle between the cone envelope and its axis as the angle of aperture. It is assumed that this angle is measured in air or corrected for this condition. The sine of this angle multiplied by the refractive index of the medium through which the beam passes, is referred to as the *aperture A*; in older books it is called the numerical aperture instead of the aperture. The maximum aperture is engraved on every microscope objective, since it determines the *resolving power* of the objective. Here, the equation

$$r = \frac{\lambda}{A} \tag{8}$$

is used to determine the resolving power, where λ is the wavelength and A the aperture. For the visible region this is simplified by using $\lambda = 500\ \text{nm} = 0.5\ \mu\text{m}$ (which corresponds to green light). For instance, an objective 40/0.65 has $40 \times$ magnification and the aperture 0.65. According to Eq. (8) its resolution is $\frac{0.5}{0.65}\ \mu\text{m} = 0.8\ \mu\text{m}$. The relevant literature sometimes quotes smaller resolution values, down to one half of the above. However, as will be shown on p. 321 ff., these details are irrelevant for microphotometry.

The microscope objective *(8)* in Fig. 6 produces, in the microscope tube, an enlarged image *(10)* of the specimen *(7)*. For more convenient observation the beam path is deflected by mirror *(9)* to one side. The eyepiece *(11)* is a magnifier for the observation of the enlarged intermediate image *(10)*. It also produces an image of the light source *(1)* and the aperture stop *(4)* at the place of the observer's pupil. The image can be visualised by placing a piece of thin paper about 1 cm above the eyepiece. Opening or closing of the aperture stop *(4)* changes the diameter of this image. An image of the luminous field stop *(3)* is produced in the plane of the intermediate image *(10)*, i.e. it can be observed through the eyepiece like the specimen. For this purpose the stop *(3)* must, of course, be closed so that its margin lies within the field of view of the eyepiece.

Because of the importance of these diaphragms and their images for the functioning of the microphotometer, it is re-iterated: the *luminous field stop (3)* is imaged on the specimen *(7)* by the condenser *(5)*, on the plane of the intermediate image *(10)* by the objective *(8)* and, finally, on the observer's retina by the eyepiece *(11)* and the eye itself. The light source *(1)* is imaged on the *aperture stop (4)* by the collector *(2)*. The aperture stop lies in the lower focal plane of condenser *(5)*. This means that the condenser produces the aperture image at infinity, the objective *(8)* projects the image in its upper focal plane, and finally the eyepiece *(11)* projects it in its exit pupil, i.e. the place of the observer's pupil.

The microscope in Fig. 6 is easily converted into a microphotometer: the mirror *(9)* is removed from the optical path, which then proceeds straight upwards. Nothing can be seen in the eyepiece, and the intermediate image *(13)* now lies within the photometer attachment. The projective *(14)* is an optical system similar to the eyepiece, with the photomultiplier *(15)* in its exit pupil. A fixed or variable *measuring diaphragm* is provided at the place of the intermediate image *(13)*; this stop occludes part of the image. All light passing through this stop is measured by the multiplier *(15)*.

Microphotometers for use outside the visible region are more complicated, especially for UV. Within the visible range, specific wavelengths can be selected by means of filters or a continuous interference filter between luminous field stop and condenser. The use of a monochromator necessitates a change in the illumination setup. Fig. 8 is the front view of a microscope, with the light coming from the side. The collector *(B)* produces an image of the high-intensity light source *(A)* at infinity. Lens *(C)* focuses this light onto the entrance slit *(D)* of the monochromator, and at the same time forms an image of the collector

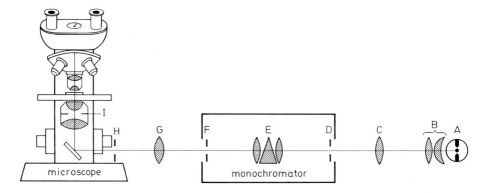

Fig. 8. Microscope and illumination arrangement with monochromator

aperture *(B)* on the prism or grating *(E)*. Following the monochromator design, the entrance slit *(D)* is imaged on the exit slit *(F)*. Lens *(G)* forms an image of *(F)* in the lower focal plane *(I)* of the condenser and, simultaneously, of prism *(E)* on the now laterally arranged luminous field stop *(H)*. The results are images of the light source and the monochromator slit in the aperture stop *(I)*, and of the prism *(E)* and the collector aperture in the luminous field stop *(H)*.

The instruments in Figs. 6 and 8 are fundamentally similar, in that the light source is imaged in the aperture stop, and the collector aperture imaged on the specimen. Due to the monochromator, however, slit *(F)* and stop *(I)* in the condenser jointly determine the effective illuminating aperture. Normally, *(I)* limits the aperture in the direction of the slit height whereas at right angles to it the aperture is limited by the slit width (Fig. 9). Only when the aperture stop *(I)* is small and slit *(F)* wide open, do conditions in an illuminating system with monochromator (Fig. 8) correspond to those in a normal setup (Fig. 6).

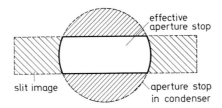

Fig. 9. Limitation of the illuminating aperture by the slit of a monochromator and the aperture stop in the condenser

The above statements refer to single-beam microphotometers, i.e. instruments using the same optical path for measurement and reference. In double-beam microphotometers a beam-splitter in the illuminator, and a recombination device behind specimen and reference, produce two separate optical paths for measurement and reference, which pass either through two microscopes or through different portions of the specimen. Double-beam instruments are more expensive because of the greater complexity. Nowadays single-beam instruments are preferred because, due to the inherent variations of his specimens, the biologist is bound to make a great number of measurements. Ease of operation is thus more important than the long-term stability offered by double-beam instruments. With modern electronics the instability of single-beam instruments can be kept so small that it becomes negligible for biologists. A detailed description of double-beam microphotometers is given by FRANÇON (1961).

Just a few remarks on "ease of operation". Whoever uses a microphotometer for the first time, will find neither the adjustment, nor the focusing and centering of luminous field stop and measuring stop an easy task. The demands for accuracy are best explained by an example. A luminous field stop with a diameter of 0.1 mm must be imaged on the specimen 100 times reduced. Multiplication of all dimensions by 10000 yields as luminous field stop a circular disk 1 m in diameter lying more than 1 km from the condenser; it is to be imaged through a 10 m thick glass wall, corresponding to the specimen slide, onto a specific spot in the specimen, having the size of a thumb nail. Considering this, it is obvious that the operation of a microphotometer requires skill, experience, and knowledge of its technical capabilities.

Spot Measurements

In biology, cells were first measured microphotometrically (CASPERSSON, 1936, 1950). Measurements in small cuvettes, for instance the micro complement fixation test, are a better introduction to microphotometry. The test, its meaning, and the automation of the measurement, are described by D'AMARO (1970), D'AMARO et al. (1970), and D'AMARO, HENSEN, and VAN ROOD (1971). This test

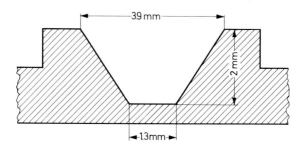

Fig. 10. Cross-section of a microcuvette in a Falcon Microtest Tissue Culture plate No. 3034

is used to determine platelet and lymphocyte antigens in man; it requires only 0.5 µl of serum and has a final reaction volume of 10.5 µl. The test is performed under liquid paraffin in Microtest Tissue Culture plates No. 3034, manufactured by Falcon Plastics. Each plate contains 60 circular microcuvettes whose cross-section is shown in Fig. 10. The volume is 11.5 µl. For the micro complement fixation test each cuvette is filled with

serum	0.5 µl
platelets	2.0 µl
complement (diluted)	2.0 µl
complement buffer	4.0 µl

the mixture is incubated before the addition of

sensitized sheep Red Cells	2.0 µl

Depending on the degree of haemolysis, the reaction product will be either clear (total haemolysis) or turbid due to an absorbing residue of red blood cells. D'AMARO's studies have shown that by measuring the transmission this test is converted into a quantitative method. Such a measurement will be described here in detail, as a useful introduction to the problems of microphotometry.

Fig. 6 shows the photometer. The terms and numbers used there will also be applied in the following. The light source *(1)* is a low-voltage filament lamp, e.g. 6 V 15 W. It is not, as usual, connected to a transformer, because the alternating current would cause periodic modulation of the light flux. A photograph of an oscilloscope screen, displaying in the lower half the lamp voltage and in the upper half the photo current coming from the photomultiplier, is shown in Fig. 11; it is obvious that the AC "shows through". The photometer lamp is thus run on DC which is stabilised by electronic regulators. These also suppress the influence of the fluctuations of the mains voltage on the light flux and the measurement.

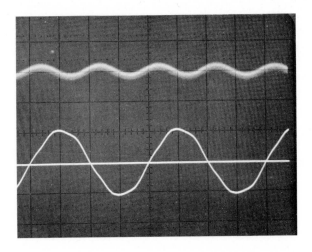

Fig. 11. Oscilloscope display of lamp voltage (below) and corresponding photocurrent (above)

The lamp coil is imaged into the condenser *(5)*. A normal microscope condenser with an iris as aperture stop is not recommended for microphotometry. The iris is easily displaced, and it is difficult to reproduce a specific aperture. Fixed aperture stops or variable diaphragms with fixed stops are preferred because of their reproducibility.

A condenser in the microscope should illuminate both large object fields (at low magnifications) and large apertures (at high magnification). Its imaging quality is a compromise between these extreme requirements. In microphotometry, accuracy and reproducibility have priority over flexibility and convenience of operation. Special condensers, generally microscope objectives, are used to illuminate the specimen. This gives a very accurate definition of the illuminating aperture, i.e. the maximum aperture engraved on the objective used as condenser, provided that the aperture is not reduced by additional stops.

Normally the areas to be measured cover only a small portion of the microscope's field of view. It then becomes necessary to illuminate only these small areas of the specimen. A microscope objective used as a condenser is the best means of producing a microscopically reduced image of the luminous field stop *(3)* in the specimen.

With the Falcon plate 3034, the bottom of a microcuvette, i.e. a circle of 1.3 mm diameter must be illuminated. The maximum diameter of the luminous field stop *(3)* in the microscope base is about 20 mm. With a $10 \times$ objective as condenser, the circle which can be illuminated in the specimen has a maximum diameter of 2 mm. With this in mind, an Achromat 10/0.22 was used as condenser for the micro complement fixation test.

Another lens between condenser and luminous field stop, for instance for centering, changes the factor with which the luminous field stop is imaged in the specimen plane. In practice, however, looking through the microscope will be enough to see whether or not the illuminated area is of the correct size. The exact reduction factor need therefore not be known.

For the micro complement fixation test an Achromat 10/0.22 could also be used as objective *(8)*; it would then produce, in the tube, a $10 \times$ magnified image of the microcuvette. This objective has 5 mm working distance, i.e. the distance between object plane and front lens mount, so that there is a danger of dipping the front lens into the paraffin. An Achromat 3.2/0.07 was therefore selected. Its working distance of 30 mm allows convenient handling of the filled Falcon plates. This objective produces a $3.2 \times$ magnified image of the bottom of the microcuvette, i.e. a diameter of 4.2 mm in the intermediate image *(10)* or *(13)*.

If the condenser aperture is larger than the objective aperture, part of the light misses the objective and is lost, an undesirable effect because of the possible errors produced by stray light. One of the basic rules of microphotometry is that the illuminating aperture should not be larger than the objective aperture. The application described above neglects this rule, the illuminating aperture being 0.22 and the imaging aperture only 0.07.

It would have been possible to meet the above requirement by using an Achromat 3.2/0.07 as condenser. However, the luminous field stop *(3)* would then have to be stopped down to a diameter of 4.2 mm to illuminate the bottom of the microcuvette alone; at such small diameters the iris is no longer circular— its lamella structure becomes obvious. To avoid having to insert and adjust a pin-hole diaphragm, the rule was ignored, and too large an illuminating aperture was accepted. The requirement can also be fulfilled by reducing the aperture of the Achromat 10/0.22 with a fixed stop to 0.07. This offers the advantages of a correct illuminating aperture and a reasonable size of the luminous field stop. If the demands on accuracy of measurement are high, such a system would have been applied. However, experiments have shown that, for the micro complement fixation test the influence of stray-light is negligible.

Suitable selection of the objectives for illuminating and imaging are of decisive importance in microphotometry. It has been described here in detail, because in normal photometry it is a problem of minor importance. With the micro complement fixation test there are no extreme requirements of resolution, image quality, working distance, transmission, and other features of the microscope optics. However, for a microphotometer to be universally applicable, an ample selection of special objectives should be available.

The essential parameters for the formation of the optical image are determined by the objectives and condensers. The next step is the adjustment of the microscope for observation of the specimen. The skilled microscopist may omit the next sections, up to Fig. 12; they are included here in order to help those who have not yet worked with a microscope. The following statements are general hints, since certain procedures depend on the design of the microscope. Instructions relating to any specific microscope are given in the operating manual.

First, a distinct part of the specimen is selected and focused with coarse and fine adjustment until at least its coarse structure becomes visible. The image quality need not be good. Then the eyepieces on the binocular tube are set to the individual pupillary distance (PD). To check the adjustment, the head must approach the eyepieces slowly; at some distance, there is only a small section of the image in each eyepiece. The nearer the observer, the larger the sections. If the PD is set correctly, the partial images coincide when the entire field of view of the eyepiece is visible. If the PD is too small, the two circular images combine before the entire image is visible; if the distance is too large, the two partial images will not coincide. Furthermore, if the PD is not correct, even a slight lateral movement of the head will cause darkening of one partial image from the rim towards the center. With some binocular tubes the PD must also be corrected on the eyepiece setting-rings, to make sure that the eyepieces are correctly focused. Eyepieces with reticules feature a ring for focusing the cross-hairs. Accommodation errors of the eye can be compensated for by adjusting the eyepieces or eyepiece sleeves. Modern, high-eyepoint, eyepieces permit work with spectacles.

The third step is the focusing and centering of the condenser. A part of the specimen which is at least partly transparent is focused. The luminous field stop (3) is then closed, and the focusing knob of the condenser is turned until the edge of the luminous field stop becomes sharp in the specimen plane. The condenser, or an auxiliary lens below the condenser, is then shifted laterally with the centering screws until the image of the luminous field stop is centered in the field of view. The luminous field stop is then opened until its rim just disappears from the field of view.

The fourth step is the centering and focusing of the lamp. An image of the light source is produced in the lower focal plane of the condenser or the objective used as condenser. It is not necessary to make a very accurate adjustment. To check the focusing, a piece of thin paper is held in front of the lower condenser opening, or one eyepiece is removed and the exit pupil of the objective is observed through the tube, because an image of the light source is produced at the exit pupil of the objective. Observation of the exit pupil is facilitated by a centering telescope which is either swung into the tube or inserted like an eyepiece. Focusing of the lamp produces an image of the light source in the entrance pupil (4) of the condenser, and thus in the exit pupil of the objective. By centering the lamp, the image of the light source is centered in both pupils and is not shifted to one side. A ground-glass collector used for uniform illumination prevents the image of the light source from being clearly visible; what can be seen is a bright spot of light with a diffuse rim. This does not give reliable criteria for focusing and centering the lamp, but larger variations are permissible because of the diffuse boundary lines. The collector in a microphotometer should not be mat. A swing-in ground glass could be provided for observation only.

After these preparations an evenly illuminated sharp image of the specimen should be visible on looking through the eyepieces. The next step is the adjustment of the beam path for photometry. The sample—for the micro complement fixation test, the bottom of a microcuvette—is brought into the center of the field of view, and the deflector (9) in Fig. 6 is switched from observation to measurement. A measuring stop must be provided in the plane of the intermediate image (13). If the total light flux through the cuvette is to be measured, a pin-hole diaphragm with a diameter of 4.2 mm must be inserted at (13) and centered so

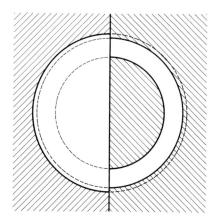

Fig. 12. Adjustment of the measuring stop (right) with regard to the image of the microcuvette (left)

that it coincides exactly with the image of the bottom of the cuvette. This would be correct for a homogeneous sample; however, in the micro complement fixation test the absorbing material is unevenly distributed in the cuvette, since some of the non-haemolized cells sink down along the inclined cuvette wall. The absorption along the walls is higher than in the center of the cuvette, as shown by the pictures in D'AMARO, HENSEN, and VAN ROOD (1971).

Measurements with a pin-hole diaphragm would therefore show a considerable distribution error (see p. 301 ff.). To increase the accuracy of measurement an annular diaphragm is used instead of the pin-hole diaphragm. This diaphragm has an inner diameter of, for example, 3 mm and an outer diameter of 4 mm. Fig. 12 illustrates the adjustment of the measuring stop with regard to the image of the cuvette. If the multiplier (15) is removed, viewing through the eyepiece (14) shows whether the annulus covers the bright bottom of the cuvette or its dark wall, but gives no indication as to whether the annulus is centered with regard to the bottom of the cuvette. The solution to the problem is the following: the luminous field stop (3) is closed until its rim becomes visible within the annulus of the measuring stop, and these two stops are then centered with regard to one another.

With a microphotometer the basic task is the precise centering of the luminous field stop, the specimen and the measuring stop with regard to each other. The unskilled microscopist will have considerable difficulties in doing this. However, when the necessary steps are carried out systematically they will very soon become routine. Those who are not familiar with the procedure should never regard the data they obtain as measuring results. The unskilled operator should practise centering by starting with simple specimens at low magnifications, and slowly go over to higher magnifications and smaller stops. Since in normal photometers the specimen is at the same time luminous field stop and measuring stop, this kind of adjustment does not exist.

Once microscope and photometer are adjusted, measurement can be started. In the micro complement fixation test a cuvette in which a haemolysis has been completed is used as reference. The bottom of the cuvette is brought into focus and the luminous field stop is closed until its image has the same diameter as the bottom of the cuvette. The Falcon plate is then moved on the microscope stage until the cuvette is centered with regard to the luminous field stop. The bottom of the cuvette alone should be illuminated but not the surrounding field, so as to reduce the influence of stray-light. Deflector *(9)* is switched from observation to measurement and the light falls on the photomultiplier *(15)*. The high voltage supply of the multiplier, and the amplification of the photocurrent are adjusted to give a reference value $T=100\%$ on the photometer. This procedure is described in detail in the operating manuals of amplifier and display units. Provided that the amplifier has no automatic zero adjustment, the path of the beam is interrupted by closing a diaphragm completely. The photocurrent due to stray-light and the dark current of the multiplier is compensated by an amplifier control. There is an interconnection between $T=100\%$ and $T=0\%$, and the setting of the controls must be repeated until both reference values remain constant. This is quite tedious and is obviously the main reason for the preference for instruments with automatic zero setting.

When the reference value $T=100\%$ has been set for a cuvette with complete haemolysis, another cuvette is selected and centered with regard to the luminous field stop. After switching over from observation to measurement the photometer displays the result, e.g. $T=53\%$ or $E=0.28$. In the micro complement fixation test any value between $T=2\%$ (no haemolysis) and $T=100\%$ may be obtained.

This detailed description may give the impression that the method is tedious and time-consuming. In the daily routine, however, the 60 cuvettes of a Falcon plate are measured within three minutes. This speed can be obtained with a motorized scanning stage for moving the plate from one cuvette to the next, and by feeding the results from the photometer into a computer (see D'AMARO, HENSEN, VAN ROOD, 1971). The computer corrects the zero-point and the reference value, so that $T=100\%$ and $T=0\%$ need not be carefully set. The outlay of a computer is justified when a great number of samples must be measured. When the instrument is operated manually, about 10 min are required to measure the 60 cells.

Because of their size no special technique is necessary for filling the cuvettes, and they are therefore suitable for applying existing photometric methods to the micro-scale. A constant path length of 2 mm can be obtained with cover glasses;

if the cuvette is filled with 15 µl of the sample 3.5 µl is displaced by the cover glass and flows into the Falcon plate.

Monochromatic light is not required for the micro complement fixation test. If other samples are measured in these cuvettes, light filters or a monochromator must be used in the illuminating beam path.

Line Scanning

The biggest problem in microphotometry is the inhomogeneity of the samples. In the micro complement fixation test (see p. 310 ff.) it was easily overcome by the use of an annular stop because of the circular cross-section of the cuvette. Generally, the only solution is to divide the sample into very small areas which are assumed to be homogeneous. Determination of the absorbance of each partial area, and summation of these values, yields the optical density of the inhomogeneous sample without distribution errors. The analysis of the RNA bases in micro-electrophoreses is given as an example.

J.-E. EDSTRÖM and V. NEUHOFF describe the RNA base analysis in chapter 6 of this book. Information on the relevant literature can also be found there.

RNA bases have their absorption maxima in the UV—uridylic acid at 263 nm, cytidylic acid at 282 nm, guanine at 247 nm, and adenine at 262 nm. They do not have measurable absorption in the visible region, so that a microphotometer for RNA base analysis must be equipped with UV illumination and UV optics. Fig. 8 shows a UV illuminator with monochromator. The light source, e.g., an XBO 150 W 1 xenon high-pressure lamp by Osram must, of course, be run on a stabilized power supply. This is independent of mains fluctuations and supplies a constant lamp current. Usually such an arc-lamp burns less stably than a filament lamp because the luminous plasma may "dance" between the electrodes. To prevent this flickering from influencing the measurement, the illumination must be very carefully adjusted so that the images of the arc fill the monochromator slit and the condenser pupil symmetrically. The lamp should be operated for longer periods since frequent ignition increases the instability of the arc. After ignition the lamp needs a warming-up time of about half an hour before the light flux is constant.

Normal microscope optics cannot be used for absorption measurements in the UV, since glass absorbs light at wavelengths below 360 nm. Mirror objectives and mirror condensers can also be used in the UV; in the microscope they each consist of two mirrors (see FRANÇON, 1961). Their disadvantage is that the central portion of the aperture cannot be used for formation of the image, so uniform illumination is difficult to achieve. Ultrafluar optics by Carl Zeiss, Oberkochen, offer the most convenient solution; Ultrafluar objectives are standard-sized microscope objectives covering the visible and UV spectral ranges (from 230 nm to 700 nm).

The micro-electrophoresis strips for RNA base analysis are about 100 µm wide. Sections of about 10 µm can be treated as being homogeneous. Ultrafluar objectives 10/0.20 are therefore used as objective and as condenser. These objectives have working distances of 7 mm. Due to the small aperture, 0.20, thicknesses of the coverglass and the specimen slide need not be considered. At $10 \times$ magnification the dimensions of measuring and luminous field stops are reasonable. The objective with the next higher power is the Ultrafluar 32/0.40. This has more convenient stop sizes, but is an immersion objective and would have made manipulation of the specimen difficult. Due to the working distance of 0.45 mm at a coverglass thickness of 0.35 mm, the thickness of the specimen slide ought not to be higher than 0.5 mm. In this case a coverglass would have been required as carrier for the specimen, with a second coverglass on top of it.

The above conditions are imperative for the successful application of microphotometry. Failure is due to a lack of consideration of the photometric requirements, the laws of optics, or the preparative techniques.

For observation, a UV microscope must be equipped with an image converter, to convert the invisible UV radiation into visible light electronically. It is attached in addition to the binocular tube, because normal observation in the visible is more convenient.

Fig. 6 shows the simplest form of microphotometer. Operation is facilitated by providing an additional viewing tube between the measuring stop and the multiplier or by imaging the measuring stop in the binocular tube by means of mirrors. The measuring stop and the centering procedure can then be observed without removing the multiplier. Such auxiliary aids are necessary when the size of the measuring stop must be changed frequently or adapted to the specimens.

The microscope is adjusted with visible light. The normal procedure with filament illumination is described on p. 313ff. The same principle holds for an illuminator with monochromator. After the condenser is centered, the image of the monochromator exit slit must be centered in the aperture stop of the condenser. This is achieved by shifting the monochromator vertically and laterally. The next step is the adjustment of the rear mirror in the lamp housing. The mirror image of the arc (plasma) must be positioned directly beneath the direct image of the arc. The pair of luminous spots is then focused onto the entrance slit of the monochromator with the aid of a mat black sheet, because light surfaces reflect so much light that the observer is dazzled. Vertical and lateral adjustment of the lamp centers the images of the light spots on the entrance slit. The correct position is observed by placing the edge of the black sheet horizontally and vertically in front of the center of the slit.

Correct illumination of the slit does not automatically ensure even illumination of the monochromator prism or grating. The axis between the lamp and the slit need not necessarily coincide with the axis of the monochromator. To obtain this condition, the lamp housing is shifted vertically and laterally, keeping the light spot images centered with regard to the entrance slit. This adjustment is made when the instrument is set up and need not be repeated. Illumination of the prism is easily checked because the prism image appears in the luminous field stop and in the plane of the object. When defocusing the lamp collector deliberately, so that the light spot images are in the prism and not on the entrance slit, faulty adjustment is indicated by excentric positions of the light spot images.

A microscope adjusted with visible light is also adjusted for UV, with one exception: when the lamp is not equipped with a mirror collector the positions of the light spot images are different for UV and visible. Focus of the lamp collector in the UV is therefore different from that in the visible range. Correct focusing is indicated by an even illumination of the field of view in the image converter.

Adjustment of the measuring stop is also easier in the visible region than in the UV. The adjustment must be done in the visible region when there is no image converter in the photometer attachment. Provided that the specimen does not display noticeable structures in the visible, the measuring stop is centered with regard to the luminous field stop image.

The micro-electrophoresis strips for the RNA base analysis have four absorption bands in the UV for each specimen. They are inhomogeneous along the strip, whereas in the transverse direction the absorption is uniform except for slight inhomogeneity at the edges of the strips. In the longitudinal direction the strip must be divided into fine measuring sections; 10 µm wide sections are sufficiently

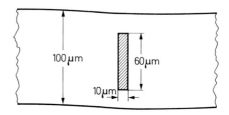

Fig. 13. Size of the field of measurement for photometry applied to micro-electrophoreses

small for most measurements. In the transverse direction the measuring areas may cover almost the entire width of the strip. The strips are not always completely straight; about $\frac{2}{3}$ of the width of the strip should therefore be taken as the height of the measuring area (Fig. 13), i.e. 60 µm to 70 µm for a strip width of 100 µm. If the height of the measuring area is increased, inhomogeneities at the edges of the strip may have a disturbing effect. Reduction of the height of the measuring area to a very small value raises the effect of contaminants in the strip on the results.

With an Ultrafluar objective 10/0.20 as condenser, the luminous field stop must be 10 times larger than the measuring field in the plane of the object, i.e. it must be 0.1 mm wide and 0.6 mm high. With these theoretical minimum dimensions it would require great skill to bring the image of the luminous field stop to coincidence with the measuring stop. Furthermore, every inhomogeneity in the specimen slide or in the thickness of the paraffin layer over the strips would cause a small yet disturbing shift of the image of the luminous field stop. The luminous field stop must therefore be larger than the theoretical minimum value, e.g. 0.3 mm wide and 1 mm high. Fundamentally, the stray-light portion in the result is larger when the luminous field stop is larger. Homogeneous specimens produce almost no stray-light, so that the luminous field stop need not be small. With scattering and strongly-absorbing specimens the luminous field stop should be optimally adapted to the measuring stop.

The size of the measuring stop is the product of image scale of the objective (10 in the example), the image scale of the intermediate optics (tube factor 1.6 in the example), and the measuring field in the specimen (10 µm or 60 µm in the example). For RNA base analysis this yields a measuring slit 0.16 mm wide and 0.96 mm high. The additions to the values of the luminous field stop should not be made for the measuring stop, because all the light is measured which passes through the measuring stop. Deviations from this rule are discussed on p. 323 ff.

With variable slit diaphragms as luminous field and measuring stops, the correct dimensions are determined with an object micrometer. After focusing the micrometer scale in the microscope, the luminous field stop is closed until it covers the desired number of scale intervals both vertically and laterally. The measuring stop is then closed while the desired scale intervals are observed through the photometer viewing tube.

Compared with the micro complement fixation test, RNA base analysis has resulted so far only in improvements and modifications. Fundamentally new is the use of the scanning stage. A micro-electrophoresis band has a length of 1 to 2 mm; measuring 10 µm sections requires 100 to 200 single measurements without interruptions. This cannot be done manually, so a scanning stage is used to automate the procedure. The scanning stage is essentially a mechanical stage capable of motion in two independent directions. It is driven by electronically controlled stepping motors which rotate at specific angular steps. The transmission between motor and stage determines the relationship between the motion of the stage and the angular step. E. Leitz, Wetzlar, and Carl Zeiss, Oberkochen, offer stages with step widths of 0.5 µm and 10 µm. An electronic unit controls the scanning stage. The simplest version is only a power supply for both motors, but is needs control pulses for each step, which may be provided by a computer. More sophisticated models include not only the power supply but also produce the pulses for the stepping motors, control sequence, and step direction of the motors, so that the stage follows fixed programs, such as lines, meander, rectangle, etc. Micro-electrophoreses can thus be scanned without computers, and the results, for instance, plotted by a recorder.

The measurement described by EDSTRÖM and NEUHOFF in chapter 6 of this book uses an on-line stage control receiving the control pulses from a PDP 12/20 computer (Digital Equipment, Maynard, Mass.). Compared with off-line operation, the on-line operation offers greater flexibility and instant data processing. The computer controls the motion of the scanning stage, stores the measured value after each step, determines the reference values, computes the optical density of each band, and derives the absolute and the relative mass of the absorbing substance. Averaging and computation of the standard deviation immediately follow the measurement.

Scanning stages are mounted on the microscope like normal microscope stages. For line scanning the preferred direction of motion is right to left. When the electrophoresis strip lies at an angle to this direction, the strip must be aligned with the aid of a rotating specimen holder. Adjustment of microscope and photometer is unaffected by the scanning stage.

To start a measurement, the microphotometer is prepared, the computer loaded with the control program and started, or the necessary parameters are set on the separate stage control. The desired wavelenght (e.g. 260 nm) and slit width (e.g. 0.25 mm) are set on the monochromator. This position should be noted for quick change between full slit width, for convenient observation, and measuring slit width. Auxiliary illumination in the visible is used to select the end of a strip, which is placed parallel to the direction of scanning. The edge of the strip, easily recognizable in visible light, is then focused. To detect the positions of the absorption bands, the illumination is switched over to UV and the monochromator slits are opened to ensure a brighter image in the image converter. The starting point is made to coincide with the crosshairs of the image converter by moving the stage. The luminous field stop is then brought into the optical path, the monochromator slit closed to the values noted before, switchover made from observation to measurement, and the measuring program of computer or separate stage control started. The measurement itself runs automatically.

At the normal measuring frequency of 50 Hz it takes 3 sec to measure a 2 mm long micro-electrophoresis band. The starting point of the next electrophoresis band on the same strip is searched for in the UV, with the monochromator slit and the luminous field stop open. For fresh alignment of the strip the light is switched over to visible to protect the specimen from being bleached by prolonged irradiation. The final checks are made in UV light, then the monochromator slit and luminous field stop are closed, and the microphotometer is ready for the next measurement.

Although all four RNA bases have absorption maxima at different wavelengths, measurement at one single wavelength, 260 nm, is sufficient for their determination. The positions of the absorption maxima and the conversion factors are determined prior to measurement. The absorbance measured at 260 nm must be multiplied by these factors to obtain the absorbance at the absorption maximum. The conversion is easy and saves three measurements, but this procedure is reasonable only when the absorption maxima are so near to each other that the correction factors lie below 3, because the unavoidable measuring errors are also multiplied by the correction factors.

Line scanning is also applied to the measurement of chromosomes. The 0.5 μm step is generally too large to resolve the band structure of genuine chromosomes. However, by scanning the positive or negative of a photograph of a chromosome, at 10 μm steps, a resolution of 0.1 μm (referred to the chromosome in the specimen) is easily obtained. The procedure is time-consuming, with additional inherent errors due to the photographic processing, yet it supplies the best results for the details of chromosome structure as far as these are detectable under an optical microscope.

Area Scanning

Microphotometric specimens are generally inhomogeneous in all directions. Cells in sections, smears, or cultures, are typical examples. Various types of instruments have been developed to measure such specimens, as well as methods to keep the distribution error as small as possible. Details and results are given by SANDRITTER and KIEFER (1965), WIED (1966), WIED and BAHR (1970a and

1970 b). As the scanning method has found widespread use due to its universality, it is assumed in the following considerations that a scanning microphotometer is used for the measurements. Technical details are given in the operating manuals of the instruments, and are not mentioned here in the interest of a detailed description of the optical and photometric conditions, which are prerequisites for successful work. Usually they are not mentioned when describing an instrument because the designer deems them unimportant, and they are omitted in the publications of results as being known already and trivial.

Microphotometric results are as good as the specimen, but most microscopic specimens are not suited for microphotometry. It is senseless to spend a fortune on equipment yet economize on the buying of good specimen slides and cover-glasses. In macrophotometry the need for perfectly ground and polished cuvette windows is unquestioned, and in micro-photometry defects become more noticeable because of the small measuring area. Ground edges and optically treated surfaces are therefore essential. For example, a 1 mm thick specimen slide for a 10 μm layer of the specimen is equivalent to a cuvette wall of 1 m thickness for 1 cm path length of the sample. Nobody would produce cuvette walls of this thickness of normal cheap window glass. The coverglasses, too, must be of top quality and of the prescribed thickness.

Utmost cleanliness and accuracy are required in making the preparation. The demands are high in photomicrography, but an unsharp dirt particle that hardly influences the information content of a photograph would falsify the photometric measurement of the spot considerably. The refractive index of the embedding medium must be optimally adapted to the specimen in order to minimise the interference by refraction. The technique of staining is important, because its reproducibility determines the reproducibility of the measurement. The structure of the specimen determines the selection of the microscope optics. Generally the demand is for the photometric resolution of structures lying near the limit of resolution of the optical microscope. According to Eq. (8) on p. 307, the resolving power of an immersion objective with 1.30 aperture for green light ($\lambda = 546$ nm) is $r = \dfrac{0.546\ \mu m}{1.3} = 0.4\ \mu m$. The literature often gives 0.25 μm as the resolving power of the optical microscope. There is no contradiction in the two statements, since they depend on the definition of the term resolution. When two points are 0.25 μm apart, it is possible to determine whether there are one or two points; whether there are really two separate points or a small dumb-bell is not answered. There are limits to the resolving power of the microscope, because of the diffraction fringes around the magnified specimen. For specimens which are large compared with the resolving power, the diffraction fringes can be neglected. Such specimens appear sharply defined and with the correct shape. Specimens near the limit of resolution have relatively large diffraction fringes. The diffraction pattern alone is visible, not the geometrical image. The presence of one or two points can be calculated from the diffraction pattern, but the geometrical structure becomes visible only when the elements forming the structure are at least 0.4 μm in size.

The next questions arising for microphotometric work are: How small can the measuring stops be? Are measuring stops of 0.25×0.25 μm of any practical use? To answer this question, square stops with edge lengths of 10 or 20 μm were

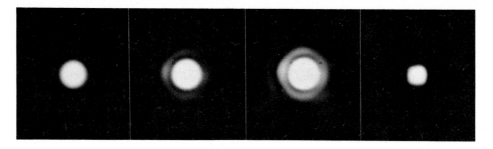

Fig. 14. Images of a square stop, 10 μm × 10 μm, taken at 0.04 aperture at the specimen. Exposure time varied (from left to right) in a ratio of 1 : 4 : 16. Image at the extreme right taken at 0.08 aperture

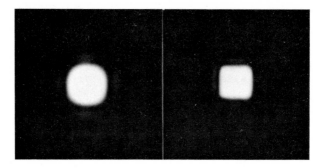

Fig. 15. Photographs of a 20 μm × 20 μm stop, the left taken at 0.04 aperture in the specimen, the right taken at 0.08 aperture

imaged at 0.04 aperture, which, according to the theory of diffraction, corresponds to imaging stops with edge lengths of 0.3 or 0.6 μm at 1.30 aperture, or edge lengths of 0.25 or 0.5 μm at 1.60 aperture.

Fig. 14 shows photographs of the 10 × 10 μm stop. The first three pictures were taken at different exposure times, at a ratio of 1 : 4 : 16. The only indication of the square shape of the stop is the uneven brightness of the diffraction rings. The picture to the right in Fig. 14 shows the same stop at the same magnification, but imaged at 0.08 aperture: there is at least a hint of the square shape. To obtain such an image of an 0.25 μm stop, the objective aperture should be above 2.50.

Fig. 15 shows pictures of the square stop with 20 μm edge length, the left taken at 0.04 aperture. This is how an image of an 0.5 μm stop looks like near the resolution limit of the microscope. The picture to the right was taken at 0.08 aperture.

Dividing the central area of the diffraction pattern by the geometrical image, gives a figure for the falsification of the effective diaphragm area by diffraction. For small stops (see Fig. 14), this factor amounts to about 3, for large stops, to 1.3. The larger the stop, the better the coincidence between geometrical and effective stop size.

In the microphotometer, the measuring stop lies in an enlarged image of the specimen, where the stop is sharply defined. In spite of this the above statements about the effective stop are applicable in this case, also. The relationship between specimen and image being reciprocal, the stop which is effective for the specimen is a considerably reduced image of the diffraction patterns shown in Figs. 14 and 15. This means that the light passing through a square in the specimen with edges of 0.25 μm can never be measured. The light passing through the diffraction pattern of the stop outside this square is also measured, but does not affect the results if the specimen is homogeneous. However, with inhomogeneous specimens the actual resolution of structural details with a 0.25 μm stop is hardly better than that with a 0.5 μm stop, as shown in Figs. 14 and 15.

In practice $\dfrac{0.5\,\mu m}{A}$ can be considered as the minimum size of the stop in the plane of the specimen, where A is the objective aperture. Stops smaller than this can only be used to recognize structural details if the photometric accuracy is disregarded. The stop sizes mentioned on p. 314 ff. are obviously over the minimum size.

The size of the stop at the plane of the specimen depends on the structure of the specimen, and the optics used. The size of the actual measuring stop in the photometer is calculated according to the following equation:

Stop size in the photometer = stop size at the specimen × magnification of objective × magnification of the intermediate optics between   (9)
the objective and the measuring stop in the photometer.

The application of Eq. (9) has already been explained by the example on p. 319.

When scanning a specimen, the size of the stop at the specimen is given by the width of the step on the scanning stage, which should not be smaller than the minimum stop size. The smallest width of step of 0.5 μm on commercial scanning stages corresponds to the minimum stop size if immersion objectives with apertures of 1.00 or more are used.

If the width of the step is of the same order of magnitude as the minimum stop, the shape of the measuring stop (circular or square) is unimportant for practical work (see Figs. 14 and 15). In actual fact, circular stops can be made more accurately, thus increasing the reproducibility of the results. The edge of a square measuring stop should be equal to the width of the step or this width converted to the plane of measurement according to Eq. (9). This guarantees, at least theoretically, complete coverage of the specimen and avoids double measurements. In practice the diffraction images of the measuring stop overlap (see Fig. 15). A step 0.5 μm wide with a circular stop of 0.6 μm diameter (referred to the plane of the specimen) will not affect the result. Even a circular stop of 0.7 μm diameter only impairs the resolution of the structures slightly, and offers double the amount of light for each single measurement as compared with a circular stop of 0.5 μm diameter; this increases the photometric accuracy. The specific problem under investigation determines what compromise is made between the resolution of the structure and the accuracy of measurement.

Using a step 0.5 μm wide and the corresponding size of stop, structures with a diameter of 1 μm or more are resolved. A dark particle with a diameter of 0.5 μm in a bright field will be detected only when it lies exactly in a measuring area. It is equally probable that it may lie on the boundary between four adjacent measuring areas, which makes measurements of the optical density of this particle impossible. The limit of resolution for scanning microphotometry is double the minimum width of step. The limits of resolution for commercial scanning stages with step widths of 0.5 or 10 μm are structures with diameters of 1 or 20 μm respectively.

Similar considerations hold for the choice of the luminous field stop. Its image at the specimen should be as large as the effective area of measurement, in order to eliminate light scattered from portions of the specimen lying outside the area of measurement. However, there are three effects which must be taken into account against this theoretical consideration: first the difficulties involved in focusing and centering the luminous field stop exactly. Second, the "migration" of the image of the luminous field stop during scanning, caused by optical in-homogeneities in the glass slide and the specimen. Third, the low resolving power of the condenser. The correction of a normal condenser is not as good as that of the microscope objective. Due to the thick specimen slides, microscope objectives used as condensers cannot be used at such high apertures as the objectives used for the formation of the image. Small illumination apertures are preferred because this reduces deviations from the Lambert-Beer law. Since Eq. (8) is applicable here, too, only large luminous field stops can be imaged with small illumination apertures. With an illumination aperture of 0.35 no image with a diameter below 1.5 μm of the luminous field stop is obtainable in the plane of the specimen.

The influence of the size of the luminous field stop on the stray-light portion of the result is called the Schwarzschild-Villiger effect. It is discussed by many authors, e.g. D. MÜLLER in SANDRITTER and KIEFER (1965). However, experiment proves to be much more useful than theoretical considerations, because the size of the effect is determined primarily by the structure of the specimen. Test measure-ments of several typical specimens are carried out with luminous field stops of different sizes. A graph of the measured total absorbance versus the size of the stop shows whether or not there is an optimal size for the luminous field stop. This would be a stop which, on reducing the diameter, does not influence the stray-light portion significantly. When the specimens scatter considerably and have high contrast, every reduction in the diameter of the stop will affect the result. As the Lambert-Beer law cannot be expected to apply to such specimens, measurements of this kind are of use primarily for comparison.

As a rule of thumb, the size of the luminous field stop at the plane of the specimen should be at least three times the diameter of the measuring field or $\dfrac{1.5\,\mu m}{A_c}$, where A_c is the illuminating aperture, whichever is the larger of the two values. This gives better reproducibility, which is of importance for purposes of comparison. If, according to test measurements, larger stops are permissible, they should be used because of the greater convenience of operation. The rule of thumb does not apply, however, to reflected-light photometry.

In the relevant literature, different opinions can be found regarding the most suitable illuminating aperture. Geometrical considerations call for small apertures, so that the differences in the path lengths of the various rays in the specimen become negligible. However, small apertures require large luminous field stops and produce more stray-light. Furthermore, small apertures reduce the amount of light reaching the multiplier, which in turn causes a "noisier" measuring signal. Therefore one uses high illuminating apertures (0.6 to 0.8) near the limits of resolution of microphotometry in order to get good reproducibility of the measurements. The compromise between accuracy and reproducibility must be decided by each worker individually. As long as results of microphotometric measurements are not accompanied by the technical data of the optical equipment used, there will be limitations on the interpretation of results obtained by different authors.

There remains little to be said about the scanning process. As a rule, the edges of the field scanned will be covered by test runs until the correct size of the field has been determined. The focus is then checked, the luminous field stop swung in, switchover made from observation to measurement, and the automatic scanning started. At high step frequencies of 50 Hz or more, the scanning stage does not come to rest and the specimen vibrates during measurement, thus impairing the resolution of structural details. The higher the demands in this respect, the longer should be the intervals between two measuring steps.

A scanning stage is a high-precision instrument. The utmost cleanliness and a regular check of all its functions is imperative for its correct working.

There are two methods of evaluating the data obtained on scanning cells. Classical, quantitative cytochemistry requires the mass of the absorbing substance, using the Lambert-Beer law and the resulting Eq. (7) given on p. 304. The total absorbance E is obtained as the sum of the absorbances for each measuring point during scanning. The measuring field, a, is given by the square of the step width. If the molar absorptivity, ε, is known, the mass can be calculated in millimol according to Eq. (7) $m = \dfrac{E \cdot a}{\varepsilon}$; examples are given on p. 304 ff. If ε is not known, the product $E \cdot a$ is calculated, as "mass in arbitrary units". The number of all absorbance values differing from zero, or of values above negligible absorbance, e.g. 0.02, is a measure of the area in the specimen which is covered by the cell. The quotient of the total absorbance and the surface gives the mean absorbance of the cell. Calculations of this kind are based on the Lambert-Beer law, which means that the photometric accuracy is of primary importance for such measurements.

From the point of view of classical quantitative cytochemistry, scanning is only a method of eliminating the distribution error of inhomogeneous cells. G.L. WIED and co-workers pointed out that data obtained by the scanning process provide not only chemical but also morphological infromation about the cell (cf. WIED and BAHR, 1970a and 1970b). The area covered by the cell is one information about the morphology of the cell. Another example is the differences between adjacent absorbance values: these supply numerical data on the homogeneity of the cell. Theoretically, any combination of measured values defines a particular feature of the specimen. However, many of such features are of little value because they are not characteristic for an entire class of similar objects. The search for, and the proof of the validity of a number of characteristic features is the basis for automation of cytological diagnoses using microphotometry. Development in this field progresses at a rapid pace. In contrast to classical cytophotometry, such considerations are not based on photometric accuracy, but on the reproducibility of the feature values. Considering the fundamental problems of photometry near the limits of resolution of the microscope, this is a pragmatic point of view.

Addendum

Remarks Regarding the Microphotometry of Autoradiographs

An example should be given to demonstrate how essential it is to pick the proper procedure from the large variety of methods in microscopy, in order to achieve the best results. This assay was performed by H. PILLER (unpublished) and refers to the microphotometry of autoradiographs. The basic problems of such measurements are described by P. DÖRMER in chapter 11 of this book. Here it will only be shown, by means of photomicrographs, how special techniques can be used to reduce or avoid disturbing effects.

The specimen is the autoradiograph of a thin section of liver, photographed with an Epiplan 40/0.65. Fig. 16 shows a picture taken with transmitted light in bright-field illumination. The silver grains appear black, but a measurement of the absorption would give completely misleading data on the density and number of silver grains, because the light is also absorbed by the stained cells of the specimen. Fig. 17 shows the same spot of the specimen in bright-field epi-illumination. The silver grains are bright because of their intense reflection, but the light is also reflected from the upper surface of the layer of gelatin and from the rear side of the slide. The background becomes bright and the photometric measurements are incorrect. The reflections from the rear side of the slide are due to the different refractive indices of glass and air, and can be largely eliminated by using an oil-chamber, a cross section of which is shown in Fig. 20. The specimen was illuminated in precisely the same manner as in Fig. 17; the result is shown in Fig. 18. A marked increase in the contrast can be noted. The best method of illumination, however, is with polarized light. The reflections from the upper surface of the layer of gelatin, as well as those from the rear side of the slide are linearly polarized and can be extinguished by the analyzer. The heavily reflecting silver grains are depolarizing, and the light which is reflected by the grains cannot be extinguished by the analyzer. The silver grains now appear bright on a black background (Fig. 19) and can be measured photometrically without stray-light.

Fig. 16. Autoradiographs of a thin section of liver, photographed with an Epiplan 40/0.65. Total magnification approx. 500 ×. Illumination: Bright-field transmitted light

Fig. 17. Same as Fig. 16, but with bright-field, epi-illumination

Fig. 18. Same as Fig. 17, but with reduced reflections, achieved by an oil chamber beneath the specimen slide

Fig. 19. Same as Fig. 17, but with polarized light; reflections reduction by crossing the polarizer and analyzer

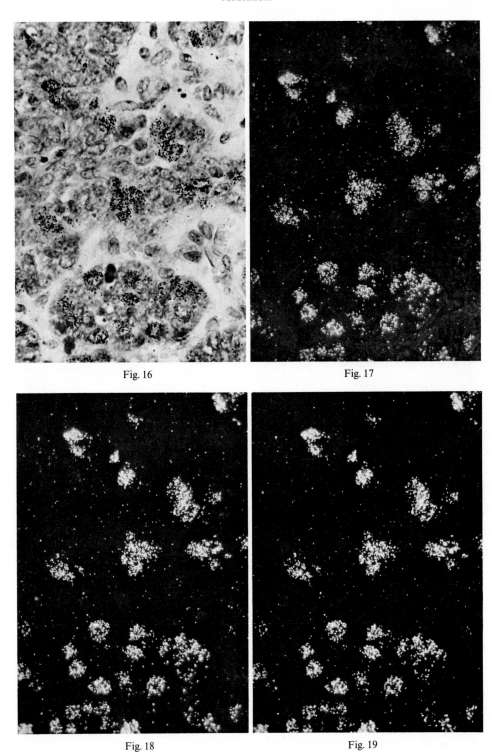

Fig. 16

Fig. 17

Fig. 18

Fig. 19

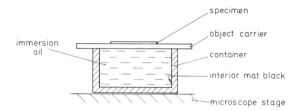

Fig. 20. Cross-section of the chamber for immersion oil for suppression of the reflections from the rear side of the slide

Literature

BEER, A.: Bestimmung der Absorption des rothen Lichts in farbigen Flüssigkeiten. Pogg. Ann. Physik **86** (2), 78–88 (1852).

BOUGUER, P.: Essai d'optique sur la gradation de la lumière. Paris: Claude Jombert 1729.

CASPERSSON, T.: Die Untersuchung der Nukleinsäureverteilung im Zellkern. Z. wiss. Mikr. **53**, 403–419 (1936).

CASPERSSON, T.: Cell growth and cell function. New York: Norton 1950.

D'AMARO, J.: The micro-complement-fixation test I. Vox Sang. (Basel) **18**, 333 (1970).

D'AMARO, J., VAN LEEUWEN, A., SVEJGAARD, A., VAN ROOD, J.J.: The micro-complement-fixation test II. Histocompatibility testing, ed. P. TERASAKI, p. 549. Copenhagen: Munksgaard 1970.

D'AMARO, J., HENSEN, E., VAN ROOD, J.J.: The micro-complement-fixation test III. Tissue Antigens **1**, 171–177 (1971).

EWING, G.W.: Instrumental methods of chemical analysis, p. 137–192. New York-Toronto-London: McGraw-Hill 1954.

EWING, G.W. (deutsch von A. MASCHKA): Physikalische Analysen- und Untersuchungsmethoden der Chemie, S. 151–217. Wien-Heidelberg: Bahmann 1964.

FRANÇON, M.: Progress in microscopy, p. 242–266. Oxford-London-New York-Paris: Pergamon 1961.

KORTÜM, G.: Kolorimetrie, Photometrie und Spektrometrie, 4. Aufl., 464 p. Berlin-Göttingen-Heidelberg: Springer 1962.

LAMBERT, J.: Photometria, sive de mensura et gradibus luminis colorum et umbrae, 1760.

MICHEL, K.: Die Grundzüge der Theorie des Mikroskops, 2. Aufl., 355 p. Stuttgart: Wissenschaftliche Verlagsgesellschaft 1964.

MÖLLRING, F.K.: Mikroskopieren von Anfang an. 66 p. Aalen: Theiss 1968.

MOELLRING, F.K.: Beginning with the microscope. 74 p. New York: Sterling 1973.

NEEDHAM, G.: The practical use of the microscope. 493 p. Springfield (Illinois): Charles C. Thomas 1958.

PESTEMER, M.: Anleitung zum Messen von Absorptionsspektren im Ultraviolett und Sichtbaren, S. 1–70. Stuttgart: Thieme 1964.

SANDRITTER, W., KIEFER, G.: Methoden und Ergebnisse der Zytophotometrie und Interferenzmikroskopie. Acta Histochem. (Jena), Suppl. VI, 459 p. (1965).

WEISSBERGER, A., ROSSITER, B.W. (editors): Physical methods of chemistry, part III B, chapt. III, p. 207–428. New York-London-Sydney-Toronto: Wiley-Interscience 1972.

WIED, G. (editor): Introduction to quantitative cytochemistry. 623 p. New York-London: Academic Press 1966.

WIED, G., BAHR, G. (editors): Introduction to quantitative cytochemistry, part II. 551 p. New York-London: Academic Press 1970(a).

WIED, G., BAHR, G. (editors): Automated cell identification and cell sorting. 403 p. New York-London: Academic Press 1970(b).

Chapter 10

Cytofluorometry

Cytofluorometry has become a valuable method for the quantitative determination of substances such as nucleic acids, amino acids, proteins, and biological amines in cells or cell organelles. Its main advantages are high sensitivity and simplicity, which make the technique especially suitable for fast analyses of large populations of objects.

For the theory of the method we refer the reader to RUCH (1970). In this chapter we shall discuss its practical use, based on the experience obtained in our laboratory during the past 10 years.

Description of Instrument

Disregarding instruments designed for special purposes, such as the determination of excitation or fluorescence spectra (LOESER and WEST, 1962; RIGLER, 1966; ROST and PEARSE, 1971; THAER, 1966), the apparatus for cytofluorometric determinations consists in principle of a fluorescence microscope combined with a photometer. Such a combination can be constructed easily, using standard parts available from optical manufacturers. These instruments may be quite useful for some purposes, but in general they do not exploit the method to its fullest extent for two reasons: firstly when working with an ordinary fluorescence microscope, the time of excitation for a measurement is in the range of seconds, and causes considerable fading due to photo-decomposition of the fluorochrome (Fig. 1).

To overcome this difficulty it is necessary to reduce the time of excitation and measurement to a fraction of a second. In the example shown in Fig. 1, an excitation time of 7 milliseconds produces practically no fading. Only after 10 measurements of the same object is a slight decrease of intensity noticeable.

Secondly the instrumental arrangements used hitherto are all more or less impractical and time-consuming, as they demand several manual operations of diaphragms, mirrors, and other parts.

Recently, Carl Zeiss, Oberkochen, in cooperation with the authors, developed a cytofluorometer for short-time excitation with program-controlled operation (RUCH and TRAPP, 1973). In this chapter, this Zeiss instrument is described; of course, most remarks will apply to other instruments also.

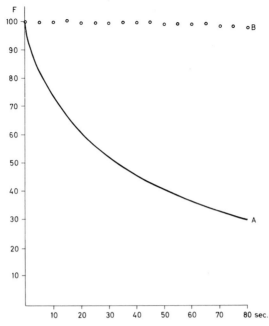

Fig. 1. Influence of photo-decomposition on the measurement of fluorescence. *A*: with continuous blue excitation; *B*: with blue excitation of 7 milliseconds duration at intervals of 5 sec. Mastocytoma cells stained with dansyl chloride. Abscissa: excitation time in seconds. Ordinate: intensity of fluorescence in arbitrary units (F)

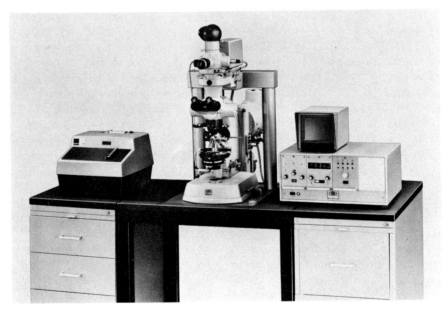

Fig. 2. Zeiss cytofluorometer. Universal microscope with photometer O 1 and AC stand. Fluorescence excitation with reflected light; TV equipment for observation of the image; modular electronic system with amplifier, decimal display unit, and operation control unit; digital printer

Basic Equipment

The basic equipment (Fig. 2) consists of a large fluorescence microscope (Universal or Photomicroscope) with excitation by reflected light, electrically operated shutters and diaphragms, and a modular electronic system for quantitative light microscopy.

Light Sources

Since the single beam system is used for measurements, the light sources have to be invariant to within $\pm 2\%$. Using highly stabilized power supplies, the mercury arc HBO 100, and the xenon arc XBO 150, both from Osram, have proved to be dependable. If excitation with the mercury lines 366, 405, 436, or 546 nm is possible, the HBO 100 is 2 to 5 times brighter (depending on the type of excitation filter) than the XBO 150. Furthermore, the HBO 100 is more practical, since it has a larger discharge arc and is therefore easier to center.

In addition to the source of light for excitation, an ordinary tungsten lamp is necessary for observing and focusing the preparation in green light to avoid fading.

Filters and Reflectors

Absorption or interference filters are most practical for the selection of the appropriate spectral range for *excitation*. A monochromator is only necessary if excitation below 360 nm or over a very narrow spectral range is required. The spectral range for the excitation depends on the fluorescent substance. If this range is unknown, it can be found as follows: the fluorescence of the specimen is inspected by eye for colour, intensity, and contrast, using a set of filters with different spectral ranges and band widths for excitation (with corresponding reflectors and barrier filters). From the filters found to be satisfactory, one is selected photometrically. The types of *reflector* and *barrier* filter are governed by the excitation filter. Therefore it is often advisable to use fixed combinations of excitation filter/reflector/barrier filter. The new fluorescence condenser III RS, from Zeiss, for reflected light, is very practical and time-saving. The filters and reflectors can be changed in combination or individually (for excitation with UV < 400 nm, violet < 450 nm, blue < 500 nm, and at 405 nm, 436 nm, 475 nm, and 546 nm). The condenser also contains a turret with different field apertures (for Koehler illumination). Supplementary filtering of the fluorescent light is necessary in order to determine the fluorescence spectrum (interference filter monochromator) or to obtain measurements over a restricted spectral range.

Objectives

Excitation with reflected light (vertical illuminator) is most advantageous for cytofluorometry, as the focusing of the condenser, necessary with transmitted light, is not required. Excitation with reflected light is most efficient when used with objectives of high numerical aperture. However, it must be remembered that the numerical aperture limits the thickness of the object which can be measured accurately. A *numerical aperture* of 1.3 limits the useful thickness of the object to about 6 μm, and one of 1.0 to about 10 μm. A relatively low numerical aperture is not a disadvantage in cytofluorometry, apart from the reduced intensity of the light, since the highest resolving power is usually not required (RUCH, 1970). Therefore, the use of objectives with the lowest useful power are recommended. Such objectives also have the advantage of being more useful for routine work than high power objectives.

It is self-evident that the lenses must have a high transmittance for short wavelengths, and must not exhibit any noticeable auto-fluorescence. The Neofluar objectives 40, 63 oil, and 100 oil, are very suitable for this purpose.

Since observation and focusing of the preparation is carried out in transmitted light, the objective must be equipped with a phase annulus; an ordinary phase condenser is used for this step.

Fluorescence Standard

The intensities of fluorescence are expressed as relative values; they allow the comparison of the amount of a substance in different objects. Apart from the object, the intensity depends on factors in the measuring instrument, such as the light source, filters, reflector, objective, and the photo-electric receiver and recorder. Since these factors are more or less variable, it is necessary to use a constant fluorescent device to standardize the instrument. A fluorescent glass plate, placed in the object plane of the objective is suitable. In the Zeiss instrument, such a standard is fixed in a magnetic disk, which can be attached to the objective quickly.

Microscope Photometer

The type *O1 microscope photometer* is specifically designed for cytofluorometry. Since it does not use a beam-splitter, the total intensity of the radiation is available either for viewing or for measuring the specimen. This photometer contains interchangeable measuring diaphragms, a viewing eyepiece, projection lenses for the connection of a photomicroscopic or TV camera, and a photomultiplier tube. The change-over (mirror) between viewing or camera and measurement is carried out either manually, with a cable release, or electrically. In order to avoid vibration during measurements the O1 photometer should be fixed to the sturdy AC stand (Fig. 2).

The *measuring diaphragm* in the plane of the image restricts the measurement to one object. The following diaphragms are available: 1) A turret with an iris diaphragm and a set of interchangeable circular diaphragms, 2) an adjustable rectangular diaphragm, and 3) a holder for electrically operated diaphragms. This last device, worked by a foot switch, allows quick in-out switching of interchangeable circular diaphragms for observation and measurement, and is therefore preferable for most routine work.

The choice of the measuring diaphragm depends on the type of object. If the object is extended, such as a microtome section for instance, and we intend to measure only a certain structure within this object, the size of the diaphragm must be adjusted exactly to the size of this structure (requiring iris or rectangular diaphragms, or even diaphragms of more complex form). This procedure is of course time-consuming and often inaccurate. It is much simpler and faster to measure isolated objects of restricted size, such as cells or organelles. Therefore, whenever possible, cell preparations of this type should be used for accurate and fast measurement. In this case the use of a circular, fixed diaphragm with a relatively large aperture is recommended. With such a diaphragm the centering of the object into the aperture is not critical and can be performed quickly. Furthermore, it is not necessary to change diaphragms for the measurement of objects of different sizes, as long as the aperture of the diaphragm is compatible with the size of the largest object in the preparation.

There are of course two limitations to the maximum size of the measuring diaphragm. One is governed by the shortest distance between objects in the preparation, and the other by the fact that the light intensity may not be zero in the clear field. Especially when working with a high sensitivity of the photoelectric receiver, excitation or fluorescent light from the glass or embedding media may be noticed. In this case it is advisable to reduce the measuring aperture to a minimum. In addition, a correction for the intensity of the measuring field may be necessary (intensity of object minus intensity in the empty field). A fixed aperture, as recommended, allows the use of a constant correction factor for all measurements. Furthermore, a fixed aperture is essential for the standardization of the instrument by means of fluorescent glass plates as described previously.

The *lamp field stop* has also to be adjusted (Koehler illumination). According to our experience, a circular lamp field stop with a diameter 2 to 3 times that of the measuring aperture is suitable. A smaller stop has little advantage, and may even be a cause of error if it is not well centered. Regarding the need of a fluorescent standard, it is most practical to use the same stop for all measurements in a research program. The centering of the field stop with the measuring aperture is carried out with the aid of a homogeneous fluorescent preparation or a fluorescent standard plate.

The viewing eyepiece may be used for the *observation of the image and the measuring aperture*. It contains cross-hairs on which the measuring diaphragm can be centered. As already mentioned this observation should be done with green light. If large series of measurements have to be carried out, the use of the viewing eyepiece is inconvenient and tiring. In this case a TV camera is preferable, and it even allows the adjustment of the contrast of the image to further facilitate routine work.

A *photomicrographic camera* may be attached to the O1 photometer, either instead of the viewing eyepiece or instead of the TV camera.

The *photomultiplier* normally used with the microscope photometer is a factory selected (Zeiss) RCA IP 28. For the measurement of red fluorescence this type must be replaced by Hamamatsu R 521. More elaborate multipliers gave no noticeable advantage for the measurement of small light intensities.

Short time excitation, of 7 milliseconds, is produced by an electrically-operated *shutter* combined with the light modulator. A second electrically-operated shutter blocks the light from the tungsten lamp during measurement. The switching on and off of the measuring diaphragm, mirror, shutters, and recording instrument, is *controlled, according to a program,* by an electronic unit worked

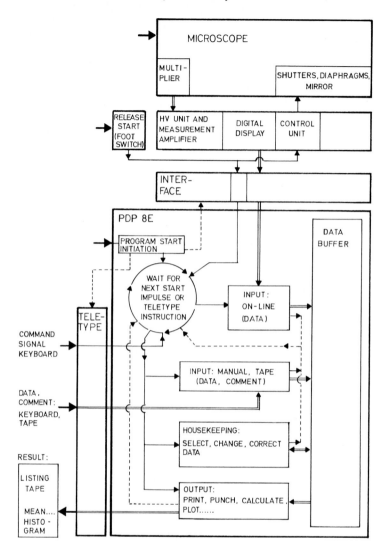

Fig. 3. On-line operation of the cytofluorometer with the PDP-8 E computer

by a foot switch. Since the duration of this sequence is less than a second, the time necessary for one measurement depends mainly on the time required for searching and centering the object. A special program is also available for the standardization of the instrument with the fluorescent glass plate.

Recording of the Intensity of Fluorescence

For this purpose the following slide-in rack-units from the Zeiss modular electronic system are necessary: the amplifier with stabilized high-voltage unit for feeding the photomultiplier tube, the decimal display unit, and the unit already mentioned for the automation of the different operations. The electronic equipment may be extended and adapted for any type of work. The data can be printed out, stored off-line on punched or magnetic tape, or processed on-line by a computer. In our laboratory a PDP-8E (4K) is connected to the Zeiss cytofluorometer. The program allows storage and immediate processing of approximately 2000 measurements; the values can be stored on punched paper tapes (Fig. 3).

Use of Instrument

Adjustment of the Instrument

Accurate measurements depend on a correctly adjusted instrument. The simplest way of *centering the light arc* into the pupil of the microscope objective is the following: A fluorescent glass plate or a sheet of white paper is placed 10 to 20 mm in front of the objective (e.g. 40:1) so that an image of the pupil with the phase annulus and the arc is projected on it. The burner is then moved vertically and horizontally until the arc is centered into the pupil. With the xenon arc, a concave mirror behind the burner should be used, positioned in such a way that the direct and the reflected images of the brightest spot of the arc lie symmetrically within the pupil. Unsymmetrical distribution of the light in the pupil (with any type of light source) gives uneven light distribution in the plane of the image in the microscope and results in inaccurate measurements of extended objects.

The *centering of the field stop and the measuring diaphragm* is carried out with the aid of the viewing eyepiece in the microscope photometer. The measuring aperture is centered on the cross-hairs of this eyepiece, then the field stop is centered on the cross-hairs, using a homogeneously fluorescent preparation for microscopy. If a TV camera is used, the center of the measuring aperture must be marked on the monitor screen; this can be done very simply, for instance by fixing two crossed threads on the monitor housing.

Measuring Procedure

The preparation is first adjusted in the microscope, under conditions of phase contrast. An inspection of objects with different sizes establishes the suitable measuring diaphragm. With this diaphragm, the microscope stage is lowered and the fluorescent standard is attached to the objective. The program switch of the electronic control unit is set to 100 (permanent excitation for standard measurement), and the voltage for the multiplier and the fine adjustment of the amplifier are varied until a value of 100.0 is shown on the decimal display unit. The control unit is then switched to automatic measurement (Fl). An object is centered on the cross-hairs (viewing eyepiece or TV camera), the measuring diaphragm is switched in, and the centering of the object checked. By a foot switch or by a cable release the following program of operation is started: closing of the shutter of the tungsten lamp—movement of the mirror to direct the light to the photomultiplier—opening of the shutter for excitation, and simultaneous light measurement—return of mirror to original position—diaphragm (automatic type) brought into place for observation of field. The duration of this program is less than one second; skilled users of the instrument may be able to adjust and measure an object within a couple of seconds.

The ratio between the values of intensity for the standard and the object is important. The latter should not be less than approximately a tenth, or more than twice the standard value of 100. If necessary a change of amplification ($10\times$ or $100\times$) can be used for adjustment. After a number of measurements of objects, the standard should be measured again and readjusted if necessary.

Some Applications

Material

Cells or isolated cell particles in suspension are the most suitable material for measurements. However, the air-drying of cells, especially if they are unfixed, should be avoided, as the uptake of the dye may be very irregular (LEEMANN, WEISS, and SCHMUTZ, 1971). The usual smear technique should, if possible, be replaced by one of the following methods:

1) Centrifugation of cells or particles onto cover slips after fixation (LEEMANN, WEISS, and SCHMUTZ, 1971). If the material is not sufficiently adherent, treatments such as acid hydrolysis, incubation with enzymes etc., may be carried out in a test tube before centrifugation.

2) The whole procedure of staining and dehydration is carried out in a test tube.

3) Filter techniques (millipore, etc.) (FROST et al., 1967).

Cytofluorometric measurements can also be made on *tissue sections*. As with absorption techniques, the physical conditions must be taken into account (thickness of sections and size of objects, fluorescence of surroundings). Frozen sections are of course suitable. Different embedding media have been employed successfully for cytochemical purposes: paraffin, polyethylene glycol, Spurr low-viscosity medium[1].

Fixation

Several different *fixatives* are suitable for cytofluorometric studies: good results were obtained with ethanol, ethanol-acetic acid mixtures, and with neutralized formalin. The use of formalin results in partial blocking of the reactive amino groups. For measurements on one kind of tissue, this did not prove detrimental; for comparison between different materials, however, caution is advised when investigating protein reactions (especially those for lysine). Ethanol and ethanol-acetic acid mixtures do not have this limitation; they may, however, give a lower dry weight per gm. wet weight than does formalin. It has to be kept

---

[1] Spurr low-viscosity embedding medium: Polyscience Inc., Paul Valley Industrial Park, Warrington, Pa. 18976, U.S.A.

in mind that all material of low molecular weight is lost despite fixation. Fixed cells (and especially isolated particles such as nuclei) may have only $\frac{2}{3}$ or even only half of the weight determined (e.g. interferometrically) for the living material.

Glutaraldehyde is not recommended because of its additional free aldehyde groups. Fixatives containing heavy metals make measurements of fluorescence impossible.

Cytofluorometry also gives good results with freeze-substituted material.

The *length of time of the fixation* for suspensions of cells is between 4 and 24 hrs.; with a shorter period the cells are not fixed properly and may decompose rapidly; longer fixation may reduce the uptake of the dye and weakens the adherence of the cells to the glass surface. Blocks of tissue should be fixed for at least 24 hrs.; only very small blocks (2 mm) should be used, to ensure thorough fixation. After fixation, the material must be rinsed carefully (with distilled water after formalin, with ethanol after mixtures containing acetic acid).

Storage

Cells or cell particles in suspension may be stored in 70% ethanol at 4° C. Some material (sperms, yeast etc.) can be kept for many months without detectable loss of substance. More delicate cells (leukocytes, mastocytoma cells etc.) should not be stored for longer than about 3 weeks, as after this period determinations of DNA and protein show decreasing amounts.

Staining Reactions

The staining reactions described in this chapter have proved useful in model experiments and in many different cellular materials (e.g. roots of plants, pollen, algae, myxomycetes, yeasts, and mammalian cells such as leukocytes, sperms, thymus, liver and kidney nuclei, and malignant cells such as ascites and mastocytoma cells). As with any other technique, however, there are limitations which must be known so as to avoid failure or misinterpretation.

The *primary fluorescence* (measured on preparations treated exactly like the stained ones except that the dye is omitted) should be considerably lower than the fluorescence of the dye. Strongly absorbant or fluorescent substances (e.g. whole liver cells, erythrocytes) debar the use of this method.

The staining reaction must be *specific and stoichiometric*. Some rare substances (e.g. the complex of tannin and nucleoprotein in some Rhoeo species) or exceptionally large amounts of, for example, strongly acidic proteins or polysaccharides, might possibly interfere, and their influence should be investigated. The specificity of the reaction can be tested on model substances, or by extractions and blocking reactions, although these are not necessarily more specific than the staining reactions. Blocking procedures often result in considerable loss of material.

The intensity of fluorescence must be proportional to the amount of dye bound. The *range of linearity* is limited by the absorption of the excitation and the emission light, i.e., by the specific absorbance of the dye and the thickness multiplied by the concentration of the object.

Since the thickness of objects may vary (this is also influenced by the technique of preparation), no limit in absolute terms can be given. To test both the stoichiometry and the linearity of the fluorescence effect, either measurements on known objects or control measurements using independent techniques (absorption, interference) are carried out (for instance, rat liver nuclei up to 16-ploid were measured, and yielded the expected 2 : 4 : 8 : 16 ratio).

Special care should be given to the choice and storage of the *dye*. Dyes procured from different suppliers do not necessarily have the same staining characteristics. Our experiences as to the stability of both the dye and the completed slides are given for each method described below. Normal laboratory standards of reagents (puriss.) and glass-ware (cleaned with detergents) are sufficient; ordinary glass slides and cover slips are used.

The stained slides, if preserved at 4° C and not exposed to light, can be *used for measurements of fluorescence* for up to about 3 weeks (except with the ninhydrin reaction). Observation and measurements of relative fluorescence on any one slide are possible for a much longer period (1–3 years).

DNA

A modified Feulgen reaction, using the fluorochrome BAO[2] [(Bis-(4-aminophenyl)-1,3,4,-oxadiazol)] instead of pararosaniline, is used for the determination of DNA (RUCH, 1970). The staining can be carried out on cells fixed with ethanol, ethanol-acetic acid, or formalin. Chromatin structures give a brilliant blue fluorescence.

Staining procedure:

rinse	— 2 min dist. water
hydrolyse	HCl (see below)
rinse	— 5 min dist. water
stain	— 2 hrs BAO solution (100 ml aqueous BAO solution 0.01 %, 10 ml 1 N HCl, 5 ml $NaHSO_3$ 10 %)
rinse	— 2 min dist. water
	— 3 × 2 min sulfite solution (180 ml dist. water, 10 ml 1 N HCl, 10 ml $NaHSO_3$ 10 %)
	— 10 min running tap water
dehydrate	— 30 sec each 50 %, 70 %, 95 %, 100 %, 100 %, ethanol
mount	— glycerol, benzyl alcohol etc. or 2 × 30 sec xylene, then Fluormount[3] for permanent slides

Excitation: UV (UG 1).

Hydrolysis: Different temperatures and concentrations of acid (as used for the regular Feulgen reaction) may be employed (e.g. 6 N HCl or 2 N HCl at room temperature, 4 N HCl at 27° C). For quantitative purposes, however, hydrolysis with 1 N HCl at 60° C is recommended. In our laboratory this has proved most reliable, and gives the highest contrast between nuclei and cytoplasm.

---

[2] BAO, ninhydrin, dansyl chloride: Fluka AG, Chemische Fabrik, 9470 Buchs, Switzerland.
[3] Fluormount: Edward Gurr Ltd., Michrome Laboratories, Coronation Road, Cressex Industrial Estate, High Wycombe, Bucks, England.

The optimal time of hydrolysis should be determined for every object (and fixation). 12 min (1 N HCl, 60° C after fixation in ethanol-acetic acid or formalin) is best for several kinds of cells (bull sperm, liver nuclei, mastocytoma cells, ascites, thymus nuclei); 8 min proved best for human leukocytes.

Stability: The 10% sodium sulfite stock solution may be kept for approximately 1 week; the BAO and sulfite solutions should be prepared freshly prior to each staining process. The solid dye decomposes slowly and should not be stored for more than about six months.

Other dyes: If the object has a noticeable blue primary fluorescence, the use of another dye instead of BAO may be advantageous: e.g. *Auramin O*[4], which yields a bright yellowish-green fluorescence (BOSSHARD, 1964). Auramin O is used in a concentration of 0.2%; otherwise the procedure is the same as with BAO.

For very small amounts of DNA (mitochondria, chloroplasts) the use of *regular Feulgen* (pararosaniline) is advisable (PLOEM, 1967). Excitation with green light minimizes the primary fluorescence. With larger amounts of DNA linearity cannot be expected because of the high absorbance of Feulgen-stained objects.

Histones

Basic proteins (mainly histones) can be determined with the acidic fluoro-chrome Sulfaflavine[5] (LEEMANN and RUCH, 1972), using a procedure similar to the Fast green FCF pH 8 reaction (ALFERT and GESCHWIND, 1953). The objects for staining must be non-dried, and fixed with formalin; DNA must be extracted, preferably by trichloroacetic acid (TCA). The chromatin structures have a yellowish-green fluorescence.

Staining procedure:

rinse	— 3 min dist. water
hydrolyse	— 3 hrs TCA 5% at 60° C
rinse	— 3 × 10 min 70% ethanol
	— 2 × 2 min dist. water
	— 3 min buffer pH 8.0 (0.01 M borate-KCl)
stain	— 30 min Sulfaflavine 0.1% in buffer pH 8.0
differentiate	— 10 min buffer pH 8.0
dehydrate	— 30 sec each 70%, 95%, 100%, 100%, ethanol
mount	— 2 × 30 sec xylene—Fluormount

Excitation: UV (UG 1) or Blue (BG 12).

DNA extraction: DNA may also be extracted by other methods (TCA 5% 15 min at 90° C, deoxyribonuclease). In our laboratory, the recipe given above yielded the best results (least loss of dry weight, good preservation of cells).

Stability: Sulfaflavine solution may be kept for about 3 weeks (at 4° C, in the dark). The dye can be stored for at least 1 to 2 years.

---

[4] Auramin O: Allied Chemical Corp., 40 Rector Street, New York 6, U.S.A.
[5] Sulfaflavine: Chroma Gesellschaft, Schmied & Co., Stuttgart-Untertürkheim, Germany.

Total Protein

Total protein is determined with Sulfaflavine[6] at pH 2.8 (LEEMANN and RUCH, 1972) by a method analogous to that for staining with Naphthol yellow S (DEITCH, 1966). Fixation with ethanol-acetic acid is recommended.

Staining procedure:

rinse	— 2 min dist. water
	— 2 min buffer pH 2.8 (citric acid-phosphate McIlvaine, $10 \times$ diluted)
stain	— 30 min Sulfaflavine 0.1 % in buffer pH 2.8
differentiate	— 10 min buffer pH 2.8
dehydrate	— 30 sec each 70 %, 95 %, 100 %, 100 %, ethanol
mount	— 2×30 sec xylene—Fluormount

Excitation: UV (UG 1) or Blue (BG 12).

For the *stability* of the dye and the staining solution see under determination of histones.

Arginine

Determination of arginine is carried out with the ninhydrin[7] fluorescence reaction (ROSSELET, 1967). For fixation ethanol-acetic acid or formalin may be used. Measurements should be made within 4–5 hrs. (reaction time). The reaction yields a greenish fluorescence.

Staining procedure:

rinse	— 2 min absolute ethanol
stain	— 4–5 hrs ninhydrin staining solution [1 part solution A (0.5 % ninhydrin in abs. methanol)+1 part solution B (1 N NaOH + anhydrous glycerol, 1:1)]
mount	— in the staining solution

Excitation: UV (UG 1).

Stability: The staining solution should be prepared freshly before each staining. It has a yellow colour which fades within a few hours; it can then no longer be used. Solution A is prepared freshly, solution B should be prepared at least 2 weeks before use, but can be kept for several months. Solid ninhydrin can be stored for months. The stability of the preparation varies according to the type of cell; due to the high concentration of NaOH some cells are destroyed within 24 hrs (e.g. leukocytes, thymus nuclei). No permanent preparations are possible.

Lysine

Lysine is determined with dansyl chloride[7] as the fluorochrome (ROSSELET and RUCH, 1968). Material fixed with ethanol-acetic acid or formalin may be used. Stained objects have a strong yellow fluorescence.

---

[6] See footnote 5, p. 339.
[7] See footnote 2, p. 338.

Staining procedure:

rinse	— 2 min in 95% ethanol
stain	— 5–12 hrs in filtered dansyl chloride solution (0.1% in 95% ethanol saturated with $NaHCO_3$)
rinse	— 30 min in filtered 95% ethanol saturated with $NaHCO_3$
dehydrate	— 2 changes of absolute ethanol
mount	— 2 × 30 sec xylene—Fluormount

Excitation: UV (UG 1).

Stability: 95% ethanol saturated with $NaHCO_3$ should be prepared at least 2 weeks prior to staining, and filtered immediately before use. The solution of dansyl chloride should be kept in the dark and for not longer than 12 hrs. Storage of the dye at 4° C is recommended.

Sulfhydryl Groups

Sulfhydryl groups are stained with mercury orange [1-(chloromercuriphenylazo)naphthol-2][8] on material fixed with acetic acid, ethanol, or TCA (5%) (BENNET and WATTS, 1958; BURNS, 1967). Reduction of the disulfide groups can be performed prior to staining of the sulfhydril groups (PEARSE, 1961). Mercury orange may also be dissolved in n-butanol, n-propanol, or n-toluol, etc. instead of in dimethyl-formamide, but yields a lower intensity of fluorescence. Air-drying of the objects before staining should be avoided; hydrogen bonding agents can be used to counteract the effect of drying (e.g. 20% phenol in the solution of mercury orange). Mercury orange has an orange-red fluorescence.

Reduction of disulfide groups:

rinse	— 2 min dist. water
reduce	— 4 hrs in thioglycollate at 37° C (47 g 98% thioglycollic acid/liter adjusted to pH 8.0 with 1 N NaOH)
rinse	— 10 min dist. water
	— 1 min acetic acid 1%
	— 5 min dist. water

Staining procedure:

dehydrate	— 30 sec each 70%, 95%, 100%, 100%, ethanol
rinse	— 2 min dimethyl-formamide
stain	— 16–48 hrs mercury orange (10 mg/100 ml dimethyl-formamide [saturated])
rinse	— 5 min dimethyl-formamide
mount	— 3 × 2 sec xylene—Fluormount

Excitation: Blue (BG 12).

Stability: The dye solution and the thioglycollate are prepared freshly before use; the dye can be kept for years.

---

[8] Mercury orange: Koch & Light Laboratories, Colnbrooks, Bucks, England.

Successive Measurements

It is often very useful to measure more than one parameter on the same cell. For this it is necessary to identify the cells on a map, for which photographic prints may be used. It is now possible to replace this rather time-consuming procedure by computer-controlled positioning of the object, using a scanning stage.

The *ratio of DNA to histone* may be measured by a combination technique analogous to the Feulgen-Fast green method, using TCA instead of HCl (BLOCH and GODMAN, 1955). Successive measurements of UV-absorption and fluorescence are, however, more reliable. Since the technique of UV-absorption requires no chemical treatment of the cells (except possibly extraction of RNA with ribonuclease), the following cytochemical reaction is in no way affected. Not only the ratio of *DNA to histone*, but that of *DNA to any protein* can be measured in this way.

Determination of the dry weight by interferometry is equally well suited for combined measurements.

With Sulfaflavine the ratio of *histone to total protein* can be determined fluorometrically (LEEMANN and RUCH, 1972) (Fig. 5). Material which is fixed with formalin and extracted with TCA is used for this procedure.

Fixation with formalin yields lower values for total protein than does fixation with ethanol-acetic acid; extraction of nucleic acids prior to the pH 2.8 staining enhances the fluorescence. Both effects roughly cancel each other out. For more extensive discussion see LEEMANN and RUCH (1972).

Procedure:

stain, differentiate, dehydrate and mount
         — as for total protein, Sulfaflavine pH 2.8
measure total protein
soak slide in xylene, remove cover slip
hydrate   — 30 sec each 100%, 100%, 95%, 70%, 50%, 30%, ethanol, dist.
             water
rinse   — 3 min buffer pH 8.0
stain, differentiate, dehydrate and mount
         — as for histone, Sulfaflavine pH 8.0
measure histone

Analysis and Evaluation of Data

The intensity of fluorescence is (under specific conditions, as discussed on p. 337 ff.) proportional to the amount of the substance measured.

A useful means of presenting and evaluating results is the *histogram* (frequency distribution) (Fig. 4). If the width of the classes is constant in the log scale, the distribution of the intensities of fluorescence for different populations can be compared directly. The width of the classes has, of course, to be chosen in relation to the distribution of values and the number of measurements. Classes

```
RAT LIVER NUCLEI, HISTONE (SULFAFLAVINE PH 8.0)

   N       10        20        30        40        50
           '         '         '         '         '
A


       X
       XX
140    XXXXX
       XXX
       XXXXXXXXXXXXXXX
       XXXXXXXXXXXXXXXXXXXXX
       XXXXXXXXXXXXXXXXXXXXXXX
200    XXXXXXXXXXXXXXXXXX
       XXXXXXX
       XXXXXX
       X
       XXXX
280    XXX
       XXXXX
       XXXXXXX
       XXXXXXXXXXXXXXX
       XXXXXXXXXXXXXXXXXXXXXXXXXX
400    XXXXXXXXXXXXXXXXXXXXXXXXXXXXXXXXXXXXXXXXXXXXXXXXXXX
       XXXXXXXXXXXXXXXXXXXXXXXXXXXXXXXXXXXXXX
       XXXXXXXXXXXXX
       XXX
       XX
560    X

       XX
       XXXX
       XXX
800    XXX
       XXXXXXXXX
       XXXX
       X

1120
```

```
RAT LIVER NUCLEI, HISTONE (SULFAFLAVINE PH 8.0)
X1=120,X2=260,              DIPLOID
N=  105        M=  184 ,4        S2=     0    596
               S=   24 ,4 =  13,2 %


RAT LIVER NUCLEI, HISTONE (SULFAFLAVINE PH 8.0)
X1=261,X2=600,              TETRAPLOID
N=  159        M=  395 ,4        S2=     2    481
               S=   49 ,8 =  12,6 %


RAT LIVER NUCLEI, HISTONE (SULFAFLAVINE PH 8.0)
X1=601,X2=1100,             OCTOPLOID
N=   26        M=  812 ,4        S2=     7    840
               S=   88 ,5 =  10,9 %
```

Fig. 4. Computer print-out of histogram, mean values, and standard deviations. N: number of nuclei, A: fluorescence in arbitrary units, X1 and X2: lower and upper limits, M: mean, S: standard deviation

which are too wide may obliterate the resolution; very narrow classes may lead to misinterpretation if the number of measurements is not large enough. The scale shown in Fig. 4, for instance, proved suitable for cytofluorometric measurements of different populations of cells (10 classes for twice the value).

The graph provides the basis for further evaluation: for instance, single populations such as diploid or tetraploid cells can be marked off for the calculation of mean, standard deviations, etc.

If more than one parameter is measured for each cell, other evaluations (correlation, linearity test etc.) (Fig. 5) may be of use.

The *reproducibility* of the fluorescence stainings and measurements is comparable to that of absorption techniques. The most accurate results, of course, are obtained from slides which are handled simultaneously. The instrument standard allows comparisons between different series of measurements over a prolonged period of time.

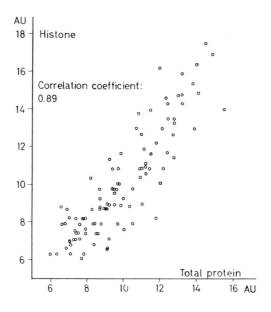

Fig. 5. Correlation between histones and total protein. Mastocytoma cells stained successively with Sulfaflavine at pH 2.8 and 8.0. AU: arbitrary units

In addition to the instrument standard, however, the use of *biological standard materials* has several advantages:

1) The unavoidable slight variations between different staining series can be eliminated.

2) Failures (e.g. due to impure reagents) are detected easily, since the approximate mean and the distribution of the standard values are already known.

3) The standard material can be calibrated, and so allows the determination of *absolute amounts of substance* (data obtained biochemically, model substances). Model experiments using protein smears, as well as different cellular objects, showed good agreement between the amounts of protein as determined fluorometrically with Sulfaflavine and by interferometry (Leemann and Ruch, 1972).

Uniform populations of cells (sperms, thymus nuclei, etc.) or populations with different but recognizable types of cells (leukocytes, liver nuclei, etc.) are suitable as standards.

The *accuracy* of the apparatus (indicated by the instrument standard) is approximately $\pm 2\%$. Minimum standard deviations obtained on uniform populations of cells amount to 6–8% for DNA, 8–10% for histone, and 10–15% for other proteins (e.g. sperms, lymphocytes, 30–50 measurements; RUCH and LEEMANN, 1972). Absorption techniques yield similar figures. It is difficult to establish how much of this deviation is caused by preparation and staining, and how much is due to biological variance.

In some instances, *correction* of the obtained values is necessary.

The *primary fluorescence* (although often negligible) is subtracted on a percentage basis. For corrections of the effect of *fluorescent surroundings* (measurement of nuclei in whole cells, tissue sections) the geometrical conditions as well as the respective intensities of fluorescence should be assessed (LEEMANN, RUCH, and STRÄULI, 1968). Measurement of the intensity per unit area (in nuclei and cytoplasm) or of intensity measured with different diaphragms (e.g.) one for the nucleus only, one for the whole cell, may be of use. In isolated material, the cytoplasm above and below the nucleus usually has little effect because of the flattening of the cells.

Unhydrolysed control preparations, stained with BAO or with Sulfaflavine at pH 8.0, should show no fluorescence. Indeed, in many cases (especially in the case of isolated nuclei), the fluorescence of unhydrolysed material is less than 10% of that obtained after optimal hydrolysis. In some instances however, control preparations fluoresce as strongly as the hydrolysed ones; the *curve for the time of hydrolysis* therefore yields no clear-cut peak. This is presumably due to the extraction of substances during hydrolysis. The optimal time for hydrolysis is determined, in this case, by observing the contrast between the nuclei and the cytoplasm.

Literature

ALFERT, M., GESCHWIND, J.I.: A selective staining method for the basic proteins of cell nuclei. Proc. nat. Acad. Sci. (Wash.) **39**, 991–999 (1953).

BENNET, H.S., WATTS, R.M.: The cytochemical demonstration and measurement of sulfhydryl groups by azo aryl mercaptide coupling, with special reference to mercury orange. In: General cytochemical methods, p. 317–374. New York: Academic Press 1958.

BLOCH, D.P., GODMAN, G.L.: A microphotometric study of the synthesis of DNA and nuclear histone. J. biophys. biochem. Cytol. **1**, 17–28 (1955).

BOSSHARD, U.: Fluoreszenzmikroskopische Messung des DNS-Gehaltes von Zellkernen. Z. wiss. Mikr. **65**, 393–408 (1964).

BURNS, J.: On increasing sensitivity of the mercury orange method for SH groups by fluorescence. Histochemie **10**, 293–294 (1967).

DEITCH, A.D.: Cytophotometry of proteins. In: Introduction to quantitative cytochemistry, vol. 1, p. 451–468. New York: Academic Press 1966.

FROST, J.K., GILL, G.W., HANKINS, A.G., et al.: Cytology filter preparations: factors affecting their quality for study of circulating cancer cells in the blood. Acta cytol. (Philad.) **11**, 363–373 (1967).

LEEMANN, U., RUCH, F.: Cytofluorometric determination of basic and total proteins with Sulfaflavine. J. Histochem. Cytochem. **20**, 659–671 (1972).

LEEMANN, U., RUCH, F., STRÄULI, P.: Characterization of the Ehrlich ascites carcinoma by means of quantitative cytochemistry. Acta cytol. (Philad.) 12, 381–394 (1968).

LEEMANN, U., WEISS, S., SCHMUTZ, E.: A centrifugation technique for cytochemical preparations. J. Histochem. Cytochem. 19, 758–760 (1971).

LOESER, C. N., WEST, S. S.: Cytochemical studies and quantitative television fluorescence and absorption spectroscopy. Ann. N. Y. Acad. Sci. 97, 329–526 (1962).

PEARSE, A. G. E.: Histochemistry, theoretical and applied, 2nd ed., p. 802. London: J. and A. Churchill, Ltd. 1961.

PLOEM, J. S.: The use of a vertical illuminator with interchangeable dichroic mirrors for fluorescence microscopy with incident light. Z. wiss. Mikr. 68, 129–142 (1967).

RIGLER, R.: Microfluorometric characterization of intracellular nucleic acids and nucleoproteins by acridine orange. Acta physiol. scand., Suppl. 267, 1–22 (1966).

ROSSELET, A.: Mikrofluorometrische Argininbestimmung. Z. wiss. Mikr. 68, 22–41 (1967).

ROSSELET, A., RUCH, F.: Cytofluorometric determination of lysine with dansylchloride. J. Histochem. Cytochem. 16, 459–466 (1968).

ROST, F. W. D., PEARSE, A. G. E.: An improved microspectrofluorometer with automatic digital data logging: construction and operation. J. Microscopy 94, 93–105 (1971).

RUCH, F.: Principles and some applications of cytofluorometry. In: Introduction to quantitative cytochemistry, vol. 2, p. 431–450. New York: Academic Press 1970.

RUCH, F., LEEMANN, U.: Human leukocytes as a standard for cytochemical protein determination. Acta cytol. (Philad.) 16, 342–348 (1972).

RUCH, F., TRAPP, L.: Zeiss Informationen 1973 (in press).

THAER, A.: Speicherbetrieb mit einem Superorthikon in der Mikrospektrographie. — Haus der Technik — Vortragsveröffentlichungen 69 (1966).

Chapter 11

Quantitative Autoradiography at the Cellular Level[1]

Introduction

In 1957 HILDE LEVI wrote that in almost no analytical method was the term "quantitative" so misused as in autoradiography. If we take the term "quantitative" to mean that absolute amounts of a substance are measured, this would seem to be true. The misuse of the term in this sense, indeed, continued after 1957, and gained ground following the introduction of tritiated thymidine for biological research.

The autoradiographic method is discussed in this review article with respect to the possibility it offers for measuring amounts of substances at the cellular level. Since so few publications according to HILDE LEVI, are indeed based on "quantitative autoradiography", it is legitimate to ask whether so much emphasis should be given to a method which is used so seldom. It is hoped, however, that the procedure for determining the rate of synthesis of DNA in individual cells, as discussed on p. 385 ff., will show the relevance and usefulness of a quantitative autoradiographic method in cell biology.

The difficulties encountered are many, so that it is quite understandable that really quantitative techniques in autoradiography have seldom been developed. One of the purposes of this review is to emphasize the details which have to be considered when a new quantitative technique is sought. The more we know about these details, the more likely are we to find suitable procedures for new applications of quantitative autoradiography. For the general aspects and commonly used methods in autoradiography, the reader is referred to a series of basic and detailed descriptions (BASERGA, 1967; BASERGA and MALAMUD, 1969; BOYD, 1955; FISCHER and WERNER, 1971; HARBERS, 1958; NIKLAS and MAURER, 1955; PRZYBYLSKY, 1970; ROGERS, 1967; SCHULTZE, 1968, 1969).

In the opinion of the present author, the greatest difficulty in using quantitative autoradiography for biological research is due to the biochemical and preparatory prerequisites. The details of the autoradiographic technique can be adapted for biological purposes by means of systematic studies. However, to what extent the biochemical prerequisites can be met, for instance, to study the synthesis and turnover of RNA in a really quantitative manner, cannot yet be predicted. This is not meant to detract from the value and significance of the qualitative autoradiographic studies of the metabolism of RNA. It means, however, that biochemical prerequisites are of primary importance for quantitative autoradiographic work.

[1] Investigations performed under the association contract EURATOM-GSF for hematology No. 031641 BIAD. Supported by the Deutsche Forschungsgemeinschaft: SFB 51/E-3.

Biochemical Aspects

The Precursor Pool

Two kinds of substances can be used as tracers: the first group comprises substances, such as labeled drugs, which the biological object did not contain at the start of the study. In this case, no unlabeled precursor has to be taken into account, so the same specific activity as in the applied compound can be assigned to the activity as determined autoradiographically. It is assumed that a stable chemical bond is formed between the isotope and the object, and that the object is not separated from its label prior to autoradiographic detection.

The second group includes all the tracers and labeled compounds that either occur normally in the biological system under study, or are synthesized or degraded to normally-occurring intermediates. All these labeled substances are diluted in their metabolic pathways by the naturally occurring unlabeled compounds. In this second group, the autoradiographically determined activity cannot be correlated directly to the specific activity of the applied compound.

All intermediates between the chemical substance applied and the end-product detected by autoradiography can be designated as the precursor pool. This pool is fed by the labeled exogenous and unlabeled endogenous precursors (Fig. 1). The flow out of the pool is caused by the synthesis of the end-product and by the degradation of intermediates. The conditions of the pool are known in most detail for the case of the metabolism of thymidine (QUASTLER, 1963a;

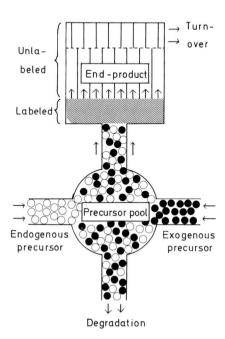

Fig. 1. Schematic representation of changes in the specific activity of a labeled precursor in the precursor pool, and of the specific activity and turnover of the end-product

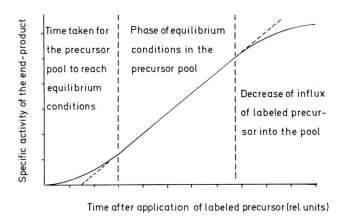

Fig. 2. Increase in specific activity in the end-product after a single application of a labeled precursor

CLEAVER, 1967; CLEAVER and HOLFORD, 1965). The following discussions therefore refer chiefly to the thymine-nucleotide pool.

When a labeled precursor is added to a system, it begins to flow into the precursor pool, and the specific activity in the pool increases from 0 to an equilibrium value. Equilibrium is reached between endogenous and exogenous precursors; the initially unlabeled precursor pool is mixed with labeled molecules, and a new pool size is attained (CLEAVER and HOLFORD, 1965). The length of time it takes to reach equilibrium is influenced largely by the size of the initial pool and the amount of the labeled precursor; this may need up to 20 min (CLEAVER, 1967). According to CLEAVER (1967), the physiological precursor pool of thymidine phosphates is of the order of 10^{-5} M and according to the investigations of FEINENDEGEN and BOND (1962), is localized in the nucleus of the cell.

As long as the specific activity in the precursor pool is changing, the increase in specific activity of the end-product is not linear (Fig. 2). Only when equilibrium is reached in the precursor pool is the rate of increase in activity of the end-product a relative measure for the rate of synthesis. After a certain time the rate of increase in activity decreases; this may be caused by reduction of the influx of the labeled precursor (Fig. 2).

For in vivo experiments it is questionable whether a period exists at all for which conditions in the precursor pool are constant. After the injection of ^{3}H-thymidine (^{3}H-TdR) into man, RUBINI et al. (1960) observed an exponential rate of clearance of the plasma. This rate was governed by two components, the half times of which were calculated to be 0.2 and 1.0 min, respectively. It is thus improbable that equilibrium would be attained in the precursor pool even for one minute. The determination of the relative rate of synthesis of the end-product, during equilibrium in the precursor pool, is therefore only possible under in vitro conditions (BRINKMANN and DÖRMER, 1972), and not under in vivo conditions.

In order to calculate the absolute rate of synthesis from the relative rate obtained under equilibrium conditions, the ratio of the influx of the unlabeled endogenous precursor to the influx of the labeled exogenous precursor must be known. This yields the specific activity of the newly-synthesized end-product. The ratio of the rates of influxes depends, among other factors, upon the molar concentration of the labeled precursor offered to the cell (Fig. 3). This holds for in vivo conditions (LANG et al., 1966, 1968) and in vitro conditions (CLEAVER and HOLFORD 1965; COOPER et al., 1966; DÖRMER and BRINKMANN, 1970 b; LANG et al., 1968). The quoted studies refer to the synthesis of DNA. However, it seems that the same conditions apply for the synthesis of RNA (MAURER and SCHULTZE, 1969).

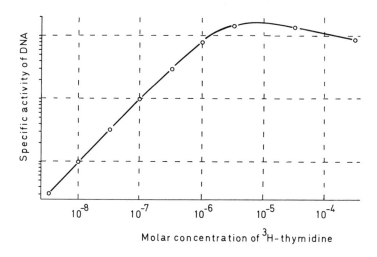

Fig. 3. Specific activity of the DNA from bone-marrow cells of rat, incubated with increasing concentrations of ^{3}H-thymidine with constant specific activity. (After DÖRMER and BRINKMANN, 1970 b)

The importance of the size of the precursor pool was made clear by irradiation experiments with Ehrlich ascites cells in vitro by HELL et al. (1960). When incorporation of ^{3}H-TdR was examined under conditions of high specific activity and low molar concentration, considerable differences in the incorporation were observed between irradiated and control cells. However, if a lower specific activity was used with a correspondingly higher molar concentration, only slight differences were found on comparison with the controls. The authors interpreted this difference as an enlargement of the precursor pool after irradiation, in which a small molar concentration of exogenous TdR is diluted to a greater extent than a large one. NEWTON et al. (1962) obtained similar results suggesting changes in the size of the pool while using ^{3}H-TdR to study HeLa cells after infection with herpes virus. When studying the influence of the size of the pool on the results, the molar concentration of exogenous TdR must not be too high, otherwise there is a risk of "toxic inhibition" of synthesis of DNA due to exogenous

TdR, described by PAINTER and RASMUSSEN in 1963 as a "pit-fall of low specific activity" (see also Fig. 3).

Fig. 3 shows that there is a saturation value for the precursor, which in the case of TdR is in the order of 5×10^{-6} M. A further increase in concentration has no influence on an increase in the specific activity of the DNA. One might think that at this concentration the precursor pool was supplied entirely from labeled exogenous TdR. However, this is not the case. Instead, amounts of unlabeled precursor, varying for different types of cells or species, are allowed into the pool (DÖRMER et al., 1972). With 10^{-5} M TdR in the incubation medium, the pro-erythroblasts of the rat obtain 77.9 %, and the polychromatic erythroblasts of man only 28.3 %, of their DNA-thymine from the exogenous supply. In order to avoid this effect during quantitative autoradiography, the endogenous pathway for the synthesis of thymidine monophosphate can be blocked by amethopterine (BRINKMANN et al., 1969) or by 5-fluorodeoxyuridine (FUdR) (BRINKMANN and DÖRMER, 1972). The specific activity of the newly-formed DNA then equals that of the labeled thymidine administered.

The pool conditions for the essential amino acids are less complicated. NIKLAS et al. (1958) studied the incorporation of ^{35}S-methionine into various organs of the rat. To obtain the rate of turnover of protein, they used the following formula:

$$\text{Incorporated amino acid activity} = A_0 \cdot \int_0^T S_t \cdot dt.$$

A_0 refers to the rate of synthesis of protein, and the integral refers to the change in specific activity of the amino acid pool, from the start (0) to the end (T) of the experiment. The specific activity of the amino acid pool should be measured at the site of protein synthesis, i.e. in the cell (MAURER, 1972). Since the cell itself cannot synthesize methionine, the specific activity of the methionine pool should be identical inside and outside the cell and, consequently, also in the blood. Active transport of amino acids into the cell (ADAMSON et al., 1966; RIGGS and WALKER, 1963) could result in temporary differences in the specific activity of these pools, but the results of CITOLER et al. (1966) indicate that no significant error is caused by such active transport under normal conditions in the rat.

Turnover of the End-Product

The length of time over which the relative rate of synthesis of the end-product is measured should, if possible, not extend beyond the point where the first labeled molecules of the end-product are about to be degraded (Fig. 1). For the synthesis of RNA this period is extremely short, as is shown when further synthesis of RNA with a labeled precursor is suddenly blocked by actinomycin D (HARRIS, 1964; TORELLI et al., 1964a, 1964b, 1965). Fig. 4 shows the rapid breakdown of newly-synthesized RNA in canine granulocytes with actinomycin D.

In view of this rapid turnover of RNA, the net synthesis of RNA in a cell is not a measure of the actual synthesis of RNA. Additional experiments, such as the one illustrated in Fig. 4, should be performed when studying the net synthesis. Unfortunately, it is not clear whether the breakdown of RNA under the influence of actinomycin D reflects physiological conditions. Since, in addition,

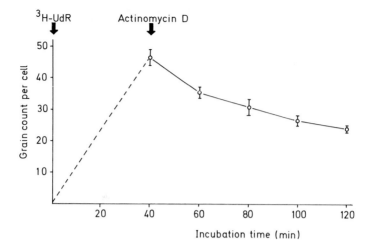

Fig. 4. Breakdown of RNA labeled with ^{3}H-uridine in canine granulocytes after the addition of actinomycin D to the incubation medium (5 μg/ml). In order to calculate the total amount of RNA synthesis, the turnover of RNA has to be added to the net synthesis

the turnover of various fractions of RNA has varying rates (AMANO et al., 1965; TORELLI et al., 1965), the applicability of quantitative autoradiography to the study of the synthesis of RNA is limited.

For the semiconservative synthesis of DNA, it can be assumed that the synthesized end-product accumulates during interphase, and is subjected to an apparent turnover only at cell division. In 1961 CRONKITE et al. calculated that, after the fifth cell division, the first unlabeled progeny from human cells, pulse-labeled with ^{3}H-TdR, were to be expected.

Studying the synthesis of proteins, NIKLAS et al. (1958), using ^{35}S-methionine, found biological half-lives for protein turnover between 2.7 hrs. (for the spleen) and many days (in striated muscle and heart muscle). According to the results of these authors, a period of 1 hr. between injection and sacrifice is suitable for studying protein synthesis. Results are not noticeably affected by degradation of the labeled proteins during this period.

Reutilization of the Labeled Compound

With long-term experiments, reutilization of labeled material formed on degradation of the labeled end-product is to be expected (BRYANT, 1963a, b; FEINENDEGEN et al., 1966; GALASSI, 1967; HEINIGER et al., 1971a, b; ROBINSON et al., 1965). This reutilization results in a slower decay of specific activity in the end-product than would be expected from the true turnover. Reutilization of labeled TdR during the formation of blood cells was demonstrated by ROBINSON et al. (1965). After an initial dose of ^{3}H-TdR they administered continuous doses of unlabeled TdR to a group of animals in order to dilute the specific activity of the recycling substances. In this group of animals, the specific activity dropped much more quickly than in animals not treated with unlabeled TdR.

The Nuclear Emulsion

Response to β-Rays

Since the majority of isotopes used in autoradiography emit β-rays, only this group will be discussed. The effect of x-rays and γ-rays on nuclear emulsions, however, is comparable to that of β-particles (BECKER, 1962; HERZ, 1969).

Nuclear emulsions are characterised by having small grain diameters, narrow distributions of grain size, and by very dense packing of grains. In Fig. 5 the distributions of grain diameters for one nuclear and two x-ray emulsions are shown. Due to the properties of the nuclear emulsions it is possible to follow the tracks of single β-particles or to count single grains. There are emulsions with different sensitivities, and a wide choice of grain size for various aspects of autoradiography. Lists were published by BASERGA (1967) and by HERZ (1969).

β-emitters have a continuous spectrum of energy ranging from values of almost zero up to a maximum energy. The latter, as well as the mean energy, varies with the isotope. For instance, carbon-14 has a maximum energy of 156 keV and a mean energy of about 50 keV, tritium has a maximum energy of 18 keV and a mean at 6 keV, and for phosphorus-32 the corresponding values are 1 710 and 680 keV. From these figures it can be seen that the distribution of the particle energies must be skewed towards higher values.

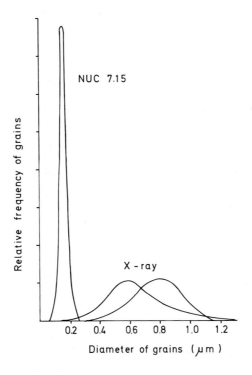

Fig. 5. Frequency distribution of grain diameters in a nuclear emulsion (NUC 7.15, Agfa-Gevaert, producer's data), and in two typical x-ray emulsions. (After HERZ, 1969)

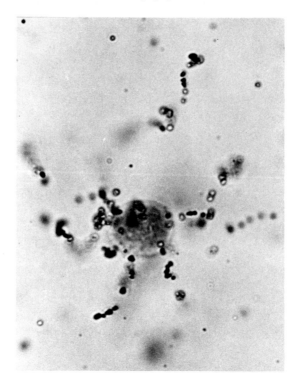

Fig. 6. Autoradiograph of a thymus lymphoblast labeled in vitro with ^{14}C-thymidine. The cells were stained in suspension prior to autoradiography. Autoradiograph of the cells suspended in Ilford G5 emulsion under conditions of 100% geometry (procedure of LEVINTHAL and THOMAS, 1957). Several ^{14}C-tracks are recognizable in the optical plane chosen

When a particle hits a crystal of silver bromide in the emulsion, traces of metallic silver are formed. These traces represent the latent image. On impact, the particle looses part of its energy. The formation of the latent image by particles or quanta has still not been completely explained. For further details, the reader is referred to the book by HERZ (1969).

Nuclear emulsions are more sensitive to β-particles with low energy than to those with high energy (DUDLEY, 1954; HERZ, 1951). This explains why the distance between two grains decreases towards the end of a track (as the kinetic energy of the particle decreases). Typical β-tracks are shown in Fig. 6. The same energy absorbed in a nuclear emulsion always results in the same degree of blackening, irrespective of the initial kinetic energy of the β-particles (GLOCKER, 1960). The energy spectrum of β-sources is not changed by absorbers (BROWNELL, 1952). This is important, since β-particles of low energy give rise to a greater number of latent images per path length than do those with high energy. Therefore, the yield of silver grains per incident electron is constant for a certain β-emitter with a certain nuclear emulsion.

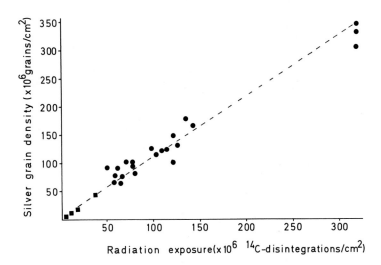

Fig. 7. Dependence of grain density on the β-flux incident on the nuclear emulsion from ^{14}C-methacrylate. Grain counts in AR.10 stripping-film were determined by microphotometry with incident light and were corrected for non-linear registration. Symbols: ■ Mean values, each calculated from 6 specimens, with a film exposure time of 11 days; ● values of individual samples with a film exposure time of 96 days. The regression line corresponds to 0.95 electrons per blackened grain. This is equivalent to 1.9 β-decays per grain

In the range of blackening used in conventional micro-autoradiography, there is a linear relationship between the β-flux through the emulsion and the resulting density of silver grains (ANDRESEN et al., 1953; TAYLOR, 1956). This relation is also valid beyond densities that can be counted visually, as is shown in Fig. 7. To show this, sections of poly-methyl-^{14}C-methacrylate were cut with varying thicknesses, and the surface activities were calculated from the data of DÖRMER and BRINKMANN (1972).

To investigate even higher grain densities, AR.10 stripping-film was exposed to x-rays, developed, and then melted by boiling. Excess gelatine was added, to give a new emulsion with a grain density in the countable range which was spread on specimen slides and counted by eye. The thickness of the new emulsion in the areas counted was determined by interference microscopy. Since the dilution factors were known, it was possible to calculate the density of silver grains in the original stripping-film specimens. Fig. 8 shows that above approximately 800×10^6 grains/cm^2 the relation to radiation exposure became non-linear.

The deviation from linearity can be explained by the coincidence of β-particles: by this we understand that more than one β-particle transmits at least part of its energy to the same crystal of silver bromide. With increasing grain density, the probability that one crystal of silver bromide will absorb energy from more than one β-particle rises. PERRY (1964) gave an exponential equation for coincidence:

$$n_{obs}/n_{max} = 1 - e^{-n_{exp}/n_{max}}.$$

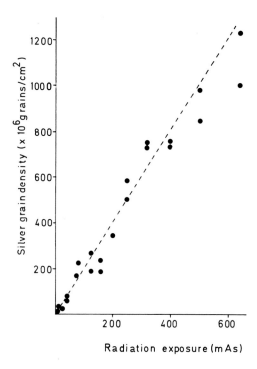

Fig. 8. Dependence of the grain density counted by eye (in a diluted emulsion), in AR.10 stripping-film developed with amidol, on the x-ray exposure (60 kV, 0.3 mmCu, exposed in plastic containers). Deviation from the linear region for grain density is seen with a radiation dose exceeding 400 mAs

n_{max} is the actual grain density at maximum blackening of the emulsion, n_{obs} is the observed, and n_{exp} the expected grain density, which would be obtained, if coincidence did not occur. It is assumed that the grain density observed is equivalent to that one actually present. The expression n_{obs}/n_{max} can be regarded as the fractional coverage of the emulsion with developed grains. As long as n_{exp} is small in relation to n_{max}, the formula may be simplified to:

$$n_{obs}/n_{max} = n_{exp}/n_{max}.$$

This shows that a linear relationship can be assumed for the radiation exposure when the fractional coverage is small. Most of the values in Figs. 7 and 8 are in the linear region, however, in Fig. 8, the transition to non-linear conditions can be seen.

In ordinary micro-autoradiography, grain densities are so low that a linear dependence on radiation exposure can be assumed. Caution is advised with tritium autoradiography; in this case n_{max} is only the maximum grain density produced by the short range of the low-energy decays of tritium. In the same nuclear emulsion, n_{max} may be lower for tritium than for carbon-14 or phosphorus-32; this means that the same grain density in a ^{3}H-autoradiograph may correspond to a higher fractional coverage than in a ^{14}C-autoradiograph.

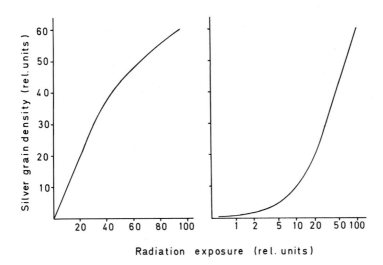

Fig. 9. Dependence of grain density on the β-flux. The scale of the radiation dose is linear (left) and logarithmic (right)

DÖRMER (1967b) and DÖRMER and BRINKMANN (1968) investigated coincidence in tritium-autoradiography, using stripping-film AR.10, and determined n_{max} as 124 grains per $10\,\mu m^2$ in one study, and as 143 in another. The higher value is more reliable, due to the experimental design.

Even with tritium-autoradiography, there is no danger of exceeding the range of linearity as long as visual grain counting is performed. If, however, higher grain densities are used for photometric evaluation, errors due to β-coincidence have to be taken into account. This is shown in Fig. 9: Beyond a certain grain density, the deviation from linearity becomes too great to be ignored. For higher grain densities, it is more convenient to assume a linear dependence on the logarithm of the radiation exposure. This second range of linear dependence, however, starts with grain densities which are too high to permit simultaneous observation of the underlying biological structures in a microscope. Micro-autoradiography, therefore, should always be performed in the lower range of linear conditions which, according to Fig. 7, should be sufficiently large. If measurements beyond the linear section are inevitable, calibrations of the radioactivity-to-grain density relationship should be performed with radioactive standard sources of varying energies. This procedure is explained on p. 380 ff.

Development

The conversion of the latent image into a visible silver grain, and the subsequent removal of the non-developed silver bromide crystals, is a procedure with many empirical factors. Development has been described in detail by, among others, JAMES and HIGGINS (1948), and JAMES and KORNFELD (1942). Due to the reducing action of the developer, silver bromide is converted to metallic silver and is deposited as a filament, starting at the site of the latent image. This process continues until the whole grain consists of a coil of silver filaments. Under the usual conditions, the developer can attack only silver bromide crystals which contain a sufficient number of silver atoms to form the latent image. All crystals not containing a latent image are dissolved by the sodium thiosulfate in the fixing bath. Simultaneously, the nuclear emulsion decreases in thickness, since it consists largely of these crystals and only a few of them are rendered developable in an autoradiograph.

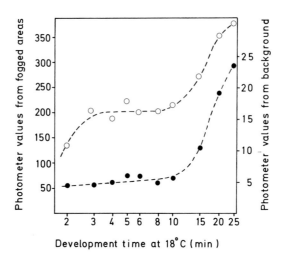

Fig. 10. Dependence of the intensity of reflectance on the time of development in D-19 developer at 18° C. Bright-field measurements with incident light, of Ilford G5 emulsions in light-fogged o and control ● areas

The number of silver grains which are developed depends on the duration and temperature of development and on the composition of the developer. Fig. 10 shows the dependence of grain density on the duration of development, for Ilford G5 emulsion. The background value, which initially increases only slightly with the time of development, shows a sharp rise after 10 to 15 min. At this stage the grain density of the light-fogged areas also increases considerably. The most suitable development time under the conditions for Fig. 10 is from 3 to 10 min.

It is possible to compare visual grain counts between autoradiographs developed in different developing batches, if constant conditions of development and fixation are observed rigorously. For this, it is advisable to study the effects of temperature, time, and composition, taking care that development is not done at the edge of the plateau conditions. At the edge of the plateau, minute and un-noticed changes in the conditions of development might cause considerable variance in the yield of silver grains. More dependable, although un-necessary for visual grain counting, is the simultaneous exposure of the nuclear emulsion to a radioactive standard source (see p. 380 ff.).

If autoradiographs are to be measured by photometry, standardization is essential. Specimens of nuclear emulsion which are irradiated simultaneously, then developed, fixed, and dried on different days may exhibit differences on photometry which cannot be tolerated.

The dependence of the photometrically-determined optical density on the chemically-determined amount of silver per unit area of a developed emulsion is called the covering power. This covering power varies with conditions of development and drying (JAMES and FORTMILLER, 1961); grain size and the structure of the filamentous silver also play a role. In our experience, it seems impossible to keep the conditions in the darkroom so invariant that no changes in the covering power occur. The simplest method of standardization is by the use of radioactive standard sources (see p. 380 ff.).

The two developers described below will be refered to frequently in this article.

1) D-19 developer:
   Add to 500 ml distilled water (50° C)
   8.8 gm hydroquinone
   5.0 gm potassium bromide
   48.0 gm anhydrous sodium carbonate
   96.0 gm sodium sulfite
   Finally, add cold distilled water to make 1 000 ml.
2) Amidol-developer (similar to D-170):
   4.5 gm amidol (diaminophenol-dihydrochloride)
   0.8 gm potassium bromide
   18.0 gm anhydrous sodium sulfite
   Add distilled water, with gentle shaking, to make 1 000 ml.

Fading of the Latent Image

The latent image, from which a silver bromide crystal becomes developable, consists of at least 4 to 10 silver atoms (HERZ, 1969). These silver atoms may revert to silver ions if chemical influences are effective during exposure. If this happens, the latent image is deleted and the silver bromide crystal becomes once more undevelopable. HERZ (1959) pointed out that some fading in nuclear emulsions is desirable in order to keep the background low. According to this author, fading can be stopped during exposure: he advises keeping the humidity in the exposure boxes low by the addition of some silica gel, and replacing the air in the boxes with an inert gas, for example CO_2.

The extent of fading during exposure, as reported by various authors, seems to vary considerably. Results have been obtained with both NTB-emulsions and AR.10 stripping-film to either confirm or disprove the occurence of fading. Considerable fading of the latent image during exposure has been found by BASERGA and NEMEROFF (1962), OJA et al. (1966), and BASERGA (1967), whereas KOPRIWA and LEBLOND (1962), APPLETON (1966), and PELC et al. (1965) did not observe the effect.

Errors are introduced by fading of the latent image during exposure, if grain densities are compared in autoradiographs exposed for varying periods. It is rather difficult to predict the degree of fading in an experiment, even if the previous experiment showed no evidence of fading at all. ROGERS (1967) suggests adding controls of light-fogged emulsion to the exposure boxes for experiments with varying exposure times. If there are no differences in the blackening of the controls for varying exposure times, fading can be ignored. However, if slight differences exist between the controls, it is risky to draw any conclusion about the fading in the autoradiographs. This is due to the fact that the degree of fading of the latent image depends, among other things, on the ionizing power of the source. The weaker the ionizing power, the more fading will occur (HERZ, 1959). Indeed, the ionizing power of light quanta is much less than that of β-particles.

The above considerations emphasize the usefulness of radioactive standard sources containing the same isotope as for the autoradiographs, and to which the nuclear emulsion is exposed simultaneously with the specimen. After varying periods of exposure, the grain densities in the autoradiographs can be referred to the grain densities for the corresponding standard sources (see p. 380ff.). The problem of negative chemography caused by the specimen has been ignored in this context, and will be discussed separately (see p. 371ff.). Of course, this effect is not revealed by the use of radioactive standards.

The Autoradiograph

Track Autoradiography

According to the definition, a track consists of 4 or more silver grains arranged linearly (ROGERS, 1967). As can be seen for the ^{14}C-tracks in Fig. 6, however, the path through the emulsion is rather curved. β-particles are electrons of nuclear origin; they have a relatively small mass and are subjected to numerous elastic collisions on passing through matter, so that the direction of the path is changed repeatedly. When the kinetic energy of the particles decreases, the degree and frequency of the changes in direction increases, so that the end of the track may be quite coiled. For kinetic energies of less than 1 MeV, the loss of energy of the β-particles increases with decreasing kinetic energy (EVANS, 1962). This decrease of energy is equivalent to an increased ionizing power. Therefore, the distances between two grains become smaller, and the grains increase in size towards the end of a track.

The principles of β-track autoradiography have been studied by ZAJAC and ROSS (1949), LEVI and HOGBEN (1955), and LEVI et al. (1963). Useful statistical procedures and details of the method can be found in the publication of LEVINTHAL and THOMAS (1957). An extensive description of β-track autoradiography is given by ROGERS (1967). LEVI et al. (1963) measured the length of β-tracks from various emitters in Ilford G5 emulsion, and counted the number of grains per track. They were able to establish some basic empirical correlations. According to their results, G, the number of grains that are developable in a track under plateau conditions is related to the initial kinetic energy, E, of the β-particle by the formula: $\log G = 1.19 \cdot \log E - 0.74$.

From this formula it can be shown that a tritium particle with the maximum energy of 18 keV is able to activate 5.7 grains in G5 emulsion. The initial kinetic energy needed to activate the 4 grains required for a track is in the order of 13.4 keV. According to the data of PERRY (1964), only 4.7% of the β-particles of tritium posses this energy. It is, therefore, obvious that tritium cannot be used for track autoradiography. In the case of carbon-14, 14% of the decays

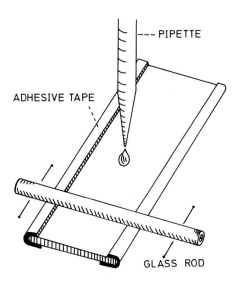

Fig. 11. A simple process for making thick layers of liquid emulsions for track autoradiography

do not possess the initial energy necessary to activate 4 grains. A β-particle of carbon-14, with the maximum energy of 156 keV, can produce 74 silver grains, and one with an average energy of 50 keV would produce 19 grains. According to the same authors, the track length, L, is related to the initial β-energy, E, by the following empirical formula: $\log L = 1.59 \cdot \log E - 1.51$.

For producing track autoradiographs thick layers of liquid emulsion are used. Ilford G5 emulsion in gel form seems to be preferred for this purpose, since it is the emulsion to which reference is made most frequently. The thickness of the emulsion depends on the energy of the isotope used. ROGERS (1967) suggests thicknesses of 15–20 μm for sulfur-35 and carbon-14. 60 μm is necessary for phosphorus-32. To prepare thick layers of emulsion, the sides of the slides can be taped with adhesive material having twice the height of the desired final thickness of the emulsion (Fig. 11).

The emulsion is melted at 42° C as usual and mixed with some distilled water containing glycerol: 4 ml of aqua dest. and 0.2 ml of glycerol are used for 20 ml of final emulsion. Glycerol acts as a plasticiser for the emulsion and reduces undesirable background effects resulting from stresses during drying. The emulsion is pipetted onto the slide, which is lying flat, and is spread with a glass rod (Fig. 11).

The slides are dried in a horizontal position at room temperature in a gentle stream of air. As long as the emulsion is not dry, it is highly probable that the tracks formed will fade, at least to some extent. This is indeed a problem since, due to the thickness of the layer, drying can last several hours. The time of exposure, however, is a great deal shorter for track autoradiographs than for grain density autoradiographs. Less than 36 hrs. exposure are often sufficient. The drying process, however, must not be accelerated, since this results in an undesirably high background (SAWICKI and PAWINSKA, 1965). The slides are exposed in light- and moisture-proof boxes with silica-gel at 4° C.

Layers of emulsion not exceeding 30 μm can be developed in the usual manner with amidol. The time for development, however, has to be extended, and results in a higher background. 10 min at 18° C with gentle agitation in amidol developer (for composition, see p. 359) is sufficient. Fixation is carried out after soaking for 2 min in a 1% solution of acetic acid. According to ROGERS (1967), no acid fixing baths should be used, since part of the silver might be dissolved out of the grains as a result of long treatment in the acid solution. 40 min fixation in a 30% solution of sodium thiosulfate is sufficient. The autoradiographs are then washed for at least two hours in running tap water.

The method of LEVINTHAL and THOMAS (1957) requires a somewhat different technique. In this case labeled particles, for instance phages labeled with ^{32}P (LEVINTHAL and THOMAS, 1957) or cells labeled with ^{14}C-TdR (DÖRMER and BRINKMANN, 1972), are suspended in the emulsion. If labeled cells are used, they are suspended in saline, fixed in methanol while in suspension, and washed in distilled water. The cells are then stained with MAYER's hemalum by adding 1 ml of the stock solution to 5 ml of cell suspension. After 10 min the cells are washed twice in distilled water and are then passed through a glass filter with pores of 45–90 μm. After one more wash, distilled water is added to make 1 ml.

The suspension of stained cells is warmed to 42° C in the darkroom and is added to 10 ml of Ilford G5 emulsion at the same temperature. 1% of glycerol is added, and the suspension is mixed carefully. Plastic rings, 0.5 mm high, inner diameter 12 mm, are taped onto slides, and are filled with the prepared emulsion from a pipette. Drying takes about 3 hrs., during which time it is likely that fading of the latent images occurs.

These thick layers of emulsion, as well as the 60 μm ones used for detecting phosphorus-32, require predevelopment at 5° C in an amidol developer of twice the normal strength. Predevelopment for 15 to 40 min allows the developer to penetrate the emulsion without exerting its reducing power. Subsequently, the autoradiographs are transferred to an amidol developer of normal concentration at 18° C. Depending on the thickness of the layers, these are developed for 15 to 40 min. Because of the long development, it is advisable to increase the concentration of KBr from the usual 0.8 gm/l to 1.25 gm/l.

The thick layers are fixed for 2 hrs. in an acid fixing bath without an intermediate stop-bath (300 gm sodium thiosulfate, 30 gm potassium bisulfite, with distilled water to make 1 l). An extremely long time is required for fixation of such thick layers if potassium bisulfite is omitted. In the author's experience, silver grains were not dissolved by the acid fixing bath. Washing for 4 hrs. in running tap water is necessary. The thick pieces of emulsion can then be removed from the rings with forceps, and transferred to distilled water. They are mounted wet, using a few drops of glycerol; it is important to mount autoradiographs prepared for track counting while the emulsion is swollen. This results in a better separation of the individual tracks and makes counting easier.

Grain Density Autoradiography

The preparation of grain density autoradiographs is much more common than that of track autoradiographs. For quantitative evaluation of grain density autoradiographs, the thickness of the emulsion must be kept constant. This is important if the path of the particles under investigation exceeds the thickness of the emulsion. Thin emulsions are useful for counting and for measuring grain densities. A stripping-film, with a thin layer of emulsion and a constant thickness, is recommended for all isotopes. Tritium is not necessarily included because of the very short range of its radiation.

If a dipping emulsion is chosen for the quantitative evaluation of tritium, it should not be diluted more than 1:1. All autoradiographs in which the emulsion is unevenly distributed should be rejected from the counting series; in areas thinly covered with emulsion the layer might be too thin even for the short range of the β-particles of tritium. Variable thickness of the emulsion is easily recognizable if the autoradiograph is stained through the emulsion, because thick layers are stained more intensely than thin ones.

Kodak stripping-film AR.10 (PELC, 1947, 1956; PELC and HOWARD, 1952; HERZ, 1951) has been most widely used. It consists of a layer of emulsion, 5 μm thick, with a carrier layer of gelatine, 10 μm thick. The processing of this stripping-film is explained in the instruction leaflets of the company and is dealt with extensively in autoradiographic texts. A technique which avoids the electrostatic effects of the stripping process and the influence of the relative humidity in the darkroom has been described by SCHMID (1965). It has been applied in our laboratory with success. The stripping plates are put into 70% ethanol for 3 min. The wet film is cut, then dehydrated in absolute ethanol for 6 min. After drying, it can be stripped from the glass plate with forceps, without electrostatic effects, and can be floated on the surface of the water.

The degree of swelling on the water surface depends a great deal on its temperature. According to our experience, the time for swelling may take up to 5 min. It is advisable to wait until swelling is complete, otherwise there is a risk of the film having an uneven thickness. If the temperature of the water varies, varying surface areas of the stripping-film result, resulting in varying thicknesses of the emulsion in the autoradiographs. Grain densities of different autoradiographs should therefore be compared with caution if the emulsion thickness is less than the maximum range of the β-rays. This, too, calls for radioactive standard sources as controls (see p. 380ff.).

β-Self-Absorption and β-Absorption

When β-particles pass through matter their energy is increasingly absorbed. There is a roughly exponential relationship between the number of absorbed β-particles and the thickness of the absorber (EVANS, 1962). The term β-self-absorption is applied if β-particles are absorbed in the source of radioactivity itself, for example in a cell nucleus labeled with ^{3}H-TdR. The term β-absorption is used if they are absorbed in a non-radioacitve material. The phenomena of β-self-absorption and β-absorption have become well known as a significant source of error in tritium-autoradiography (ADA et al., 1966; FALK and KING, 1963; FITZGERALD et al., 1951; LAJTHA and OLIVER, 1959; MAURER and PRIMBSCH, 1964; OJA et al., 1966; PELC and WELTON, 1967; PERRY, 1964).

The experiments of MAURER and PRIMBSCH (1964), from which Fig. 12 is taken, have given us the best insight into these phenomena. In the absence of β-self-absorption a linear relation between the thickness of the section and grain density would be expected (dotted lines). However, it is obvious that there is strong β-self-absorption for the tritium in all components of the cell. Depending

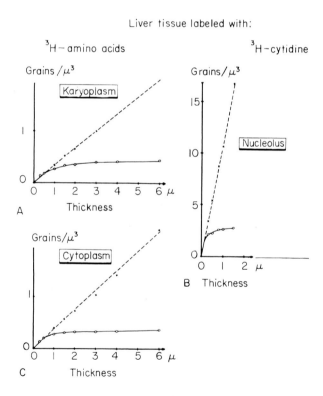

Fig. 12. Grain density in AR.10 stripping-film, depending on the thickness of sections of various components of liver cells labeled with tritiated precursors. The dotted lines show the results expected in the absence of β-self-absorption. The solid lines record the actual observations. (From MAURER and PRIMBSCH, 1964)

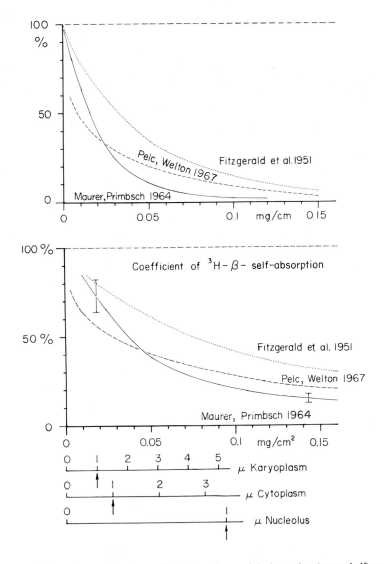

Fig. 13. Compilation of several analyses of the dependence of β-absorption (upper half) and β-self-absorption (lower half) on dry-weight per unit area in tritium-autoradiography. Scales for the thickness of individual cell components converted to the scale of dry weight per area are shown below. (From Maurer and Schultze, 1969)

on the varying dry weights, there are differences in the degree of β-self-absorption for 1 μm sections from nucleolus, nucleus, and cytoplasm. Data published by 3 authors have been compiled by Maurer and Schultze (1969) (Fig. 13). Maurer and Primbsch (1964) stressed the strong β-absorption of tritium decays in the cytoplasm above labeled nuclei in smeared cells. A layer of cytoplasm with 0.5 μm (wet thickness) absorbs 50% of the tritium radiation passing through it.

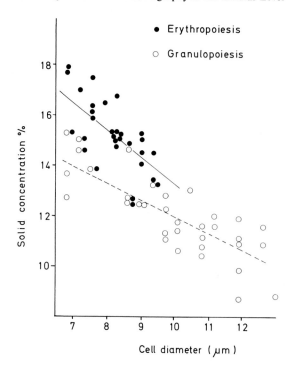

Fig. 14. Interference microscopy measurements of the concentration of solids in bone-marrow cells from mouse fixed in methanol and suspended in destilled water. The concentration of solids is plotted against the diameters of the cells. The smaller the diameters of the cells, the more advanced are they in the maturation process. The early precursors, therefore, lie to the right of the diagram, while the later ones are on the left

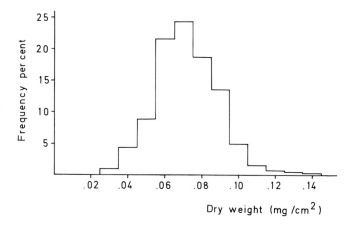

Fig. 15. Frequency distribution of the dry weight per unit area of smeared nuclei from regenerating liver of rat. Interference microscopy measurements of 817 nuclei fixed in methanol. (After DÖRMER and BRINKMANN, 1970a)

For the quantitative evaluation of tritium by autoradiography, histological sections with an effectively infinite thickness are recommended (MAURER and PRIMBSCH, 1964). By this, we mean that the thickness of the sections exceeds the maximum range of the β-rays of tritium. Fig. 12 shows that such an "infinite" thickness can be assumed for sections greater than 3 μm. Grain density, i.e. the grain count per unit of area, is then a measure of the radioactivity per unit dry weight of the cell or cell component. However, it can only be used for the comparison of concentrations of radioactivity in different cells if these cells have the same dry weight per unit volume. This condition does not apply, for instance, to the case of bone marrow cells (Fig. 14). The dry weight per unit volume differs from erythropoietic to granulopoietic cells and from one maturation stage to another. For tritium-autoradiographs of such cells sectioned at infinite thickness, grain density only gives a relative measure of radioactivity per unit dry weight. The grain count per cell or nucleus in tritium-autoradiographs is a generally unsuitable parameter for comparison.

Fig. 15 shows the distribution of the dry weight per area of smeared cells. Even if the error due to β-absorption in the overlying cytoplasm can be disregarded, conditions are unsuitable for tritium-autoradiography, because β-self-absorption cannot be neglected, and the conditions of an infinite thickness are not fulfilled. In a study of protein synthesis in cultured mouse fibroblasts, ZETTERBERG and KILLANDER (1965) avoided these difficulties by interference microscopy. They determined the regions of the cells which satisfied the conditions of infinite thickness. In green light with air as a medium, such regions have an optical path

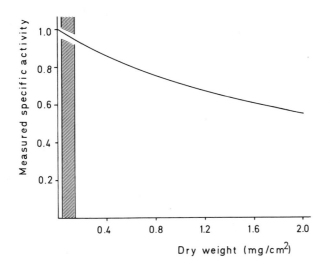

Fig. 16. Dependence of the measured specific activity of ^{14}C-methacrylate samples of constant specific activity, on the weight per area of the samples. Autoradiographic determination at low grain densities. No allowance has been made for β-back-scattering in the thin samples mounted on glass. If this is disregarded, the observed specific activity indicates the extent of β-self-absorption. The dry weights per unit area of the smeared nuclei from Fig. 15 have been included as a hatched column.
(After DÖRMER and BRINKMANN, 1972)

difference of at least 1 wavelength. DÖRMER et al. (1968) determined, first, the dry weight per unit area of nuclei from smears of liver cells and second, the grain density. This was then corrected for β-self-absorption. This is a timeconsuming process which can be replaced satisfactorily by ^{14}C-autoradiography.

Conditions for β-absorption and β-self-absorption are much more advantageous in ^{14}C-autoradiography (Fig. 16). The weight per unit area of the smeared liver cells shown in Fig. 15 is given as a column in Fig. 16. It can be seen that reduction of the observed specific activity due to by β-self-absorption in smeared cells labeled with ^{14}C is negligible (DÖRMER and BRINKMANN, 1972).

Difficulties are encountered in ^{14}C-autoradiographs from sectioned material, since the saturation thickness is only reached at 3–4 mg/cm^2, corresponding to a section thickness of 23–31 μm. The activity of sectioned cells can hardly be estimated, because it is uncertain which portion of a cell is present in the section. Also, the problem of cross-fire has to be considered (see p. 368ff.). In conclusion, quantitative tritium-autoradiography is suitable for sectioned material, with some reservations. ^{14}C-autoradiography, on the other hand, is useful for smeared cells.

Back-Scattering

The number of β-particles registered that have been scattered back from the support of the radioactive source onto the nuclear emulsion depends on several factors. It increases with decreasing thickness of the radioactive source (GLENDENIN and SOLOMON, 1950; YANKWICH and WEIGL, 1948) and with increasing atomic number of the back-scattering element (EVANS, 1962; GLENDENIN and SOLOMON, 1950; ODEBLAD, 1952; YANKWICH and WEIGL, 1948); it is more pronounced when the surface of the support is rough (LIBBY, 1957).

As for labeled cells on a slide, the conditions of 50% geometry for the autoradiographic detection of radioactivity are complicated by β-back-scattering. This causes an error if the radioactivity of thin sources is compared with that of thick ones. Comparison between cells labeled with ^{14}C, in a smear, and an infinitely thick reference source is an example of this (DÖRMER and BRINKMANN, 1972). In infinitely thin sources mounted on a glass support, an amount of β-back-scattering, amounting to 12–17% of all registered decays, has been observed (ODEBLAD, 1952; YANKWICH and WEIGL, 1948).

Geometric Factors

In an autoradiographic image of a radioactive source there is a spread of blackening which corresponds to the range of the rays. Depending on the isotope used, haloes of decreasing grain density can be found around cylindrical sources, as for example, labeled nuclei in a smear (Fig. 17). According to ROBERTSON et al. (1959), the dose, d, per radioactive decay with an initial dose d_0, measured at a distance ρ from a point source is given by the formula:

$$d = d_0 \cdot \frac{1}{\rho^2} \cdot e^{-\mu\rho}.$$

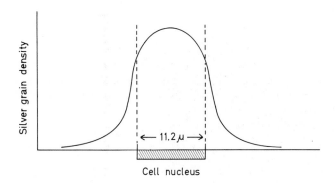

Fig. 17. Scanning line of the grain density in AR.10 stripping-film of a smeared erythroblast labeled with ^{14}C-thymidine. (From DÖRMER, 1972)

μ is a constant for the absorbing matter. The authors point out that for isotopes with β-energies equal to or larger than that of ^{14}C ($E_{max} = 156$ keV), the square of the distance (the $1/\rho^2$ component) governs the decline in the radioactive dose. This means that geometric factors are the main causes of the decline, and not the energy of the β-emitter. However, in the case of tritium, the exponential component is of greater importance.

DONIACH and PELC (1950), and LAMERTON and HARRISS (1954) showed (Table 1) that the spread of the autoradiographic image depends on 3 factors:

1) The thickness of the sample, for instance, of a radioactive cylinder,
2) the thickness of the nuclear emulsion, and
3) the gap between emulsion and sample.

The gap is very important in contact autoradiography, where the film is pressed against the sample during exposure; for technical details, see BOYD (1955), EVANS (1948), GROSS et al. (1951). This procedure has been generally abandoned in favor of the stripping- and dipping-techniques. In these techniques the gap between the emulsion and the specimen is so small as to be of little significance.

Table 1. Influence of geometric factors on the spread of the autoradiographic image, measured by the distance to half the mean grain density at the edge of cylindrical sources. (After LAMERTON and HARRISS, 1954)

Conditions	1 µm radius of cylinder	5 µm radius of cylinder	10 µm radius of cylinder
2 µm specimen 2 µm emulsion	1.02 µm	1.94 µm	2.6 µm
5 µm specimen 5 µm emulsion	1.92 µm	3.28 µm	4.4 µm
2 µm specimen 5 µm emulsion	1.36 µm	2.52 µm	3.6 µm

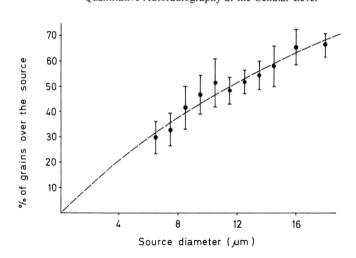

Fig. 18. Dependence of the percentage of silver grains in AR.10 stripping-film projected onto ^{14}C-labeled nuclei, on the nuclear diameter. The dotted line corresponds to an exponential function. (From DÖRMER, 1972)

If the radioactivity of different sources, e.g. labeled cells, is to be compared, either all the grains produced by the sources must be measured, or at least the same percentage thereof. The percentage of grains projected on the source depends a great deal on the size of the source (Fig. 18). The results illustrated in Fig. 18 were obtained from autoradiographs with AR.10 stripping-film, of smears of various types of cells labeled with ^{14}C-TdR. An empirical correlation can be established between the diameter, d (μm), of the source and the fraction f of grains projected on the source (DÖRMER and BRINKMANN, 1972):

$$f = 1 - e^{-ad}.$$

The exponential coefficient a has a value of 0.0606.

The total number of silver grains in AR.10 stripping-film produced by a nucleus labeled with ^{14}C-TdR can thus be calculated if the grain count over the nucleus and the nuclear diameter are known. It is advantageous to use a counting procedure corresponding to the above formula, because the radii of the grain distributions around labeled cells frequently overlap each other. If the emulsion over a nucleus is free from the effects of cross-fire from the neighboring cells, the total grain count of this nucleus can be calculated.

Since cells labeled with ^{14}C in a smear can be regarded as almost infinitely thin (see Fig. 16), the geometric error due to the thickness of the specimen can be disregarded. A gap between emulsion and nucleus of not more than 0.5 μm can also be disregarded (LAMERTON and HARRISS, 1954). The thickness of cytoplasm overlying smeared nuclei is, according to electron microscopic determinations, less than 0.1 μm (DÖRMER et al., 1968). The only source of error thus remaining, when applying the above formula, is the variable thickness of the emulsion of the stripping-film (see Table 1).

Several attempts, both theoretical and practical, have been made to solve the problems of geometry in quantitative autoradiography (BLACKETT et al., 1959; BLEECKEN, 1961; DONIACH and PELC, 1950; GROSS et al., 1951; HERRMANN et al., 1961; LAMERTON and HARRISS, 1954; LOEVINGER, 1956; MARSHALL et al., 1959; NADLER, 1951). However, no simple solution has been found which meets all demands. This emphasizes the advantages of using tritium. In tritium-autoradiography, it is easy to include all grains produced by the source in the measurement. In addition, the thickness of the emulsion can be regarded as infinite when using AR.10 stripping-film.

Chemography

BOYD (1955) introduced the term (histo-)chemography for all properties of the specimen which affected the nuclear emulsion chemically. Such action may result in either leaching of the latent images (negative chemography) or enhancement of them (positive chemography). BOYD (1955) listed numerous substances with chemographic activity.

Generally speaking, negative chemography can occur whenever a specimen has been insufficiently fixed or has been prepared without fixation, for example, by freeze-drying (see p. 376ff.). Positive chemography is known to occur particularly after prestaining of the specimens with various dyes prior to autoradiography. Examples of such substances are orcein (STÖCKER and MÜLLER, 1967), celestin blue (DEUCHAR, 1962), and methylene blue. "Autochemographs" of unlabeled cells stained with methylene blue were made in our laboratory, and were hardly distinguishable from autoradiographs with ^{3}H-amino acids. Exceptions to the general rule are easily found. Cholesterol, for example, will produce positive chemography if not removed from a specimen on fixation (KORR et al., 1970).

Because of chemographic side-effects, there are only a few dyes suitable for pre-staining of specimens. One of these is the Feulgen stain; however, it cannot be used in quantitative autoradiographic work, because part of the label is removed during hydrolysis (LANG and MAURER, 1965). The present author has found MAYER'S hemalum to be suitable; after staining, a rinse for 20 min in running tap water and, subsequently, in distilled water is recommended.

Methods for detecting negative chemographic effects easily need to be developed. ROGERS (1967) suggests exposing stripping-film to light and then covering unlabeled specimens with it. The result of such an experiment is shown in Fig. 19. The effects have apparently been caused by dust, since they were located in the "autochemograph" in regions remote from any smeared material. There are some reservations concerning this method (see p. 360ff.), since the negative chemographic effects produced are more pronounced than they would be in an autoradiograph.

A more reliable method of detecting chemographic artifacts is to expose a nuclear emulsion with a thin unlabeled specimen on a radioactive support. Plates of poly-methyl-^{14}C-methacrylate can be used for this purpose. Such plates, with a specific activity of 10 μCi/gm polymer, and dimensions of 3×3 cm, are available from Amersham, England. They are cut into several pieces with a sharp knife (cut the lines first and then break along them). After cleaning the pieces in methanol

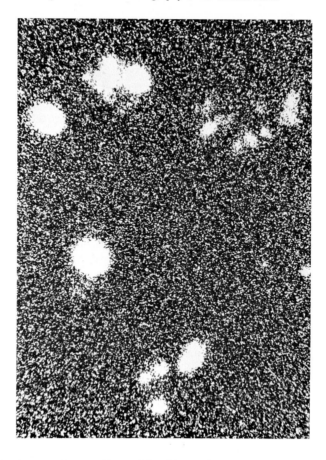

Fig. 19. Negative chemography caused by particles of dust or dirt on a smeared slide in an area without cells. After covering the sample with light-fogged stripping-film AR.10, it was dried for 3 hrs. and then developed in amidol developer

and distilled water they are glued to slides. Unlabeled specimens are transferred to the radioactive supports. For smeared cells, or tissues sectioned at 3 μm, the influence of β-absorption can be disregarded (DÖRMER and BRINKMANN, 1972). The unlabeled specimens are fixed and prepared for autoradiography in the usual manner and are then covered with emulsion. Exposure for 5 days with AR.10 stripping-film is sufficient.

A detail of such a sample is shown in Fig. 20. The smeared bone marrow cells were fixed for 20 min in methanol. Reduction of the grain density is recognizable around and over erythrocytes. In another experiment, when the cells were fixed for 48 hrs. in methanol, no such effects were observed (DÖRMER and BRINKMANN, 1972).

Some positive chemography inevitably results from the use of D-19 developer (for composition, see p. 359). This phenomenon is shown by the fact that the silver grains over the specimens become larger than the grains of the background

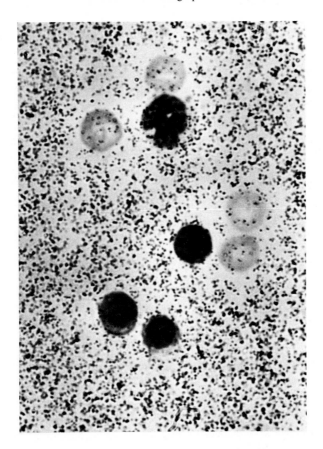

Fig. 20. Negative chemography in an area of unlabeled cells, especially erythrocytes, that were smeared onto a ^{14}C-methacrylate plate and then fixed in methanol for 20 min. Exposure with AR.10 stripping-film, development in amidol developer. After the same cells had been fixed in methanol for 48 hrs., no further chemographic effects were detectable

(MERTZ, 1969; ROGERS, 1967; ROGERS and JOHN, 1969). In photometric evaluations of autoradiographs this effect may result in gross errors (see p. 384 ff. and Fig. 25).

Background

There is no nuclear emulsion which is not affected by environmental irradiation, leading to the continuous formation of latent images. If an autoradiograph is found to have no or few background grains, fading of the latent images must be suspected. The addition of 0.05 % potassium bromide to the water from which the stripping-film is picked up has been suggested. It is true that the background level will be reduced satisfactorily, but this is caused by fading of the latent images, as experiments in our laboratory have shown. This procedure, therefore, should not be applied in quantitative autoradiographic work.

The background becomes a severe problem in quantitative autoradiography if the lowest grain density attributable to the label is comparable with the highest grain density of the background. This problem may arise in visual grain-counting work, since here low grain densities are preferred. The simplest way to avoid this is to use longer times of exposure in order to obtain a more favorable autoradiographic label-to-background ratio. Sometimes, however, this results in grain densities which can only be counted photometrically (p. 382 ff.).

Under suitable conditions, a mean value of background grain density can be subtracted from all counts over a labeled specimen. The grain density of the background should always be determined over an unlabeled part of the specimen which is otherwise identical to the labeled part. In experiments with precursors of RNA or protein, in which all cells may be labeled, it is advisable to add unlabeled material of the same kind to the slides.

If accurate determinations have to be carried out with low grain densities, allowance for the background can be made according to the formulae of STILLSTRÖM (1963, 1965). For a given number of silver grains the following relationship holds:

$$pl = (pm - pk)/(1 - pk)$$

where pl is the fraction of cells which are truly labeled, pk the fraction of cells which are unlabeled but positive, and pm the fraction of all cells which are positive. This formula thus gives the fraction of cells which are truly labeled. The true mean grain count can be calculated in a similar way. These calculations, however, must be done individually for every grain count; the use of a computer is recommended (MOFFATT, 1971).

In view of the remaining errors in quantitative autoradiography, it is questionable as to whether this exact elimination of background influence will give essentially better results. In the case of tritium-autoradiography, the errors due to β-self-absorption and, eventually, to β-absorption are so large that the error incurred by subtracting a mean background value from all counts may be disregarded.

Preparing the Samples

Fixation

Fixation of tissues and cells should fulfill a number of conditions:
1) The specimens should be converted to a stable state.
2) The fixing agent should not produce chemographic effects.
3) The chemographic power of the unfixed specimen should be stopped.
4) All labeled precursors that have not been incorporated into the end-product should be removed from the specimen.
5) No artificial bonds between the labeled precursors and the specimen should occur on fixation.
6) The labeled end-product should not be dissolved and removed from the specimen.

The first condition is met adequately by all the fixing agents used in histology and cytology. The second condition is more restrictive. Fixing agents containing mercury, e.g. ZENKER's fluid, cause positive chemography (BOYD, 1955). After fixation with formalin, fogging has been observed (WILLIAMS, 1951), as well as leaching of the emulsion (ROGERS, 1967). ROGERS (1967) also occasionally observed leaching after fixation with glutaraldehyde. In general, the effects of formalin on the nuclear emulsion are not observed if the fixed specimens are washed for a sufficient length of time prior to autoradiography.

Chemographic activity of the specimen itself is a source of considerable errors in quantitative autoradiography. It is dealt with separately on p. 371ff. Here, it should be noted that most chemographic effects caused by specimens can be suppressed by appropriate fixing techniques.

Table 2. Effect of fixation in methanol on the removal of the acid-soluble pool from smeared thymocytes labeled with ^{14}C-thymidine (20 min incubation with ^{14}C-TdR). Additional treatment of the cells with cold 5% TCA, a safe method for removing acid-soluble substances, has no further effect

Treatment	Number of autoradiographs	^{14}C-TdR per labeled cell (moles $\times 10^{-18}$)
20 min methanol / 20 min dist. water	5	54.9 ± 2.6 [a]
20 min methanol / 5 hrs. 5% TCA (+4° C) / 5 hrs. dist. water (+4° C)	5	53.0 ± 1.6 [a]

[a] Standard error of the mean.

The fourth requirement, of removing all labeled molecules which are not parts of the end-product, can be fulfilled, as illustrated by Table 2. This gives the results of fixation for 20 min in methanol, followed by a 20 min wash in distilled water: the entire pool of acid-soluble material is readily removed from smeared thymocytes labeled with ^{14}C-TdR without any acid agent. The results of BASERGA and KISIELESKI (1963), and the extensive studies of ANTONI et al. (1965) lead to the same conclusion. It is demonstrated that all the free precursors of the synthesis of DNA and protein are removed by fixing the samples in formalin or methanol; this probably applies for precursors of RNA also. Such a conclusion can be drawn, since even parts of the labeled RNA are lost on treatment with practically all fixatives (see below).

The fifth condition, that no artificial binding of the precursors should occur, is emphasized by the investigation of PETERS and ASHLEY (1967, 1969). These authors showed that, during fixation with glutaraldehyde and osmic acid, free ^{3}H-amino acids were bound to liver tissue, and that removal of the activity subsequently by various washing procedures was not possible. Formalin also had a similar effect. This was, however, of no quantitative significance, since less than 1% of the free amino acids were bound to the liver tissue.

The most serious problem of fixation is the requirement that no components of the labeled end-product should be lost. BOYD (1955) surveyed the effects of fixation on the loss of phosphorus-32 from the tissues. Because of the heterogeneity of the ^{32}P-labeled end-products, the degree of loss depended largely on the fixing procedure; losses of up to 80% of the total activity could occur. The data of HARBERS (1958) agree with these findings; in addition, they show that the loss also depends on the time between application of the label and sacrifice.

Of special importance is the loss of labeled RNA. Several studies gave contradictory results. KOPRIWA and LEBLOND (1962), comparing grain counts of ^{3}H-cytidine labeled liver after fixation in either formalin, CARNOY's or BOUIN's fluid, were unable to detect appreciable differences in the losses incurred. SANDRITTER and HARTLEIB (1955) used micro-spectrophotometry in the UV to investigate the problem, and detected no differences between samples fixed in formalin or CARNOY's fluid, when compared to freeze-dried tissue. However, FEINENDEGEN et al. (1960) showed that varying fractions and amounts of labeled RNA were dissolved on fixing in either CARNOY's solution, methanol, or 70% ethanol; as a consequence, different patterns of grain density over the nucleolus and nucleus were produced. SCHNEIDER and MAURER (1963) compared the fixing effects of CARNOY's fluid and methanol with that of formalin in various tissues labeled with ^{3}H-cytidine. With CARNOY's fluid reduction in grain counts was 30 to 80%, with methanol it amounted to 78%.

Using biochemical methods, ANTONI et al. (1965) studied the effect of fixation on the losses of DNA, RNA, and proteins. Simultaneously, the loss of activity was measured by scintillation counting. After fixation of liver in formalin, 25% of the quantity of RNA and 25% of its activity were lost. After fixation in methanol, only 10% of the quantity of RNA was lost; however, 60% of the activity could no longer be detected. Comparable results were obtained on fixation with CARNOY's fluid. The duration of the procedures for this set of experiments was 24 hrs. If, however, fixation for 1 hr. was performed, the losses of amounts and activity from ascites tumor cells remained within the range of error. Compared with the results of the fixation of RNA, processing of samples for studies of DNA and proteins posed minor problems.

In view of the ambiguous results for the fixation of RNA, parallel biochemical determinations of the quantity and specific activity are strongly recommended when performing quantitative autoradiographic studies of RNA. The problem remains that this procedure will not indicate the loss of special fractions, as for instance, from the nucleolus or nucleus. No generally accepted technique for fixation of RNA is yet available which is also suitable for quantitative studies of turnover; further investigation of this problem is certainly needed. The fixation of DNA and proteins offers fewer problems. With proteins, caution should be exercised when using fixing agents of the aldehyde type. In any case, the fixing procedure should be checked for chemography (see p. 371ff.).

Autoradiography of Soluble Compounds

For this type of autoradiography three conditions must be met:
1) No activity should be lost from the sample, from the time of removal of the tissue until the end of the autoradiographic processing.

2) Localization of the label at the cellular level should not be disturbed, for example by diffusion processes.

3) The specimen should not exert a chemographic effect on the nuclear emulsion.

Various techniques for the autoradiography of soluble compounds have been described, and have been reviewed by FEINENDEGEN (1968).

In 1969 STUMPF and ROTH reviewed their technique in detail (ROTH, 1969; STUMPF and ROTH, 1964, 1966, 1967, 1969). This technique apparently meets all the conditions outlined above. Samples are cut at $-60°$ C or lower in a cryostat; they are freeze-dried without thawing and then brought into contact with the dry nuclear emulsion. This technique is superior to that of APPLETON (1964, 1966, 1968), which dispenses with freeze-drying. ROGERS and JOHN (1969) showed that, using the technique of STUMPF and ROTH, negative chemography was more easily suppressed than with the technique of APPLETON. ROGERS and JOHN (1969) stated that "carefully planned controls for positive and negative chemography should similarly become a routine part of every experiment if we wish to claim that autoradiography is a serious scientific technique ... ". This condition should be applied not only to the autoradiography of soluble compounds, but to all types of quantitative autoradiography.

Quantitative Determination of Substances

In this section it is assumed that the specific activity of the labeled compound is known at the site of synthesis of the end-product. The difficulties inherent in this assumption are discussed on p. 348 ff. If the specific activity is known, the amount of substance can be calculated from the radioactivity. This section, therefore, deals with the quantitative determination of radioactivity.

Track Counting

The number of tracks produced during exposure is a measure of the radioactivity. For example, $1 \mu Ci$ corresponds to 3.7×10^4 disintegrations per second. The geometry of the source is of cardinal importance for the ratio of tracks to disintegrations. If the source lies on a support, for example, a slide, and is covered with nuclear emulsion, a 50% ratio can be assumed, because, on a statistical basis, only every second decay is directed towards the emulsion; the counted tracks must be multiplied by a factor of 2 to give the number of disintegrations.

Due to back-scattering (p. 368 ff.), decays reflected from the glass support may pass through the emulsion. For specimens which approach infinite thinness, the portion of back-scattering from the slide is of the order of 12 to 17% of all the observed decays (p. 368 ff.). Assuming a 50% geometry, this means that results are overestimated by 24 to 34%. Since the degree of back-scattering depends on the atomic number of the support for the source, it is advisable to use supports made from plastic material (BOSTRÖM et al., 1952). ODEBLAD (1952) showed that, with methacrylate as a support, the effect of back-scattering amounted to only a few percent. For autoradiographic techniques with 100% geometry (see p. 360ff.), the problem of back-scattering does not exist.

Since the formation of a track requires at least 4 silver grains, some β-decays with low energies are unable to produce tracks. In the case of carbon-14, these amount to 14% of all decays (LEVI et al., 1963). The number of tracks has to be corrected for this fraction of unobserved decays.

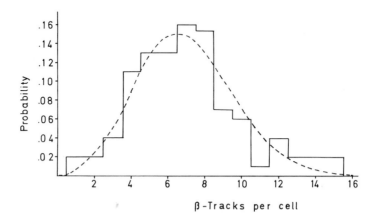

Fig. 21. Frequency distribution of β-tracks per cell with conditions of 100% geometry. Thymus lymphocytes were labeled with ^{14}C-thymidine, fixed and stained in suspension, and then suspended in Ilford G5 emulsion (procedure according to LEVINTHAL and THOMAS, 1957). The dotted curve corresponds to a Poisson distribution of 7 events. The mean radioactivity per cell can be calculated from the mean number of 7 decays during the time of exposure, giving an average value of 54.3×10^{-18} moles for the ^{14}C-thymidine incorporated per cell. (After DÖRMER and BRINKMANN, 1972)

The length of time required for the emulsion to dry is a further problem. For very thick emulsions, 3 hrs. has been assumed to be the period during which pronounced fading of the latent images occurs (DÖRMER and BRINKMANN, 1972; LEVINTHAL and THOMAS, 1957). In one of our experiments the exposure time was only 18 hrs. This means that the initial period of time when fading takes place is a considerable source of error. In track autoradiography, therefore, labeling with lower activities than are used in grain density autoradiography is recommended, so as to give reasonably long times for exposure.

Only a very few quantitative experiments have been performed by track autoradiography, obviously because counting tracks is much more tedious than counting grain densities. Nevertheless, in a quantitative study, ROGERS et al. (1966) were able to determine the number of molecules of cholinesterase per motor end-plate of muscle; the results were verified by independent methods. It should be emphasized that track autoradiography is a very direct method for the determination of extremely small quantities of radioactivity (Fig. 21) which deserves more attention than it has as yet received.

Grain Yield

For every combination of an isotope with an emulsion under the same geometric conditions, the yield of silver grains per disintegration is constant. Numerous determinations of the yield of grains, especially for tritium, can be found in the literature (Table 3); from this table it can be seen that all isotopes having the energy level of carbon-14 or higher are characterized by a grain yield of 1 to 2. The experiments summarized in Fig. 7 indicate a yield corresponding to 0.95 incident electrons per grain for ^{14}C, corresponding to 1.9 disintegrations per grain rendered developable.

Results for tritium vary considerably. Some authors have measured specimens with differing degrees of β-self-absorption, and have recorded the maximum and minimum values of the yield. This indicates that calculations of the radioactivity of tritium using grain yield will be too inaccurate because of β-self-absorption. Theoretically, calculations of this nature are possible for isotopes of higher energy. However, not a single study is known to the author in which radioactivity is calculated by such a constant value for the grain yield, although, in many descriptions of quantitative autoradiography, emphasis is given to this procedure. This may be because of the following technical difficulties:

1) Geometric conditions have to be constant, including the thickness of the layer of emulsion (p. 368 ff.).

2) Either the extent of fading must be known, or fading must not occur at all (p. 359 ff.).

3) Processing of the autoradiographs must be completely reproducible (p. 357 ff.).

Table 3. Silver grain yield (number of electrons necessary to render 1 grain developable) taken from various estimates and determinations in the literature

Isotope	Emulsion	Grain yield	Authors
^{125}I	L4	2.3–4.3	ADA et al. (1966)
^3H	L4	8.4–33	
^{14}C	AR.10	1–2	ANDRESEN et al. (1953)
^3H	AR.10	19	CLEAVER and HOLFORD (1965)
^{14}C	AR.10	1.9	DÖRMER (present study)
^{131}I	AR.10	1.8	CORMACK (1955)
^{32}P	AR.10	0.8	
^3H	SINO	18	FRIESER et al. (1962)
^{131}I	AR.10	1.9	HERZ (1951)
^{32}P	AR.10	0.8	
^3H	AR.10	5–20	HUGHES et al. (1958)
^3H	NTB-3	5.8	HUNT and FOOTE (1967)
^3H	AR.10	200	KISIELESKI et al. (1961)
^{14}C		1.8–2	LAJTHA and OLIVER (1959)
^{35}S	AR.10	1.8	LAMERTON and HARRISS (1954)
^{14}C	AR.10	2	LEVI (1957)
^3H	AR.10	16	MAURER and PRIMBSCH (1964)
^3H	AR.10	7–42	OJA et al. (1966)
^3H	AR.10	10.9	WIMBER et al. (1960)
^3H	NTB	19.3	

Radioactive Standard Sources

The use of standard sources of radioactivity is widespread (MANN, 1956; SELIGER, 1956). In quantitative autoradiography, the nuclear emulsion is exposed to the radioactive specimen and the radioactive standard source simultaneously. An example of an infinitely thick ^{14}C-standard, consisting of poly-methyl-^{14}C-methacrylate, is shown in Fig. 22. The actual yield of grains on an individual autoradiograph can be determined from the surface activity of the standard source and the density of grains over it. This value can be used for calculating the radioactivity of single cells by counting the overlying grains. It follows that:

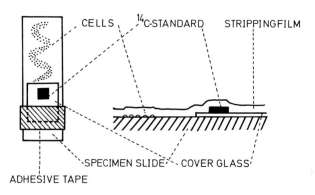

Fig. 22. Method of application of infinitely thick ^{14}C-standard sources to radioactive samples for quantitative ^{14}C-autoradiography using stripping-film AR.10. (After DÖRMER and BRINKMANN, 1972)

1) the thickness of the emulsion can vary from one autoradiograph to the next;

2) the degree of fading need not be known;

3) development can be varied deliberately.

Some precautions, however, must be observed:

1) Because of β-self-scattering (KARNOVSKI et al., 1955), the sample and the standard source should consist of similar materials.

2) Corrections have to be made if the geometric conditions of the standard source and the sample are different. For instance, if infinitely thin specimens are compared with infinitely thick standard sources, allowance has to be made for β-self-absorption in the standard source and for β-back-scattering in the sample.

3) The thickness of the emulsion over the standard source and the sample must be the same.

4) Chemography, and fading of the latent images, must be the same for the sample and the standard source. This is of special importance in the negative chemography encountered with insufficiently or unsuitably fixed biological objects (see p. 371 ff.). An example of differences in positive chemography between the sample and the standard source is given on p. 384 ff. (see also Fig. 25).

Several determinations of absolute radioactivity by means of radioactive standard sources have been performed in autoradiography. Some authors have also proposed methods for preparing autoradiographic standard sources. BARNARD and MARBROOK (1961) suggested the acylation of frog erythrocytes with ^{3}H-acetic anhydride in pyridine-acetonitrile. The reference activity must be determined in a scintillation counter before use. MAMUL (1956) suggested the preparation of radioactive wedges consisting of either ^{14}C or ^{35}S in gelatine between two inclined glass plates. Such a wedge has an advantage over a standard source having only one constant surface activity. The wedge can be regarded as a series of standard sources with increasing radioactivity, so measurements of grain density can be extended into the range of non-linearity with the radioactive dose (see p. 355 ff.).

Some experiments have been conducted with standard sources containing ^{45}Ca in plaster of Paris (DUDLEY and DOBYNS, 1949; JOWSEY, 1966; JOWSEY et al., 1965). ^{35}S has been formed into a radioactive standard with shellac (AXELROD and HAMILTON, 1947), PVC (KUTZIM, 1963), or plaster of Paris (DOMINGUES et al., 1956). In most of these experiments, a series of standards with different activities were taken for one autoradiograph. RITZEN (1967a, b, 1968) employed poly-butyl-^{3}H-methacrylate as infinitely thick standards.

An indirect method consists of obtaining calibration curves with radioactive standard sources, and using these to calculate the radioactivity in samples exposed independently (ANDRESEN et al., 1953; BOSTRÖM et al., 1952; ODEBLAD, 1952; WASER and LÜTHI, 1962). This method takes a mid-position between calculations with the aid of the grain yield and the simultaneous exposure of the emulsion to standard source and specimen. The study of ODEBLAD (1952) illustrates the painstaking steps necessary for calibration when using this indirect method. For an example of application see the section on p. 385 ff., which deals with the direct method for the determination of the absolute incorporation of ^{14}C-TdR into single cells.

Manual and Automatic Methods of Grain Counting

Visual Counting

In visual grain counting of histological sections, an average of 30 grains per labeled cell is considered a good compromise, if fine-grained nuclear emulsions are used. At higher grain densities individual grains are no longer distinguishable. At lower counts, there is an increasing probability that in a portion of labeled cells the grain density does not exceed the background level. Assuming a grain yield of 1.0 and a count of 30 grains, the standard error due to the randomness of radioactivity is $100/\sqrt{30} = 18\%$. This error is increased by subtracting a background value (PERRY, 1964). Therefore, narrow peaks can not be expected in frequency distributions of grain density when the counts are done manually.

The accuracy can be increased by counting on photographs (OSTROWSKI and SAWICKI, 1961) or on a screen onto which the image is projected from a microscope (MICOU and GOLDSTEIN, 1959). Before such complicated methods are applied, the question should be asked as to whether the reduction by approximately

5–8 % in the error in counting yields more information concerning the biological problem. In most cases, the answer to this question will be negative, because of the extent of the remaining errors. A further means of improving the conditions of counting is to use incident, instead of transmitted, light (DÖRMER, 1972; ROGERS, 1967).

Silver Grain Photometry

In principle, the best method of photometry is densitometry. Blackening, as measured by densitometry, is defined as $\log_{10}(I_0/I)$, where I_0 is the flux of light incident on the autoradiograph, and I the flux transmitted. As is demonstrated in Fig. 23, blackening is proportional to grain density (ARENS *et al.*, 1931; NUTTING, 1913; KLEIN, 1958). In the lower range of blackening there is a linear relationship to the radioactive dose (BECKER, 1962; BOTHE, 1922; BOYD, 1955; DUDLEY, 1954; HERZ, 1969; PODDAR, 1955). The advantage of the linear relationship to the radioactive dose is offset by the fact that, for densitometric measurement, the cells have to be separated from the nuclear emulsion. This implies that densitometry can be applied only in exceptional cases of single cell autoradiography.

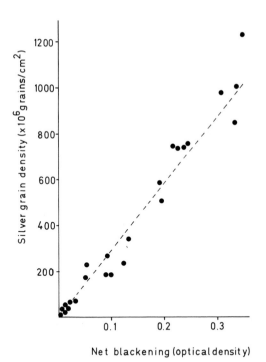

Net blackening (optical density)

Fig. 23. Relationship between actual grain density and blackening as determined by densitometry. There is a linear correlation over the entire range of measurement. The experiment was that referred to in Fig. 8. From comparison of the two figures, the following conclusion can be drawn: in the range of grain densities applicable to autoradiography with AR.10 stripping-film, there is a linear correlation between actual grain density and blackening, however, the linear relationship between actual grain density and radiation exposure does not extend over the whole of this range

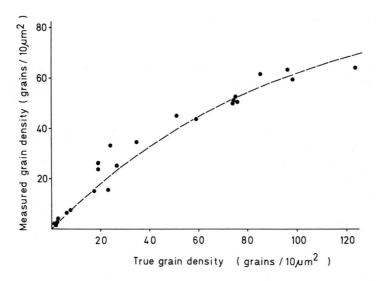

Fig. 24. Relationship between the apparent grain density as determined by incident-light micro-photometry and the true grain density. Same experiment as in Figs. 8 and 23. While blackening is related linearly to grain density, the reflectance, obtained by incident light photometry, is an exponential function (dotted line). (From DÖRMER, 1972)

The various photometric techniques for quantitative autoradiography have been reviewed recently (DÖRMER, 1972; GOLDSTEIN and WILLIAMS, 1971). For measurements at the cellular level, photometry with incident-light bright-field illumination is most suitable (DÖRMER, 1967a, 1972; DÖRMER and BRINKMANN, 1968; DÖRMER et al., 1966; ROGERS, 1961, 1967). Silver grains are illuminated by the same objective as is used for observation and measurement, and the amount of reflected light is registered. In the range of grain densities countable by eye, this amount is proportional to the grain density. With higher grain densities saturation becomes obvious (Fig. 24). This can be expressed in terms of an exponential function (DÖRMER, 1972):

$$n_{\mathrm{phot}} = n_{\mathrm{sat}}(1 - e^{-n_{\mathrm{true}}/n_{\mathrm{sat}}}),$$

where n_{phot} is the grain density obtained from the photometric measurements, n_{sat} is the apparent grain density in the range of photometric saturation, and n_{true} is the actual grain density of a field. For AR.10 stripping-film developed in an amidol-developer, n_{sat} has been found to be 95 grains per 10 μm² (DÖRMER, 1972).

The same experiment gave the data for Fig. 24 as well as for Fig. 8 and Fig. 23. On comparing the three diagrams, it is apparent that the relationship between the true grain density and the grain density as measured by incident-light photometry deviates increasingly from linearity, while there is still a good proportionality between the true grain density and the radiation exposure. The error incurred by

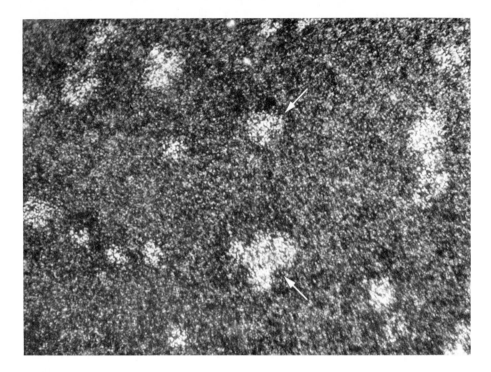

Fig. 25. Microphotograph of reflectance of grains of silver in incident light dark-field (Ultropak). Unlabeled cells were smeared on a plate of ^{14}C-methacrylate and fixed in methanol for 20 min. After exposure, stripping-film AR.10 was developed with D-19 developer. The smeared cells were found under the brightly reflecting areas. The higher reflectivity of the grains over the unlabeled cells is caused by the larger surface area of these grains. The upper arrow points to a single cell, the lower one to a cluster of cells

not correcting the photometric data by means of the above formula can be seen from Fig. 24; it depends on the measured grain density. The formula given above should be used to correct the data whenever processing with a computer is possible.

If standard sources of radioactivity are used, and grain counting done with a photometer, it is essential that the grains over the standard source and over the specimen are of the same size. This does not occur when using D-19 developer (for composition, see p. 359). Probably due to a positive chemographic effect, the grains over the specimen are larger than those over the standard source (Fig. 25) (DÖRMER and BRINKMANN, 1972). The difference in size is not observed when amidol is used as a developer (for composition, see p. 359); it is known, however, that with amidol as a surface developer, negative chemography is more likely to occur than when a hydroquinone developer is used (ROGERS, 1967; ROGERS and JOHN, 1969). Therefore, whenever amidol-developer is used in quantitative autoradiography, the effects of negative chemography and fading have to be excluded rigorously (see p. 359ff. and p. 371ff.).

If grain densities are to be assessed by photometry with a microscope, the investigator should be familiar with the conditions of this procedure. This applies to the preparation of the autoradiographs as well as to the photometric principles and design. An extensive description cannot be given here; the reader is referred to review articles by DÖRMER (1972) and by GOLDSTEIN and WILLIAMS (1971). Instructions for the preparation of autoradiographs for photometric evaluation are found on p. 386ff.

Application. The Rate of Synthesis of DNA in Individual Cells

Until recently the application of quantitative autoradiography has been quite limited. This is understandable, considering all the biochemical and preparative problems, and those of processing and evaluation. However, the method of determining the rate of synthesis of DNA in single cells has already been applied so extensively (BRINKMANN and DÖRMER, 1972; DÖRMER, 1973a, b; DÖRMER and BRINKMANN, 1970a, b, 1972; DÖRMER et al., 1972), that a more detailed description will be given here.

The following types of cells are suitable for these investigations: a) cells brought into suspension easily, such as those of bone marrow, spleen or thymus; b) culture cells grown on slides.

The procedure for cells of group a): The cells are brought into suspension, for which mechanical mincing of the tissue may be necessary. TC-medium 199 (Difco, U.S.A.) can be used for suspension. The cells are washed twice; 10^{-5} M 5-fluoro-2'-deoxyuridine (FUdR) and 10^{-5} M deoxycytidine (CdR) are added to the medium for the last wash. The suspension is passed through a glass filter with pore size of 45–90 μm. After filtration, the suspension is diluted to 10000 to 20000 cells per mm^3, with medium containing the aforementioned additives. The entire procedure is performed at room temperature. 0.95 ml of suspension is pipetted into each of a series of test tubes which are then put in a shaking water bath at 37° C with a frequency of 150 stroke/min. The cells are preincubated with FUdR and CdR for 6 min, then 0.05 ml of a solution of ^{14}C-TdR (specific activity approx. 60 mCi/mmole) is added, to give a final concentration of 10^{-5} M TdR. Some of the tubes are incubated with ^{14}C-TdR for 3 min, and others for 7 min. To stop the reaction, the tubes are transferred to ice water. After centrifugation at 0° C the supernatant is discarded, and the cells are resuspended at 0° C in a few drops of a 10% solution of bovine albumin and smeared on slides with a coverslip.

The procedure for cells of group b): 3 100 ml flasks are used, as are commonly employed for the vertical staining of slides. 80 ml of TC-medium 199 containing 10^{-5} M FUdR and 10^{-5} M CdR are pipetted into the first two flasks; the second flask also contains 10^{-5} M ^{14}C-TdR. The third flask contains methanol. The first two flasks are placed in a water bath, and warmed to 37° C. Slides with the adhering culture cells are placed individually into flask No. 1 for 6 min, then into No. 2 for either 3 or 7 min, and finally, into flask No. 3. Fixation in methanol lasts for 48 hrs., with one change of methanol. Further treatment for both types of cells is the same.

After fixation, the backs of the slides are coated as usual with chromalum-gelatin. A standard source of ^{14}C is attached to the lower third of the slide, as shown in Fig. 22. The cells and the standard source are covered with AR.10 stripping-film. The samples with cells incubated for 7 min are exposed for 25 ± 5 days, while those of 3 min are exposed for 58 ± 12 days. The autoradiographs are developed for 6 min in amidol (for composition, see p. 359). A stopping bath is not used. The autoradiographs are placed for 20 min into an acid fixing bath with some chromalum added for hardening (sodium thiosulfate 300 gm, potassium bisulfite 30 gm, chromalum 1 gm, made up to 1 l with distilled water). After a rinse for 2 hrs. in running tap water, the autoradiographs are immersed in a 1% aqueous solution of chromalum for 3 min. The wet film is then eased back gently with a soft brush from the standard source and the coverglass: these are removed, and the film is returned to its original position. The treatment with chromalum hardens the film so that it is not stretched when it is moved. This prevents the emulsion layer blackened by the standard source from becoming thinner than that blackened by the cells. The autoradiographs are then allowed to dry vertically in air.

Evaluation of grain densities can be performed most conveniently with an incident light microphotometer (for technical details, see DÖRMER, 1972). Staining which is compatible with photometric measurements can be carried out with a modified Giemsa-stain: the autoradiographs are immersed in a filtered solution of 5 volumes Giemsa stock solution (Merck Company, Darmstadt) and 95 volumes phosphate buffer pH 6.4 for 40 min. They are then dried for 24 hrs. in a gentle stream of air. The emulsion and carrier layer are cleared at 15° C in a solution of 35 volumes absolute alcohol and 65 volumes phosphate buffer pH 6.4. The autoradiographs are air-dried without further rinsing. Mounting with a coverslip is not recommended. For photometry, adjustment of the size of the measuring aperture to the nuclei, and calculation of the total grain count, as described on p. 370 ff., should be performed. For this purpose a computer is very useful.

The amount of TdR incorporated per cell is calculated from the formula:

$$\text{TdR [moles]} = C \cdot \frac{E \cdot F}{D \cdot G},$$

where

C is the grain count per cell;

D is the mean grain count per unit area of the ^{14}C-standard source;

E is the volume of the ^{14}C-standard source, with the same surface area as for D, and giving silver grains unaffected by β-self-absorption. In the case of ^{14}C-methacrylate, a thickness of 17.94 µm, without β-self-absorption, is equivalent to an infinitely thick source of the same type, subject to β-self-absorption;

F is the radioactivity per volume of the standard source;

G is the specific activity of the ^{14}C-TdR employed.

The rate of incorporation of ^{14}C-TdR between the third and seventh minutes of incubation is constant (BRINKMANN and DÖRMER, 1972). During this period, the rate of incorporation of ^{14}C-TdR equals the rate of synthesis of DNA, since the endogenous synthesis of thymidylate is blocked by FUdR. Therefore, the rate of synthesis of DNA can be calculated for the individual cells incubated for

Myelocytes of a healthy individual (n = 287)

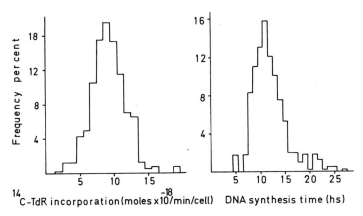

Fig. 26. Frequency distributions of the rate of synthesis of DNA (left) and time of synthesis of DNA (right) for single myelocytes of a healthy human. Measurements made by Dr. BRINKMANN

7 min (Fig. 26). Since the amount of thymine contained in a diploid cell is known (man: $6\,300 \times 10^{-18}$ moles; rat: $5\,270 \times 10^{-18}$ moles) (DÖRMER and BRINKMANN, 1970a, b), the length of time that it would take a cell to synthesize the entire amount of thymine from exogenous ^{14}C-TdR can be calculated. This time corresponds to the time of synthesis of DNA for the cell (Fig. 26).

The results obtained for human myelocytes, shown in Fig. 26, correspond to those derived by the technique of labeled mitoses for the same type of cell (STRYCKMANS et al., 1966). According to this and other comparable results, the method seems to be suitable for investigating the kinetics of cell proliferation. In order to calculate the mean values of the time of synthesis of DNA from the values for the rate of incorporation, one must apply the harmonic mean (DÖRMER, 1973b; QUASTLER, 1963b).

Literature

ADA, G. L., HUMPHREY, J. H., ASKONAS, B. A., McDEVITT, H. O., NOSSAL, G. J. V.: Correlation of grain counts with radioactivity (^{125}I and tritium) in autoradiography. Exp. Cell Res. **41**, 557–572 (1966).

ADAMSON, L. F., LANGELUTTIG, S. G., ANAST, C. S.: Amino acid transport in embryonic chick bone and rat costal cartilage. Biochim. biophys. Acta (Amst.) **115**, 345–354 (1966).

AMANO, M., LEBLOND, C. P., NADLER, N. J.: Radioautographic analysis of nuclear RNA in mouse cells revealing three pools with different turnover times. Exp. Cell Res. **38**, 314–340 (1965).

ANDRESEN, N., CHAPMAN-ANDRESEN, C., HOLTER, H., ROBINSON, C. V.: Quantitative autoradiographic studies on the amoeba chaos chaos with ^{14}C. C. R. Lab. Carlsberg, Sér. chim. **28**, 499–528 (1953).

ANTONI, F., KÖTELES, G. J., HEMPEL, K., MAURER, W.: Über die Eignung verschiedener Fixationen und perchlorsäurehaltiger Lösungen für autoradiographische Untersuchungen des RNS-, DNS- und Proteinstoffwechsels. Histochemie **5**, 210–220 (1965).

APPLETON, T. C.: Autoradiography of soluble labelled compounds. J. roy. micr. Soc. **83**, 277–281 (1964).

APPLETON, T.C.: Resolving power, sensitivity and latent image fading of soluble-compound auto-radiographs. J. Histochem. Cytochem. **14**, 414–420 (1966).

APPLETON, T.C.: The application of autoradiography to the study of soluble compounds. Acta histochem. (Jena), Suppl. **8**, 115–131 (1968).

ARENS, H., EGGERT, J., HEISENBERG, E.: Zusammenhang zwischen Schwärzung, Silbermenge, Deck-kraft, Kornzahl und Korndimension entwickelter photographischer Schichten. Z. wiss. Photogr. **28**, 356–366 (1931).

AXELROD, D.J., HAMILTON, J.G.: Radio-autographic studies of the distribution of lewisite and mustard gas in skin and eye tissues. Amer. J. Path. **23**, 389–398 (1947).

BARNARD, E.A., MARBROOK, J.: Quantitative cytochemistry using directly applied radioactive reagents. Nature (Lond.) **189**, 412–413 (1961).

BASERGA, R.: Autoradiographic methods. In: H. BUSCH (ed.), Methods in cancer research, vol. I, p. 45–116. New York-London: Academic Press 1967.

BASERGA, R., KISIELSKI, W.E.: Effects of histologic and histochemical procedures on the intensity of the label in radioautographs of cells labeled with tritiated compounds. Lab. Invest. **12**, 648–655 (1963).

BASERGA, R., MALAMUD, D.: Autoradiography. Techniques and application. New York-Evanston-London: Harper & Row Publishers 1969.

BASERGA, R., NEMEROFF, K.: Factors which affect efficiency of autoradiography with tritiated thymidine. Stain Technol. **37**, 21–26 (1962).

BECKER, K.: Filmdosimetrie. Berlin-Göttingen-Heidelberg: Springer 1962.

BLACKETT, N.M., KEMBER, N.F., LAMERTON, L.F.: The measurement of radiation dosage distribution by autoradiographic means with reference to the effect of bone-seeking isotopes. Lab. Invest. **8**, 171–178 (1959).

BLEECKEN, S.: Untersuchung des autoradiographischen Auflösungsvermögens mit Strahlungsquellen verschiedener Betaenergien. Atompraxis **9**, 321–324 (1961).

BOSTRÖM, H., ODEBLAD, E., FRIBERG, U.: A quantitative autoradiographic study of the incorporation of S^{35} in tracheal cartilage. Arch. Biochem. Biophys. **38**, 283–286 (1952).

BOTHE, W.: Über photographische β-Strahlenmessung. Z. Physik **8**, 243–250 (1922).

BOYD, G.A.: Autoradiography in biology and medicine. New York: Academic Press 1955.

BRINKMANN, W., DÖRMER, P.: In vitro-Verfahren zur Bestimmung der DNS-Synthese-Dauer ein-zelner Zellen. Biochemische Voraussetzungen und Ergebnisse. Histochemie **30**, 335–343 (1972).

BRINKMANN, W., DÖRMER, P., MUSCHALIK, P.: Eine neue Methode zur Bestimmung der DNS-Synthesegeschwindigkeit von Knochenmarkzellen in vitro. Blut **19**, 529–536 (1969).

BROWNELL, G.L.: Interaction of phosphorus-32 beta rays with matter. Nucleonics **10**, 30–35 (1952).

BRYANT, B.J.: In vivo reutilization of the DNA thymidine of necrotized liver cells by cells of testis and intestine. Exp. Cell Res. **32**, 209–212 (1963a).

BRYANT, B.J.: Reutilization of lymphocyte DNA by cells of intestinal crypts and regenerating liver. J. Cell Biol. **18**, 515–523 (1963b).

CITOLER, P., CITOLER, K., HEMPEL, K., SCHULTZE, B., MAURER, W.: Autoradiographische Unter-suchungen mit zwölf H^3- und fünf C^{14}-markierten Aminosäuren zur Größe des nuclearen und cytoplasmatischen Eiweißstoffwechsels bei verschiedenen Zellarten von Maus und Ratte. Z. Zell-forsch. **70**, 419–448 (1966).

CLEAVER, J.E.: Thymidine metabolism and cell kinetics. Amsterdam: North-Holland Publ. Comp. 1967.

CLEAVER, J.E., HOLFORD, R.M.: Investigations into the incorporation of ^{3}H-thymidine into DNA in L-strain cells and the formation of a pool of phosphorylated derivatives during pulse labelling. Biochim. biophys. Acta (Amst.) **103**, 654–671 (1965).

COOPER, R.A., PERRY, S., BREITMAN, T.R.: Pyrimidine metabolism in human leukocytes. I. Contribu-tion of exogenous thymidine to DNA-tymine and its effect on thymine nucleotide synthesis in leukemic leukocytes. Cancer Res. **26** (I), 2267–2275 (1966).

CORMACK, D.V.: The beta-ray sensitivity of autoradiographic stripping film. Brit. J. Radiol. **28**, 450 (1955).

CRONKITE, E.P., GREENHOUSE, S.W., BRECHER, G., BOND, V.P.: Implication of chromosome structure and replication on hazard of tritiated thymidine and the interpretation of data on cell proliferation. Nature (Lond.) **189**, 153–154 (1961).

DEUCHAR, E. M.: Staining sections before autoradiographic exposure: Excessive background graining caused by celestin blue. Stain Technol. **37**, 324–328 (1962).

DÖRMER, P.: Erfahrungen mit der photometrischen Silberkornzählung in der Autoradiographie. Leitz Mitt. Wiss. Techn. **4**, 74–78 (1967a).

DÖRMER, P.: Auflichtphotometrische Untersuchungen zur Größe der Koinzidenz in der Autoradiographie mit Tritium. Histochemie **8**, 1–8 (1967b).

DÖRMER, P.: Photometric methods in quantitative autoradiography. In U. LÜTTGE (ed.), Microautoradiography and electron probe analysis, p. 7–48. Berlin-Heidelberg-New York: Springer 1972.

DÖRMER, P.: Kinetics of proliferation in normal hemopoietic and leukemic cells. In: E. GERLACH, K. MOSER, E. DEUTSCH, W. WILMANNS (eds.), Erythrocytes, thrombocytes, leukocytes, p. 356–362. Stuttgart: G. Thieme 1973a.

DÖRMER, P.: Kinetics of erythropoietic cell proliferation in normal and anemic man. A new approach using quantitative autoradiography. Progr. Histochem. Cytochem. 1973b (in press).

DÖRMER, P., BRINKMANN, W.: Silberkornzählung mit dem Auflicht-Mikroskopphotometer. Ein Beitrag zur quantitativen Autoradiographie. Acta histochem. (Jena), Suppl. **8**, 163–169 (1968).

DÖRMER, P., BRINKMANN, W.: Auflichtphotometrie von Mikroautoradiogrammen für quantitative Einbaustudien an Einzelzellen. Z. analyt. Chem. **252**, 84–89 (1970a).

DÖRMER, P., BRINKMANN, W.: Estimation of the DNA synthesis rate of bone marrow cells after administration of labelled thymidine in vitro. Proceedings of the IAEA symposium on "In vitro procedures with radioisotopes in medicine". Wien: IAEA press 1970b.

DÖRMER, P., BRINKMANN, W.: Quantitative ^{14}C-Autoradiographie einzelner Zellen. Histochemie **29**, 248–264 (1972).

DÖRMER, P., BRINKMANN, W., DAHR, P., JR.: Thymidylat-de novo-Synthese, salvage pathway und DNS-Synthese während der Differenzierung von Erythroblasten. Blut **25**, 185–189 (1972).

DÖRMER, P., BRINKMANN, W., STIEBER, A., STICH, W.: Automatische Silberkornzählung in der Einzelzell-Autoradiographie. Eine neue photometrische Methode für die quantitative Autoradiographie. Klin. Wschr. **44**, 477–482 (1966).

DÖRMER, P., REICHART, B., HUHN, D.: Kerntrockengewicht und Beteiligung von ^{3}H-Thymidin an der DNS-Synthese in Einzelzellen der regenerierenden Rattenleber. Z. Zellforsch. **86**, 559–570 (1968).

DOMINGUES, F. J., SARKO, A., BALDWIN, R. R.: A simplified method for quantitation of autoradiography. Int. J. appl. Radiat. **1**, 94–101 (1956).

DONIACH, I., PELC, S. R.: Autoradiograph technique. Brit. J. Radiol. **23**, 184–192 (1950).

DUDLEY, R. A.: Photographic detection and dosimetry of beta rays. Nucleonics **12**, 24–31 (1954).

DUDLEY, R. A., DOBYNS, B. M.: The use of autoradiographs in the quantitative determination of radiation dosages from Ca^{45} in bone. Science **109**, 327–342 (1949).

EVANS, R. D.: Interactions of α and β particles with matter. In: T. F. DOUGHERTY, W. S. S. JEE, C. W. MAYS, and B. J. STOVER (eds.), Some aspects of internal irradiation. Oxford-London-New York-Paris: Pergamon Press 1962.

EVANS, T. C.: Selection of radioautographic technique for problems in biology. Nucleonics **3**, 52–59 (1948).

FALK, G. J., KING, R. C.: Radioautographic efficiency for tritium as a function of section thickness. Radiat. Res. **20**, 466–470 (1963).

FEINENDEGEN, L. E.: Autoradiographie der wasserlöslichen Substanzen. Acta histochem. (Jena), Suppl. **8**, 107–114 (1968).

FEINENDEGEN, L. E., BOND, V. P.: Differential uptake of ^{3}H-thymidine into the soluble fraction of single bone marrow cells, determined by autoradiography. Exp. Cell Res. **27**, 474–484 (1962).

FEINENDEGEN, L. E., BOND, V. P., HUGHES, W. L.: Physiological thymidine reutilization in rat bone marrow. Proc. Soc. exp. Biol. (N.Y.) **122**, 448 (1966).

FEINENDEGEN, L. E., BOND, V. P., SHREEVE, W. W., PAINTER, R. B.: RNA and DNA metabolism in human tissue culture cells studied with tritiated cytidine. Exp. Cell Res. **19**, 443–459 (1960).

FISCHER, H. A., WERNER, G.: Autoradiography. Berlin-New York: Walter de Gruyter 1971.

FITZGERALD, P. J., EIDINOFF, M. L., KNOLL, J. E., SIMMEL, E. B.: Tritium in radiography. Science **114**, 494–498 (1951).

FRIESER, H., HEIMANN, G., RANZ, E.: Einwirkung radioaktiver Nuklide auf photographische Schichten. Phot. Korresp. **98**, 131–140 (1962).

GALASSI, L.: Delayed and direct labeling after a systemic injection of thymidine-^{3}H. J. Histochem. Cytochem. **15**, 565–572 (1967).

GLENDENIN, L. E., SOLOMON, K. A.: Self-absorption and backscattering of β-radiation. Science 112, 623–626 (1950).

GLOCKER, R.: Das photographische Schwärzungsgesetz für Elektronenstrahlen verschiedener Energie. Z. Physik 160, 568–572 (1960).

GOLDSTEIN, D. J., WILLIAMS, M. A.: Quantitative autoradiography: An evaluation of visual grain counting, reflectance microscopy, gross absorbance measurements and flying spot microdensitometry. J. Microscopy 94, 215–239 (1971).

GROSS, J., BOGOROCH, R., NADLER, M. J., LEBLOND, C. P.: The theory and methods of the radioautographic localization of radioelements in tissues. Amer. J. Roentgenol. 65, 420–458 (1951).

HARBERS, E.: Autoradiographie als histochemisches Untersuchungsverfahren. In: W. GRAUMANN und K. NEUMANN (Hrsg.), Handbuch der Histochemie, Bd. I/1, S. 400–598. Stuttgart: G. Fischer 1958.

HARRIS, H.: Breakdown of nuclear ribonucleic acid in the presence of actinomycin D. Nature (Lond.) 202, 1301–1303 (1964).

HEINIGER, H. J., FEINENDEGEN, L. E., BÜRKI, K.: Reutilization of thymidine in various groups of rat bone marrow cells. Blood 37, 340–348 (1971a).

HEINIGER, H. J., FRIEDRICH, G., FEINENDEGEN, L. E., CANTELMO, F.: Reutilization of 5-^{125}I-iodo-2'-deoxyuridine and ^{3}H-thymidine in regenerating liver of mice. Proc. Soc. exp. Biol. (N.Y.) 137, 1381–1384 (1971b).

HELL, E., BERRY, R. J., LAJTHA, L. G.: A pitfall in high specific activity tracer studies. Nature (Lond.) 185, 47 (1960).

HERRMANN, W., HARTMANN, G., BRUST, R.: Das Auflösungsvermögen mikroautoradiographischer Aufnahmen in Abhängigkeit von der Energie der verwendeten β-Strahlung (Teil I). Atompraxis 9, 315–320 (1961).

HERZ, R. H.: Photographic fundamentals of autoradiography. Nucleonics 8, 24–39 (1951).

HERZ, R. H.: Methods to improve the performance of stripping emulsions. Lab. Invest. 8, 71–75 (1959).

HERZ, R. H.: The photographic action of ionizing radiations. New York-London-Sidney-Toronto: John Wiley & Sons Inc. 1969.

HUGHES, W. L., BOND, V. P., BRECHER, G., CRONKITE, E. P., PAINTER, R. B., QUASTLER, H., SHERMAN, F. G.: Cellular proliferation in the mouse as revealed by autoradiography with tritiated thymidine. Proc. nat. Acad. Sci. (Wash.) 44, 476–483 (1958).

HUNT, W. L., FOOTE, R. H.: Efficiency of liquid scintillation counting and autoradiography for detecting tritium in spermatozoa. Radiat. Res. 31, 63–73 (1967).

JAMES, T. H., FORTMILLER, L. J.: Dependence of covering power and spectral absorption of developed silver on temperature and composition of the developer. Phot. Sci. Eng. 5, 297–304 (1961).

JAMES, T. H., HIGGINS, G. C.: Fundamentals of photographic theory. New York: John Wieley & Sons, Inc. 1948.

JAMES, T. H., KORNFELD, G.: Reduction of silver halides and the mechanism of photographic development. Chem. Rev. 30, 1–32 (1942).

JOWSEY, J.: Densitometry of photographic images. J. appl. Physiol. 21, 309–312 (1966).

JOWSEY, J., LAFFERTY, W., RABINOWITZ, J.: Analysis of distribution of Ca^{45} in dog bone by a quantitative autoradiographic method. J. Bone Jt Surg. A 47, 359–370 (1965).

KARNOVSKY, M. L., FOSTER, J. M., GIDEZ, L. I., HAGERMAN, D. D., ROBINSON, C. V., SOLOMON, A. K., VILLEE, C. A.: Correction factors for comparing activities of different carbon-14-labeled compounds assayed in flow proportional counter. Analyt. Chem. 27, 352–354 (1955).

KISIELESKI, W. E., BASERGA, R., VAUPOTIC, J.: The correlation of autoradiographic grain counts and tritium concentration in tissue sections containing tritiated thymidine. Radiat. Res. 15, 341–348 (1961).

KLEIN, E.: Die Beziehung zwischen der Schwärzung und der Größe der entwickelten Silberaggregate. Z. Elektrochem. 62, 993–999 (1958).

KOPRIWA, B. M., LEBLOND, C. P.: Improvements in the coating techniques of radioautography. J. Histochem. Cytochem. 10, 269–284 (1962).

KORR, H., FISCHER, H. A., SEILER, N., WERNER, G.: Cholesterinkristalle in Kryostat-Gefrierschnitten von nervösem Gewebe als Ursache von Artefakten in Autoradiogrammen. Histochemie 23, 138–143 (1970).

KUTZIM, H.: Die quantitative Bestimmung der Verteilung von S^{35}-Sulfat bei der Maus mittels Autoradiographie. Nucl.-Med. (Stuttg.) 3, 39–50 (1963).

LAJTHA, L. G., OLIVER, R.: The application of autoradiography in the study of nucleic acid metabolism. Lab. Invest. **8**, 214–224 (1959).

LAMERTON, L. F., HARRISS, E. B.: Resolution and sensitivity considerations in autoradiography. J. Phot. Sci. **2**, 135–144 (1954).

LANG, W., MAURER, W.: Zur Verwendbarkeit von feulgen-gefärbten Schnitten für quantitative Autoradiographie mit markiertem Thymidin. Exp. Cell Res. **39**, 1–9 (1965).

LANG, W., MÜLLER, D., MAURER, W.: Prozentuale Beteiligung von exogenem Thymidin an der Synthese von DNS-Thymin in Geweben der Maus und in HeLa-Zellen. Exp. Cell Res. **49**, 558–571 (1968).

LANG, W., PILGRIM, CH., MAURER, W.: Prozentualer Anteil von ^{3}H- oder ^{14}C-Thymidin an der DNS-Synthese von Zellarten der Maus. Naturwissenschaften **53**, 210 (1966).

LEVI, H.: A discussion of recent advances towards quantitative autoradiography. Exp. Cell Res., Suppl. **4**, 207–221 (1957).

LEVI, H., HOGBEN, A. S.: Quantitative beta track autoradiography with nuclear track emulsions. Dan. Mat.-Fys. Medd. No 9 (1955).

LEVI, H., ROGERS, A. W., WEIS BENTZON, M., NIELSEN, A.: On the quantitative evaluation of autoradiograms. Kgl. Danske Videnskab. Selskab. Mat.-Fys. Medd. **33**, No 11 (1963).

LEVINTHAL, C., THOMAS, C. A., JR.: Molecular autoradiography: The β-ray counting from single virus particles and DNA molecules in nuclear emulsions. Biochim. biophys. Acta (Amst.) **23**, 453–465 (1957).

LIBBY, W. F.: Simple absolute measurement technique for beta radioactivity. Analyt. Chem. **29**, 1566–1570 (1957).

LOEVINGER, R.: The dosimetry of beta sources in tissue. The point-source function. Radiology **66**, 55–62 (1956).

MAMUL, YA. V.: Quantitative autoradiography using a radioactive wedge. Int. J. appl. Radiat. **1**, 178–183 (1956).

MANN, W. B.: The preparation and maintenance of standards of radioactivity. Int. J. appl. Radiat. **1**, 3–23 (1956).

MARSHALL, J. H., ROWLAND, R. E., JOWSEY, J.: Microscopic metabolism of calcium in bone. II. Quantitative autoradiography. Radiat. Res. **10**, 213–233 (1959).

MAURER, W.: Methodisches zur autoradiographischen Untersuchung des Eiweiß-Stoffwechsels. Acta histochem. (Jena), Suppl. XII, 65–72 (1972).

MAURER, W., PRIMBSCH, E.: Größe der β-Selbstabsorption bei der ^{3}H-Autoradiographie. Exp. Cell Res. **33**, 8–18 (1964).

MAURER, W., SCHULTZE, B.: Problems in autoradiographic studies of DNA, RNA, and protein synthesis. In: L. J. ROTH and W. E. STUMPF (eds.), Autoradiography of diffusible substances, p. 15–28. New York-London: Academic Press 1969.

MERTZ, M.: Bestimmung der Silberkorngröße in Autoradiogrammen bei Auflicht und Durchlicht. Histochemie **17**, 128–137 (1969).

MICOU, J., GOLDSTEIN, L.: A simple method to reduce the strain in manual grain counting of autoradiographs. Stain Technol. **34**, 347–348 (1959).

MOFFAT, D. J., YOUNGBERG, S. P., METCALF, W. K.: The validity of autoradiographic labeling. Cell Tiss. Kinet. **4**, 293–295 (1971).

NADLER, M. J.: Some theoretical aspects of radioautography. Canad. J. med. Sci. **29**, 182–194 (1951).

NEWTON, A., DENDY, P. P., SMITH, C. L., WILDY, P.: A pool size problem associated with the use of tritiated thymidine. Nature (Lond.) **194**, 886–887 (1962).

NIKLAS, A., MAURER, W.: Autoradiographie. In: Hoppe-Seyler/Thierfelder, Handbuch der physiologisch- und pathologisch-chemischen Analyse, 10. Aufl., Bd. II, S. 734–773. Berlin-Göttingen-Heidelberg: Springer 1955.

NIKLAS, A., QUINCKE, E., MAURER, W., NEYEN, A.: Messung der Neubildungsraten und biologischen Halbwertszeiten des Eiweißes einzelner Organe und Zellgruppen bei der Ratte. Biochem. Z. **330**, 1–20 (1958).

NUTTING, P. G.: On the absorption of light in heterogeneous media. Phil. Mag. **26**, 423–426 (1913).

ODEBLAD, E.: Contributions to the theory and technique of quantitative autoradiography with P^{32} with special reference to the granulosa tissue of the Graafian follicles in the rabbit. Acta radiol. (Stockh.), Suppl. **93**, 1–123 (1952).

OJA, H. K., OJA, S. S., HASAN, J.: Calibration of stripping film autoradiography in sections of rat liver labelled with tritium. Exp. Cell Res. **45**, 1–10 (1966).

OSTROWSKI, K., SAWICKI, W.: Photomicrographic method for counting photographic grains in autoradiograms. Exp. Cell Res. **24**, 625–628 (1961).

PAINTER, R. B., RASMUSSEN, R. E.: A pitfall of low specific activity radioactive thymidine. Nature (Lond.) **201**, 409–410 (1963).

PELC, S. R.: Autoradiograph technique. Nature (Lond.) **160**, 749–750 (1947).

PELC, S. R.: The stripping-film technique of autoradiography. Int. J. appl. Radiat. **1**, 172–177 (1956).

PELC, S. R., APPLETON, T. C., WELTON, M. E.: State of light autoradiography. In: C. P. LEBLOND and K. B. WARREN (eds.), The use of radioautography in investigating protein synthesis, p. 9–21. New York and London: Academic Press 1965.

PELC, S. R., HOWARD, A.: Techniques of autoradiography and the application of the strippingfilm method to problems of nuclear metabolism. Brit. med. Bull. **8**, 132–135 (1952).

PELC, S. R., WELTON, M. G. E.: Quantitative evaluation of tritium in autoradiography and biochemistry. Nature (Lond.) **216**, 925–927 (1967).

PERRY, R. P.: Quantitative autoradiography. In: D. M. PRESCOTT (ed.), Methods in cell physiology, vol. I, p. 305–326. New York: Academic Press 1964.

PETERS, T., JR., ASHLEY, C. A.: An artefact in radiography due to binding of free amino acids to tissues by fixatives. J. Cell Biol. **33**, 53–60 (1967).

PETERS, T., JR., ASHLEY, C. A.: Binding of amino acids to tissues by fixatives. In: L. J. ROTH and W. E. STUMPF (eds.), Autoradiography of diffusible substances, p. 267–278. New York-London: Academic Press 1969.

PODDAR, R. K.: On the quantitative relation between isotopic beta radiation and its photographic response. Indian J. Phys. **29**, 189–198 (1955).

PRZYBYLSKI, R. J.: Principles of quantitative autoradiography. In: G. L. WIED and G. F. BAHR (eds.), Introduction to quantitative cytochemistry-II, p. 477–505. New York-London: Academic Press 1970.

QUASTLER, H.: Effects of irradiation on synthesis and loss of DNA. In: M. HAISSINSKY (ed.), Actions chimiques et biologiques des radiations, p. 147–186. Paris: Masson & Cie. 1963a.

QUASTLER, H.: The analysis of cell population kinetics. In: L. F. LAMERTON and R. J. M. FRY (eds.), Cell proliferation, p. 18–36. Oxford: Blackwell Sci. Publ. 1963b.

RIGGS, T. R., WALKER, L. M.: Some relations between active transport of free amino acids into cells and their incorporation into protein. J. biol. Chem. **238**, 2663–2668 (1963).

RITZEN, M.: Mast cells and 5-HT. Uptake of labelled 5-hydroxytryptamine (5-HT) and 5-hydroxytryptophan in relation to storage of 5-HT in individual rat mast cells. Acta physiol. scand. **69**, 1–12 (1967a).

RITZEN, M.: A method for the autoradiographic determination of absolute specific radioactivity in cells. Exp. Cell Res. **45**, 250–252 (1967b).

RITZEN, M.: A simple method for determination of absolute "specific activity" in individual cells. Acta histochem. (Jena), Suppl. **8**, 275–278 (1968).

ROBERTSON, J. S., BOND, V. P., CRONKITE, E. P.: Resolution and image spread in autoradiographs of tritium-labeled cells. Int. J. appl. Radiat. **7**, 33–37 (1959).

ROBINSON, S. H., BRECHER, G., LOURIE, I. S., HALEY, J. E.: Leukocyte labeling in rats during and after continuous infusion of tritiated thymidine: Implications for lymphocyte longevity and DNA reutilization. Blood **26**, 281–295 (1965).

ROGERS, A. W.: A simple photometric device for the quantitation of silver grains in autoradiographs of tissue sections. Exp. Cell Res. **24**, 228–239 (1961).

ROGERS, A. W.: Techniques of autoradiography. Amsterdam-London-New York: Elsevier Publishing Company 1967.

ROGERS, A. W., DARZYNKIEWICZ, Z., BARNARD, E. A., SALPETER, M. M.: Number and location of acetylcholinesterase molecules at motor endplates of the mouse. Nature (Lond.) **210**, 1003–1006 (1966).

ROGERS, A. W., JOHN, P. N.: Latent image stability in autoradiographs of diffusible substances. In: L. J. ROTH and W. E. STUMPF (eds.), Autoradiography of diffusible substances, p. 51–68. New York-London: Academic Press 1969.

ROTH, L. J.: Autoradiography of diffusible substances. In: L. J. ROTH and W. E. STUMPF (eds.), Autoradiography of diffusible substances, p. 1–13. New York-London: Academic Press 1969.

RUBINI, J.R., CRONKITE, E.P., BOND, V.P., FLIEDNER, T.M.: The metabolism and fate of tritiated thymidine in man. J. clin. Invest. **39**, 909–918 (1960).

SANDRITTER, W., HARTLEIB, J.: Quantitative Untersuchungen über den Nukleinsäureverlust des Gewebes bei Fixierung und Einbettung. Experientia (Basel) **11**, 313–314 (1955).

SAWICKI, W., PAWINSKA, M.: Effect of drying on unexposed autoradiographic emulsion in relation to background. Stain Technol. **40**, 67–68 (1965).

SCHMID, W.: Autoradiography of human chromosomes. In: J.J. YUNIS (ed.), Human chromosome methodology, p. 91–110. New York-London: Academic Press 1965.

SCHNEIDER, G., MAURER, W.: Autoradiographische Untersuchung über den Einbau von H-3-Cytidin in die Kerne einiger Zellarten der Maus und über den Einfluß des Fixationsmittels auf die H-3-Aktivität. Acta histochem. (Jena) **15**, 171–181 (1963).

SCHULTZE, B.: Die Orthologie und Pathologie des Nukleinsäure- und Eiweißstoffwechsels der Zelle im Autoradiogramm. In: H.W. ALTMANN, F. BÜCHNER, H. COTTIER, G. HOLLE, E. LETTERER, W. MASSHOFF, H. MEESSEN, F. ROULET, G. SEIFERT, G. SIEBERT, und A. STUDER (Hrsg.), Handbuch der allgemeinen Pathologie, Bd. II/5, S. 466–670. Berlin-Heidelberg-New York: Springer 1968.

SCHULTZE, B.: Autoradiography at the cellular level. In: A.W. POLLISTER (ed.), Physical techniques in biological research, 2nd ed., vol. III, part B. New York-London: Academic Press 1969.

SELIGER, H.H.: The application of standards of radioactivity. Int. J. appl. Radiat. **1**, 215–232 (1956).

STILLSTRÖM, J.: Grain count corrections in autoradiography. Int. J. appl. Radiat. **14**, 113–118 (1963).

STILLSTRÖM, J.: Grain count corrections in autoradiography-II. Int. J. appl. Radiat. **16**, 357–363 (1965).

STÖCKER, E., MÜLLER, H.-A.: Zur chemischen Induktion von Silberkörnern im Stripping Film bei der Orcein-Quetsch-Technik. Histochemie **11**, 167–170 (1967).

STRYCKMANS, P., CRONKITE, E.P., FACHE, J., FLIEDNER, T.M., RAMOS, J.: Deoxyribonucleic acid synthesis time of erythropoietic and granulopoietic cells in human beings. Nature (Lond.) **211**, 717–720 (1966).

STUMPF, W.E., ROTH, L.J.: Vacuum freeze drying of frozen sections for dry-mounting, high-resolution autoradiography. Stain Technol. **39**, 219–223 (1964).

STUMPF, W.E., ROTH, L.J.: High resolution autoradiography with dry mounted, freeze-dried frozen sections. Comparative study of six methods using two diffusible compounds ^{3}H-estradiol and ^{3}H-mesobilirubinogen. J. Histochem. Cytochem. **14**, 274–284 (1966).

STUMPF, W.E., ROTH, L.J.: Freeze-drying of small tissue samples and thin frozen sections below −60° C. J. Histochem. Cytochem. **15**, 243–251 (1967).

STUMPF, W.E., ROTH, L.J.: Autoradiography using dry-mounted freeze-dried sections. In: L.J. ROTH and W.E. STUMPF (eds.), Autoradiography of diffusible substances, p. 69–80. New York-London: Academic Press 1969.

TAYLOR, J.H.: Autoradiography at the cellular level. In: G. OSTER and A.W. POLLISTER (eds.), Physical techniques in biological research, vol. III, p. 545–580. New York: Academic Press 1956.

TORELLI, U., GROSSI, G., ARTUSI, T., EMILIA, G., ATTIYA, I.R., MAURI, C.: RNA and protein metabolism in normal human erythroblasts and granuloblasts. Acta haemat. (Basel) **32**, 271–279 (1964a).

TORELLI, U., GROSSI, G., ARTUSI, T., EMILIA, G., ATTIYA, I.R., MAURI, C.: RNA turnover rates in normal peripheral mononuclear leucocytes. Exp. Cell Res. **36**, 502–509 (1964b).

TORELLI, U., ARTUSI, T., GROSSI, G., EMILIA, G., MAURI, C.: An unstable ribonucleic acid in normal human erythroblasts. Nature (Lond.) **207**, 755–757 (1965).

WASER, P.G., LÜTHI, U.: Über die Fixierung von ^{14}C-Curarin in der Endplatte. Helv. physiol. pharmacol. Acta **20**, 237–251 (1962).

WILLIAMS, A.I.: A method for prevention of leaching and fogging in autoradiographs. Nucleonics **8**, 10–14 (1951).

WIMBER, D.E., QUASTLER, H., STEIN, O., WIMBER, D.: Analysis of tritium incorporation into individual cells by autoradiography of squash preparations. J. biophys. biochem. Cytol. **8**, 327–331 (1960).

YANKWICH, P.E., WEIGL, J.W.: The relation of backscattering to self-absorption in routine beta-ray measurements. Science **107**, 651–653 (1948).

ZAJAC, B., ROSS, M.A.S.: Calibration of electron-sensitive emulsions. Nature (Lond.) **164**, 311–312 (1949).

ZETTERBERG, A., KILLANDER, D.: Quantitative cytophotometric and autoradiographic studies on the rate of protein synthesis during interphase in mouse fibroblasts in vitro. Exp. Cell Res. **40**, 1–11 (1965).

Chapter 12

Micro-Dialysis

Micro-Dialysis Chamber

The purification of very small quantities of protein solutions by dialysis is a frequent problem. Fig. 1 illustrates two dialysis chambers made of Dynal (Polyoxymethylene) which are suitable for the dialysis of 100 µl or 15 µl respectively. A 4 mm, or 2 mm, thick disc (external diameter ca. 20 mm) with a central hole which may measure between 1 mm and 6 mm according to the volumes to be dialysed (15, 30, 50, 100, 250, 500 µl) is first fitted on one side with a suitable semi-permeable membrane, and a similarly bored rubber washer[1] and a sealing cap are screwed on in such a way that the central hole is sealed only by the permeable membrane. The membrane is moistened in a suitable buffer before use.

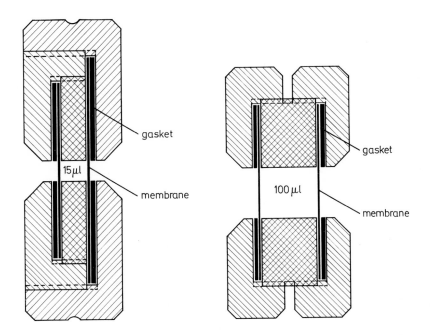

Fig. 1. Scheme for micro-dialysis chambers

---

[1] The rubber washers can be prepared by punching pieces with the right diameter out of operation gloves.

The solution to be dialysed is placed in the chamber, the opening is closed with a semi-permeable membrane, another bored-out rubber washer laid on top, and the apparatus is screwed shut with a similar sealing cap. The dialysis chamber is fitted into a holder made of non-rusting spring steel, on a bar magnet which is enclosed in plexiglass. Fig. 2 shows a complete dialysis apparatus[2] of this type for 15 μl, which can be placed in a beaker of a suitable external medium, and which is turned by a magnetic stirrer during dialysis. For volumes of less than 100 μl the inner chamber is only 2 mm thick, so that the thickness of the liquid layer, and thus the diffusion path, is as small as possible. In the vessels with volumes of 30 μl or 15 μl respectively, the outer sealing caps are made so

Fig. 2. Micro-dialysis chamber in situ

that one cap carries the screw for the dialysis chamber on the inside and that for the second sealing cap on the outside. In order to be able to fasten these narrow vessels securely, the sealing caps have two small-bore holes to take a studded key. After dialysis, the liquid is removed from the chambers by piercing through a membrane with a suitable glass capillary.

Micro-Electrodialysis

Fig. 3 shows an apparatus for the electro-dialysis of small volumes for the simple removal of ions from protein solutions. A plexiglass tube with an inner diameter of 2 mm and a length of 15 or 20 mm (for 20 or 30 μl, respectively) is equipped with screw threads at both ends. First one end is fitted with a semi-permeable membrane and a bored-out rubber washer, and closed with a corre-

---

[2] Obtainable from E. Schütt jr., Göttingen, Germany.

spondingly larger plexiglass tube. After insertion, taking care not to introduce air bubbles into the sample to be dialysed, the open end of the tube is also fitted with a membrane and washer, and screwed up. After introducing the electrode buffer, again taking care not to introduce air bubbles, platinum electrodes are inserted into the buffer vessels.

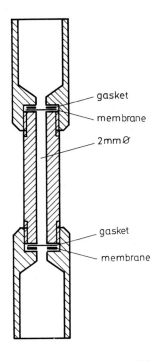

Fig. 3. Diagram of a chamber for micro-electrodialysis

Equilibrium Dialysis

The equilibrium dialysis of volumes of about 30 µl is possible with the apparatus[2] shown in Fig. 4. A hemispherical cavity is bored and polished in each of two identical plexiglass blocks. Each of these cavities is surrounded by a groove to receive a silicone rubber ring washer. A suitable semi-permeable membrane is put in place and the two plexiglass blocks are then screwed on to each other so that the two dialysis chambers are separated from each other by the membrane. A fine hole is bored through each plexiglass block so that it just reaches the tip of the dialysis cavity. The solutions to be dialysed against each other can be introduced through these holes with a fine glass capillary, and known volumes can be removed at the desired times by the same route, using a microlitre syringe. It is not necessary to seal the external openings of these holes during the equilibrium dialysis, since, because of the tiny diameter and the length of the channels, no liquid can escape from the chamber. The

Membrane gasket

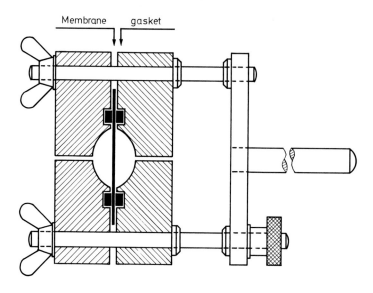

Fig. 4. Diagram of a micro equilibrium dialysis chamber. (Figs. 1, 3 and 4 from NEUHOFF and KIEHL, 1969)

plexiglass block is also fitted with a holder which can be clamped into the bit of a suitable stirrer. If it is necessary that the contents of the chambers be stirred thoroughly during the equilibrium dialysis, a small glass bead may be placed in each half of the chamber before assembly.

Literature

NEUHOFF, V., KIEHL, F.: Dialysiergeräte für Volumen zwischen 10 und 500 µl. Arzneimittel-Forsch. (Drug Res.) **19**, 1898–1899 (1969).

Chapter 13

Micro-Homogenisation

When working with biological material, homogenisation is frequently a prerequisite for analysis. To homogenise minute amounts of tissue samples, e.g. isolated cells or very small core cylinders or biopsy samples, homogenisation by repeated freezing and thawing is frequently employed. Since this is often insufficient for the complete disintegration of the tissues, micro homogenisers must be used. EICHNER (1966) has devised the micro-homogeniser shown in Fig. 1, made from a loop of Nikrothal wire (AB Kanthal, Halsthammer, Sweden); the wire has a diameter of 60 or 35 μm. To make the homogeniser, a piece of wire is bent into a loop, inserted into a thin polyethylene tube, and fixed there with quick-setting cement (e.g. Harvard glass wax[1]). The top of the loop is then bent to a fine point using a fine watchmaker's forceps, and the whole loop is twisted once round in the process. The loop is connected to a motor (dentist's drill) by two polyethylene tubes which fit suitably into each other. One is pushed firmly onto a small milling cutter in the handle of the drill to form a mounting, and the wire loop is inserted into the other (see Fig. 1). Homogenisation is performed under a strong lens or a stereomicroscope, so that the hand piece of the homogeniser can be fixed to a firm support in the position desired. The sample of tissue in the capillary can be held firmly on the stereomicroscope table by means of a simple holder (e.g. plasticine). The high elasticity of the wire makes centering of the individual components unnecessary. The heat generated on homogenisation is unimportant if one does not use too high a speed.

The homogenised material may either be manipulated further in the capillary (see also capillary centrifugation chapter 5) or removed from it without loss of material. Particles adhering to the wire loop are retained in the liquid owing to the high surface tension at the liquid meniscus in the capillary. This wire loop homogeniser is particularly suitable for homogenisation in Drummond microcaps. If so little material is available that homogenisation must be carried in 1 or 2 μl capillaries, the wire loop homogeniser described is the only method available.

For 5 and 10 μl capillaries, the technically simpler homogenisation is performed with dentist's nerve channel drills. Fig. 2 shows two drills of this type (Beutelrock drills, diameter 0.25 and 0.45 mm) which are well suited for 5 μl and for 10 μl Drummond microcaps respectively. They can be fitted into the hand piece of a dentist's drill (Fig. 3), which is best driven via an infinitely variable foot treadle. It is important to reverse the direction of the drill, so that the drill

---

[1] Richier u. Hoffmann, Harvard Dental Ges., W.-Berlin, Germany.

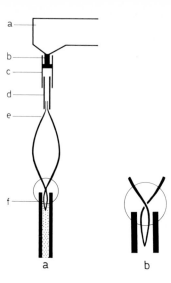

Fig. 1. *a* Schematic general view of the micro-homogeniser. *a* Handpiece of a dentist's drill, *b* small milling head, *c* polyethylene tube, *d* polyethylene tube, *e* wire loop, *f* capillary with material *b* Enlarged section from (*a*) shows the position of the twist at the tip of the wire loop. The tip serves primarily for the easier introduction of the instrument into the capillary; the homogenisation is carried out by the bulging section of the loop above the twist. (From EICHNER, 1966)

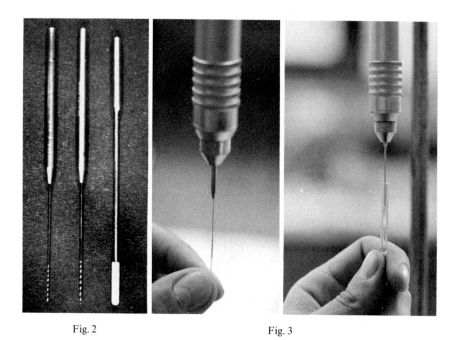

Fig. 2   Fig. 3

Fig. 2. Nerve canal drills for homogenisation in 5 and 10 µl capillaries, and a pestle for homogenisation in capillaries with inner diameter of 2 mm

Fig. 3. Homogenisation with a nerve canal drill in a 5 µl capillary (left) and with a pestle in a pyrex glass tube (right)

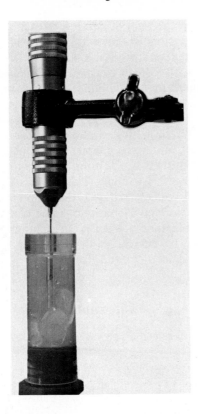

Fig. 4. Micro-homogeniser with ice cooling. (From NEUHOFF, 1968)

turns to the left, the reverse of its normal rotation. This results in the cutting head of the drill turning in the opposite sense from usual, and the liquid in the capillary is not thrown out, but is directed down towards the sealed bottom of the capillary. When using the drills, attention should be paid to the fact that they take up a larger volume than the wire loop, and consequently the capillaries should not be filled completely. It is recommended to test, before the actual homogenisation of a sample, how far a capillary may be filled when homogenising with the drill. Because of the larger surface area of the drill, the liquid can become hot during high-speed homogenisation. If temperature-sensitive solutions are to be homogenised, the capillary must therefore be cooled in ice water.

Fig. 4 shows a micro-homogeniser with ice cooling, for a volume of 50 µl, made out of plexiglass. The pestle fits into the hand piece of the dentist's drilling apparatus, and is tipped by a piece of Dynal of 2 mm diameter, which is screwed to a carrier made of stainless steel (cf. Fig. 2). These micro-homogenisers are obtainable for various volumes[2]. After homogenisation, the samples must be removed for further work-up, e.g. centrifugation. This procedure can be further

[2] E. Schütt jr., Göttingen, Germany.

simplified if, as shown on the right of Fig. 3, homogenisation is carried out in a glass tube which has been carefully sealed at the bottom by annealing, and which fits directly into the capillary rotor of a capillary centrifuge (see chapter 5). Cooling is carried out with a small beaker containing iced water, if necessary. Allowance must be made for the volume displacement of the pestle during homogenisation. Pestles of this type can be produced easily for various diameters of glass tube, and have proved to be very suitable for micro-homogenisation.

Another method of micro-homogenisation consists of adding some splinters of razor blade (for preparation see chapter 6) to the sample and solution in the glass tube, and then shaking the tube on a whirlmixer. All tissues are completely fragmented by this method. This process can be intensified if the tissue is first frozen, and then shaken with the splinters of blade after thawing out. It is not necessary to remove the splinters of blade before centrifugation, and they do not hinder pipetting out of the supernatant with a fine pipette after centrifugation. However, if removal of the fragments is required, this is best effected by drawing a magnet along the glass tube.

Literature

EICHNER, D.: Ein Mikro-Homogenisator. Experientia (Basel) 22, 620 (1966).
NEUHOFF, V.: Mikro-Disc-Elektrophorese von Hirnproteinen. Arzneimittel-Forsch. (Drug Res.) 18, 35–39 (1968).

Chapter 14

Wet Weight Determination in the Lower Milligram Range

When using micro methods it is frequently necessary to obtain quantitative values for the wet weight of very small tissue samples. For dry weight determination, a number of sensitive methods (quartz fibre balance [see chapter 8], electron microscopy [see chapter 7], interferometry etc.) exist. For fresh weight determinations it is essential to avoid the loss of water when the sample is being transferred to the balance. Using the procedure described by NEUHOFF (1971), determination of the fresh weight of tissue samples in the range between 1 mg and 0.01 mg can be made easily.

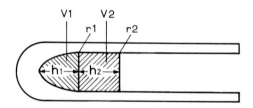

Fig. 1. Diagram of a heat sealed 5 µl capillary filled with a sample for wet weight determination

A 5 µl (or for larger samples, a 10 µl or bigger) Drummond microcap is heat-sealed carefully at one end by rotation in the flame of a spirit burner. It is necessary to check with a magnifying-glass that the seal is paraboloid in shape (see Fig. 1) as the calculation of the volume is based on this form. After filling the capillary completely with a suitable buffer solution, a minute amount of fresh tissue is transferred to the top of the capillary and centrifuged (15 min, 15000 rpm) in a centrifuge equipped with a special capillary rotor (see chapter 5). It is possible to fill the capillary with the sample directly, by punching the open end of the buffer-filled capillary into a lightly-frozen tissue section of about 1 mm thickness.

For determination of the wet weight of the tissue, the capillary is surrounded by water to avoid the magnifying effect of the glass walls. It is then observed through a microscope and the shadow of the part of the capillary containing the sample is projected onto graph paper (see Fig. 17 in chapter 6). The volume is calculated from a sketch of the sample using the formula for a paraboloid of revolution; a correction factor for the magnification is also included:

$$V_1 = \frac{\pi}{2} \cdot \left(\frac{r}{\text{magn.}} \right)^2 \cdot \frac{h_1}{\text{magn.}}.$$

For determining the magnification, an object micrometer is put under the microscope; without changing the optical system, the projection of the magnified micrometer is then sketched on the graph paper.

If the specific gravity of the tissue is known (for routine measurements this is assumed to be 1.0) it is possible to calculate the wet weight of minute amounts of tissue. If the paraboloid section of a heat-sealed 5 µl capillary is just filled, the amount of tissue is approximately 0.01–0.03 mg. For the determination of smaller quantities of tissue, it is necessary to use capillaries with smaller diameters (2 µl or 1 µl Drummond microcaps).

If more than the paraboloid part of the capillary is filled with the sample, it is necessary to calculate this volume separately, by using the formula of a cylinder which has also been corrected for the magnification:

$$V_2 = \pi \cdot \left(\frac{r}{\text{magn.}} \right)^2 \cdot \frac{h_2}{\text{magn.}},$$

$V_1 + V_2$ gives the total amount of tissue.

It is possible to homogenise the samples in a suitable buffer, directly in the capillary (see chapter 13) to be used for the micro electrophoresis of proteins (see chapter 1) or for the determination of free amino acids (see chapter 2). For the determination of phsopholipids, it is necessary to replace the water phase by chloroform/methanol first (see chapter 3).

Literature

NEUHOFF, V.: Wet weight determination in the lower milligram range. Analyt. Biochem. **41**, 270–271 (1971).

Chapter 15

Micro-Magnetic Stirrer

A small piece of iron wire (diameter 0.5–1 mm, length 1–5 mm) is sealed into a polythene tube of suitable diameter, so that water cannot enter. In order to do this, the polythene tube is first sealed at one end by squeezing it together with a heated forceps, whose points have been bent outwards (see Fig. 1). It is often difficult to obtain a watertight seal if forceps with sharp instead of with rounded ends are used, since the polythene tube at the ends of the wire can be damaged when it is compressed with the heated forceps. It is important to heat the forceps to the correct temperature: if they are too cold the seal is imperfect, while overheated forceps may easily set the polythene tube on fire. The wire is introduced into the open end of the polythene tube so that about 3 mm of polythene tube projects beyond the end of the wire. The other end is then sealed with the heated forceps. Finally the surplus polythene at both ends is removed with scissors.

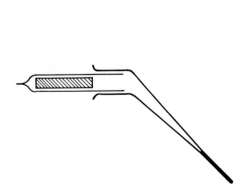

Fig. 1. Sketch for the production of the micro-magnetic stirrer

Fig. 2. Arrangement for stirring solutions with spring steel forceps

Fig. 2 shows a simple arrangement for stirring solutions, which may be employed if fragile pieces of agarose, or even acrylamide gels, are to be rapidly destained after staining. A stirrer in such solutions could easily damage the samples. For this, a soft pair of forceps made of spring steel is fixed in a clamp by means of two small pieces of wood, dipped in the solution to be stirred, and placed on a magnetic stirrer. When the magnetic stirrer is switched on the arms of the tweezers vibrate, and agitate the solution in the beaker. The intensity of the stirring motion can be varied by adjusting the magnetic stirrer.

Chapter 16

Production of Capillary Pipettes

Pyrex glass tubing (soft glass is not suitable!) of 4 mm external diameter and 2 mm inner diameter, is cut into 10 cm lengths. A piece of tubing is held between thumb and first finger and held at the point of the flame of a pressure burner. (Switching-on sequence for the burner is: gas, O_2, compressed air; switching-off in reverse order.) The length of the flame is adjusted to 3–4 cm as shown in Fig. 1.

Capillaries with extra-fine lumina, which are essential for many micro-techniques (e.g. for loading Drummond microcaps or for the application of samples for micro-chromatography) can be made by first pre-pulling the tube as shown in Fig. 2. After cooling and re-heating the pre-pulled section, the fine capillary is pulled out. This is achieved by rotating the tube evenly in the flame,

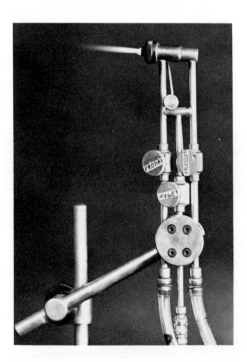

Fig. 1. Burner for preparing capillary pipettes mounted on an adjustable stand. (Obtainable from E. Schütt jr., Göttingen, Germany)

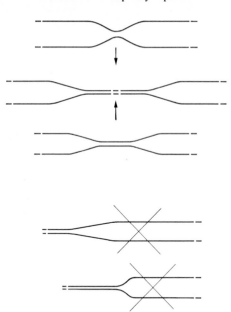

Fig. 2. Diagram for pulling capillary pipettes. The degree of pre-pulling of the pyrex tube determines the final thickness of the capillary

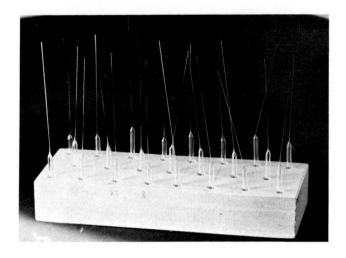

Fig. 3. Stand for a set of capillary pipettes

bringing the middle of the pre-pulled piece of tubing to red heat. The glass is then *removed from the flame* and pulled out quickly with a moderate and even force. The further the capillary is pulled out, the finer will be the point. It is important to leave the capillary to cool before separating the ends, as the glass reaches temperatures in excess of 1000° C. A set of capillaries can be held in a wooden stand until needed (Fig. 3). A capillary of the required length is prepared be breaking off the capillary tip just befor use (thereby guaranteeing that the capillary tip is clean). Only capillaries of the type sketched in Fig. 2 should be used; it is imperative that the flame is adjusted correctly and that the heating of the glass tube is strictly localized. If capillaries with a slightly bent tip are required (e.g. for applying samples to micro-chromatoplates) the ends of the glass tube are held at the required angle while the hot capillary is being pulled.

For further details in the preparation of different micro-pipettes, reference to the following books may be made: O.H. LOWRY and J.V. PASSONNEAU (1972), H. MATTENHEIMER (1966) and H.M. EL-BADRY (1963).

Literature

EL-BADRY, H.M.: Micromanipulators and Micromanipulation. In: Monographien aus dem Gebiet der qualitativen Mikroanalyse, Bd. 3, hrsg. von A.A. BENEDETTI-PICHLER. Wien: Springer 1963.
LOWRY, O.H., PASSONNEAU, J.V.: A flexible system of enzymatic analysis. New York-London: Academic Press 1972.
MATTENHEIMER, H.: Mikromethoden für das klinisch-chemische und biochemische Laboratorium, 2. Aufl. Berlin: Walter de Gruyter & Co. 1966.

Subject Index

Boldface numbers refer to headings, italic numbers to figures

Molecular Biology, Biochemistry and Biophysics